Veterinary Pharmaceuticals in the Environment

ACS SYMPOSIUM SERIES **1018**

Veterinary Pharmaceuticals in the Environment

Keri L. Henderson, Editor
Iowa State University

Joel R. Coats, Editor
Iowa State University

Sponsored by the
ACS Division of Agrochemicals

American Chemical Society, Washington DC

Library of Congress Cataloging-in-Publication Data

Veterinary pharmaceuticals in the environment / editors, Keri L. Henderson, Joel R. Coats.

p. cm. -- (ACS symposium series ; 1018)

Includes bibliographical references and index.

ISBN 978-0-8412-6962-0

1. Veterinary drugs--Environmental aspects. 2. Veterinary drug residues. 3. Environmental risk assessment. I. Henderson, Keri L. II. Coats, Joel R. III. Series: ACS symposium series ; 1018.

SF917.V492 2009
628.5'2--dc22

2009034963

The paper used in this publication meets the minimum requirements of American National Standard for Information Sciences—Permanence of Paper for Printed Library Materials, ANSI Z39.48–1984.

Distributed by Oxford University Press

PRINTED IN THE UNITED STATES OF AMERICA

Foreword

The ACS Symposium Series was first published in 1974 to provide a mechanism for publishing symposia quickly in book form. The purpose of the series is to publish timely, comprehensive books developed from the ACS sponsored symposia based on current scientific research. Occasionally, books are developed from symposia sponsored by other organizations when the topic is of keen interest to the chemistry audience.

Before agreeing to publish a book, the proposed table of contents is reviewed for appropriate and comprehensive coverage and for interest to the audience. Some papers may be excluded to better focus the book; others may be added to provide comprehensiveness. When appropriate, overview or introductory chapters are added. Drafts of chapters are peer-reviewed prior to final acceptance or rejection, and manuscripts are prepared in camera-ready format.

As a rule, only original research papers and original review papers are included in the volumes. Verbatim reproductions of previous published papers are not accepted.

ACS Books Department

Contents

1. **PREFACE: Veterinary Pharmaceuticals in the Environment 1**
Dr. Keri Henderson, Dr. Joel Coats

2. **Veterinary pharmaceuticals in the environment: an introduction......... 3**
Keri L.D. Henderson, Joel R. Coats

Fate and bioavailability

3. **A Review of the Fate of Manure-Borne, Land-Applied Hormones...... 11**
H. Hakk, F.X.M. Casey, Z. Fan, G. L. Larsen

4. **Fate of Chlorate Salts Excreted from Animals...................................... 27**
D. J. Smith, C. E. Oliver, H. Hakk, R. C. Anderson, G. L. Larsen

5. **Influence of MnO_2 on the transformation of oxy- and chlortetracycline in pond water.. 45**
Kennedy F. Rubert IV, Curtis J. Hedman, Tan Guo, Joel A. Pedersen

6. **Comparative Biotransformation of Fluoroquinolone Antibiotics in Matrices of Agricultural Relevance.. 67**
Heinz-Georg Wetzstein, Josef Schneider, Wolfgang Karl

7. **Environmental fate and chemistry of a veterinary antibiotic—tylosin 93**
Dingfei Hu, Keri L.D. Henderson, Joel R. Coats

8. **An overview of the fate and effects of antimicrobials used in aquaculture.. 105**
Yong-Hak Kim, Carl E. Cerniglia

9. **Fate and bioavailability of sulfamethazine in freshwater ecosystems 121**
Keri L.D. Henderson, Thomas B. Moorman, Joel R. Coats

Assessing potential environmental impacts

10. Livestock Hormones in Aquatic Ecosystems **135**
Yue-wern Huang, Dev K. Niyogi, Paul K. Nam, Janet M. Bandeff

11. Characterization of Chlortetracycline-induced Glutathione *S*-Transferase to Conjugate Chloroacetanilide and Chlorotriazine Herbicides **153**
Michael H. Farkas, James O. Berry, Diana S. Aga

12. Targets, Effects and Risks in Aquatic Plants Exposed to Veterinary Antibiotics **169**
Richard A. Brain, Keith R. Solomon, Bryan W. Brooks

13. Risks of agricultural pharmaceuticals in surface water systems and soils **191**
Keith R. Solomon, Derek G. Hillis, Linda Lissemore, Paul K. Sibley

14. Risk assessment considerations for veterinary medicines in aquatic ecosystems **205**
Bryan W. Brooks, Richard A. Brain , Duane B. Huggett, Gerald T. Ankley

Indexes

Author Index **227**

Subject Index **229**

Chapter 1

PREFACE

Veterinary Pharmaceuticals in the Environment

Dr. Keri Henderson and Dr. Joel Coats

Department of Entomology Iowa State University Ames, IA 50011

An important symposium was presented on "Veterinary Pharmaceuticals in the Environment" at the 233rd National Meeting of the American Chemical Society on March 25-26, 2007, in Chicago, IL. The excellence and timeliness of the presentations indicated the need for an ACS Symposium Series book addressing the environmental chemistry and toxicology of this group of emerging contaminants.

The purpose of this symposium was to bring together scientists from academia, government, and industry to discuss and present data relevant to the significance of veterinary pharmaceuticals in the environment. A broad range of topics was covered, including environmental chemistry studies focusing on transport, mobility, sorption, persistence, and bioavailability of the compounds, as well as development of analytical techniques relevant to detection of the pharmaceuticals in environmental matrices, discussion of ecotoxicological studies of veterinary pharmaceuticals, and information relevant to ecological risk assessments. The primary classes of drugs addressed herein are veterinary antibiotics and synthetic hormones. The antibiotics include some that are also used in human medicine and others that are unique to animal medicine. A number of the most widely used antibiotics are employed two different ways: therapeutically to cure or prevent diseases, or as growth promoters. Synthetic hormones are utilized to influence the development of animals, in particular in regards to weight gain and body composition. Many of the environmental concerns over veterinary pharmaceuticals arose from runoff and wastes from confined animal feeding operations (CAFOs), where tens of thousands of meat animals are confined in a small area. In such situations, the drugs that are excreted intact or partially metabolized can still be biologically active when they enter the environment; even natural hormones from the animals constitute a significant portion of the overall input of bioactive chemicals.

To date, there have been numerous monitoring studies that have confirmed the existence of very low concentrations of animal pharmaceuticals and hormones, especially in surface waters. The next important question is whether those residues are of any significance in the environment, including questions about how long they persist, what analytical methods should be used to quantify them, where they come from, how they move, whether they accumulate in environmental sinks, what kinds of biological effects they have, the species affected, and possible prevention or mitigation strategies. One of the most prominent controversies is the one generated around the possibility of antibiotics inducing or sustaining antibiotic-resistant microorganisms in the environment. Another focus is on the hormonal effects on fish and wildlife, especially as the hormones (synthetic or natural) influence reproduction and sexual development in wild mammals, birds, fish, and amphibians.

We believe that this book addresses some pressing scientific questions, including the monitoring of the drugs in the environment, the development of analytical chemical and immunochemical methods of quantifying low levels of the residues, their mobility, biological availability, uptake, effects on organisms, as well as perspectives on ecological risk assessment of these types of chemicals.

Acknowledgments

We appreciate the time and effort that the authors contributed toward this book. We thank the ACS Division of Agrochemicals for providing financial assistance and the venue for the symposium on which this book is based. John J. Johnson's assistance was also invaluable, as Program Chair of the AGRO Division. We are grateful to all of the peer reviewers of the chapters for their expertise and their effort, and we appreciate the support and assistance from Bob Hauserman, Jessica Rucker, and ACS Books which culminated in the publishing of this volume. Keri Henderson dedicates this book to her daughter Reese, and Joel Coats dedicates it to his grandchildren, Leola, Chloe, Katherine, Ivy, and Grace.

Chapter 2

Veterinary pharmaceuticals in the environment: an introduction

Keri L.D. Henderson[1] and Joel R. Coats[1]

[1]Department of Entomology, Iowa State University, Ames, IA, USA

The production of swine, cattle, and poultry raised for human consumption represents a significant portion of the U.S. agricultural economy. To maximize production, livestock producers regularly use pharmaceuticals and hormones as supplements in animal feed and water to increase rates of weight gain, and prevent or treat diseases among their livestock. For example, in swine it is estimated that antibiotics are used for disease prevention and growth promotion in more than 90% of starter feeds, 75% of grower feeds, 50% of finishing feeds, and 20% of sow feeds. Equally relevant numbers are seen in beef cattle production (*1-4*). It has been well documented that measurable quantities of pharmaceuticals are excreted, often in original form, in feces and urine of livestock. Livestock waste, containing pharmaceuticals and hormones, is often used as fertilizer for farm fields or pastures and may result in nonpoint source pollution of ground or surface waters (*5*). Field application of manure often involves injection or incorporation of the waste, which may bring nutrients, hormones, and pharmaceutical residues spatially closer to tile drainage systems. These drainage systems often flow directly into nearby streams, rivers, or other waterways, and may therefore act as a source of waterway contamination. A national reconnaissance study performed by Kolpin and colleagues (*6*) was among the first to bring widespread attention to the issue of pharmaceuticals in the environment. Many studies have examined environmental chemistry and ecotoxicology of human drugs, however fewer have examined veterinary pharmaceuticals.

Although pharmaceutical residues have been studied extensively in tissues and excrement, relatively little is known about the environmental fate of pharmaceuticals, hormones, and their metabolites once the excreta reaches soil and water environments (*7,8*). Sorption of chemicals onto solid phases, such as soil or sediment, or degradation is extremely important because it could affect the fate and impact of these substances in that environment.

Recently, veterinary pharmaceuticals were found in 48% of 139 stream waters tested in 30 states, according to the United States Geological Survey (*6*). Recent sediment monitoring studies by Kim and Carlson (*9,10*) detected extractable antibiotic residues up to 0.1 mg/kg in sediment from an impacted river; typical concentrations were 0.001 to 0.03 mg/kg. The significance of low concentrations of veterinary pharmaceuticals in lagoon water, soil, and surface water ecosystems is not well understood, particularly when so few data are available regarding sorption and bioavailability of hormones and pharmaceuticals in environmental matrices. Simply detecting the compounds does not provide adequate information on the ecological significance. These concentrations appear to be too low for significant activity; however ecotoxicity data exist for only a small number of compounds. Low or transient concentrations of pharmaceuticals could have several effects; in addition to the obvious potential effects based on mode of action, they may have secondary modes of action causing harm to non-target populations in aquatic and terrestrial habitats.

Even at low environmental concentrations, hormones and pharmaceuticals and their metabolites may be biologically active in environmental matrices; these compounds are designed to elicit specific effects, and while those effects may not lead to mortality, the may significantly alter physiology or behavior, thus impacting organismal fitness. For example, antibiotics entering the environment could potentially alter bacterial populations and their activity in soil, sediment, and water, thus affecting biodegradation, nutrient cycling, and water quality. In addition, there is concern that antibiotics in the environment may induce antibiotic resistance, resulting in adverse human health effects. Certainly, there is significant evidence for development of antibiotic resistance within animals and in the excretion of antibiotic-resistant bacteria in manure (*11-13*). Much less is known about the ability of low concentrations of antibiotics to induce resistance in the environmental microbial population or to provide selective pressure for maintenance of antibiotic resistance genes among microorganisms, although the transfer of antibiotic-resistance from agricultural settings to humans has been reported (*14*). Similarly, low concentrations of hormones have also demonstrated potential biological activity (*15-20*).

There are concerns over potential direct and indirect effects of pharmaceuticals on non-target eukaryotic organisms in important habitats found across agricultural landscapes. In particular, small ponds and wetlands that serve as key breeding sites for amphibians and support invertebrate communities that provide food for migrating birds (*21*) can receive significant amounts of contaminated agricultural runoff, which could contain pharmaceutical residues (*22,23*). The aquatic risk of these chemicals has not been extensively characterized (*24,25*); even less is known about potential implications for terrestrial habitats.

Studies on the fate, bioavailability, and effects of veterinary pharmaceuticals in terrestrial and aquatic systems will lead to improved management practices through the understanding of the potential for risk to the environment. Studies presented in this book provide information that is essential to understanding the environmental mobility and degradation of veterinary pharmaceuticals and hormones. Bioavailability studies serve as a crucial link to understanding the likelihood of environmental residues impacting species or microbial communities in a negative way. Published reports have confirmed that very low levels of residues are detected in water and sediment monitoring studies (*6,9,10,26,27*), but the significance of those residues will depend, to a great extent, on their biological availability. Ecotoxicological studies herein point to possible impacts of pharmaceutical and hormone residues on wildlife and plants; these studies are put in a risk assessment context, and importantly suggest unique considerations for ecological risk assessments of pharmaceuticals. The synthesis of these data into this book will provide a more comprehensive evaluation of the fate and potential effects of these contaminants in the environment, leading to more informed management decisions.

The purpose of this symposium was to bring together scientists from academia, government, and industry to discuss and present data relevant to the potential significance of veterinary pharmaceuticals and hormones in the environment. Topics presented in the subsequent chapters include fate studies examining sorption, mobility, degradation, and persistence in soil and aquatic systems, bioavailability in aqueous systems, ecotoxicological studies, and recommendations for risk assessment of veterinary pharmaceuticals.

Acknowledgements

We would like to thank all of the authors who contributed chapters to this book, as well as those who participated in the symposium.

References

1. Hayes, D.J.; Jensen, H.H.; Backstrom, L.; Fabiosa, J. Economic impact of a ban on the use of over-the-counter antibiotics. Center for Agricultural and Rural Development, Iowa State University, Ames, IA. Staff Report 99-SR 90. **1999.** http://www.card.iastate.edu/publications/texts/99sr90.pdf.
2. USDA, Swine 2000 survey. Accessed June 20, 2002. http://www.aphis.usda.gov/vs/ceah/cahm/Swine/Swine2000/antibiotics2.PDF
3. USDA APHIS Cattle on Feed Evaluation. Accessed February 2, 2004. http://www.aphis.usda.gov/vs/ceah/cei/antiresist.antibiouse.pdf; http://www.aphis.usda.gov/vs/ceah/cahm/Beef_Feedlot/Cofdes1.pdf
4. Bush, E.J.; Biehl, L.G. Use of antibiotics and feed additives by U.S. pork producers. *Proceedings of U.S. Animal Health Association*. Nov. 7, 2001. Hershey, PA.
5. Loke, M.L.; Ingerslev, F.; Halling-Sorensen, B.; Tjornelund, J. Stability of tylosin A in manure containing test systems determined by high performance liquid chromatography. *Chemosphere* **2000,** *40*, 759-65.
6. Kolpin, D.W.; Furlong, E.T.; Meyer, M.T.; Thurman, E.M.; Zaugg, S.D.; Barber, L.B.;Buxton, H.T. Pharmaceuticals, hormones and other organic wastewater contaminants in US Streams, 1999-2000: A national reconnaissance. *Environ. Sci. Technol.* **2002,** *36*, 1202-11.
7. Tolls, J. Sorption of veterinary pharmaceuticals in soils: a review. *Environ. Sci. & Technol.* **2001**, *35*, 3397-3406.

8. Coats, J.R.; Metcalf, R.L.; Lu, P.Y.; Brown, D.D.; Williams, J.F.; Hansen, L.G. Model ecosystem evaluation of the environmental impacts of the veterinary drugs phenothiazine, sulfamethazine, clopidol, and diethylstilbestrol. *Environ. Health Perspect.* **1976,** *18*, 167-79.
9. Kim, S.C.; Carlson, K. Quantification of human and veterinary antibiotics in water and sediment using SPE/LC/MS/MS. *Anal. Bioanal. Chem.* **2007,** *387*, 1301-15.
10. Kim, S.C.; Carlson, K. Temporal and spatial trends in the occurrence of human and veterinary antibiotics in aqueous and river sediment matrices. *Environ. Sci. Technol.* **2007,** *41*, 50-7.
11. Beaucage, C.M.; Fox, J.G.; Whitney, K.M. Effect of long-term tetracycline exposure (drinking water additive) on antibiotic-resistance of aerobic gram-negative intestinal flora of rats. *Am. J. Vet. Res.* **1979**, *40*, 1454-7.
12. Aarestrup, F.M.; Nielsen, E.M.; Madsen, M.; Engberg, J. *Antimicrob. Agents Chemother.* **1997,** 41, 2244-50.
13. Kelley, T.R.; Pancorbo, O.C.; Merka, W.C.; Barnhart, H.M. Antibiotic resistance of bacterial litter isolates. *Poultry Sci.* **1998**, *77*, 243-7.
14. Oppegaard, H.; Steinum, T.M.; Wateson, Y. Horizontal transfer of a multi-drug resistance plasmid between coliform bacteria of human and bovine origin in a farm environment. *Appl. Environ. Microbiol.* **2001**, *67*, 3732-3734.
15. Purdom, C.E.; Hardimon, P.A.; Bye, V.J.; Enụ, N.C.; Tyler, C.R.; Sumpter JP. Estrogenic effects of effluents from sewage treatment works. *Chem. Ecol.* **1994**, *8*, 275-285.
16. Harries, J.E.; Sheahan, D.A.; Jobling, S.; Mattiessen, P.; Neall, P.; Sumpter, J.P.; Tylor, T.; Zaman, N. Estrogenic activity in five United Kingdom rivers detected by measurement of vitellogenesis in caged male trout. *Environ. Toxicol. Chem.* **1997**, *16,* 534-542.
17. Desbrow, C.; Routledge, E.J.; Brighty, G.C.; Sumpter, J.P.; Waldock, M. Identification of estrogenic chemicals in STW effluent, 1. Chemical fractionation and in vitro biological screening. *Environ. Sci. Technol.* **1998,** *32,* 1549-1558.
18. Jobling, S.M.N.; Tyler, C.R.; Brighty, G.; Sumpter, J.P. Widespread sexual disruption in wild fish. *Environ. Sci. Technol.* **1998**, *32*, 2498-2506.
19. Orlando, E.F.; Kolok, A.S.; Binzcik, G.A.; Gates, J.L.; Horton, M.K.; Lambright, C.S.; Gray, L.E.; Soto, A.M.; Guillette, L.J. Endocrine-disrupting effects of cattle feedlot effluent on an aquatic sentinel species, the fathead minnow. *Environ. Health Persp.* **2004**, *112*, 353-358.
20. Huang, Y.W.; Hunter, L.D.; Phillips, J.R. Human exposure to medicinal, dietary, and environmental estrogens. *Toxicol. Environ. Chem.* **2007**, *89*, 141-160.
21. Reynolds, R.E. Breeding duck population, production, and habitat surveys, 1979-1985. *Transactions of the North American Wildlife and Natural Resource Conference* **1987,** *52*,186-205.
22. Maltby, L.; Farrow, D.M.; Boxall, A.B.A.; Calow, P.; Betton, C.L. The effects of motorway runoff on freshwater ecosystems .1. Field study. *Environ. Toxicol. Chem.* **1995,** *14*, 1079-92.

23. Lahr, J. Ecotoxicology of organisms adapted to life in temporary freshwater ponds in arid and semi-arid regions. *Arch. Environ. Contam. Toxicol.* **1997**, *32*, 50-57.
24. Baguer, A.J.; Jensen, J.; Krogh, P.H. Effects of the antibiotics oxytetracycline and tylosin on soil fauna. *Chemosphere* **2000**, *40*, 751-7.
25. Wollenberger, L.; Halling-Sorensen, B.; Kusk, K.O. Acute and chronic toxicity of veterinary antibiotics to *Daphnia magna*. *Chemosphere* **2000**, *40*, 723-730.
26. Diaz-Cruz, M.S.; de Alda, M.J.L.; Barcelo, D. Environmental behavior and analysis of veterinary and human drugs in soils, sediments and sludge. *Trends Anal. Chem.* **2003**, *22*, 340-51.
27. Beausse, J. Selected drugs in solid matrices: a review of environmental determination, occurrence and properties of principal substances. *Trends Anal. Chem.* **2004**, *23*, 753-61.

Fate and Bioavailability

Chapter 3

A Review of the Fate of Manure-Borne, Land-Applied Hormones

H. Hakk[1], F.X.M. Casey[2], Z. Fan[2], and G. L. Larsen[1]

[1]USDA ARS, Biosciences Research Laboratory, 1605 Albrecht Blvd, Fargo, ND 58105-5674
[2]Department of Animal and Range Sciences, North Dakota State University, Fargo, North Dakota 58105

The presence of naturally-occurring reproductive hormones in the environment, even at part-per trillion concentrations, may raise significant concerns regarding the health of aquatic organisms. The hormones that have been shown to alter the endocrine systems of animals at low ppt levels are 17β-estradiol and testosterone, which are classified as endocrine disrupting compounds. Most hormones are produced and released into the environment through the excreta of humans and of domestic and wild animals at low ppm levels. Because of the potency and the large environmental inputs, the fate and transport of 17β-estradiol and testosterone in soil-water systems need to be elucidated. Research projects were conducted to determine the fate, transport, and sorption of these hormones in soil-water systems, and to begin to develop comprehensive models to describe these various processes. Column studies revealed that both hormones are rapidly sorbed to soil, although 17β-estradiol had significantly higher partitioning coefficients than testosterone. Very little testosterone and no 17β-estradiol were observed in effluent water of the soil column experiments. Both hormones were redistributed through the soil columns, although the majority of the hormone was sorbed to the top few cm of soil. Degradation of hormones was predominantly a biotic process, and testosterone was degraded more readily than 17β-estradiol during incubation in a wet soil. Fate and transport processes were discerned from these laboratory studies, and were quantified using mathematical models. Despite studies showing that 17β-estradiol and testosterone had minimal mobility and were readily degraded, other research has shown that they are regularly detected in surface and ground water. Therefore, hypotheses need to be developed and tested to explain the consistent appearance of 17β-estradiol and testosterone in ground and surface waters.

Reproductive Hormones in the Environment

The presence of low levels of biologically active compounds in the environment has become a concern in the United States and throughout the world (e.g. World Health Organization, United Nations). A congressional bill (H.R. 1712) was submitted to the 106th US Congress that proposed to amend the Federal Water Pollution Act to authorize an endocrine disruptor compounds (EDC) screening program (*1, 2*). Kolpin et al. (*3*) have completed an extensive reconnaissance of U.S. surface waters in which 139 streams were sampled across 30 states, and they observed reproductive hormones in approximately 40% of the samples. Wildlife, particularly aquatic, is known to be impacted by exposure to reproductive hormones, especially because they can produce adverse effects at remarkably low concentrations (1-10 ng L^{-1}; *4, 5, 6*). For example, 17β-estradiol has been found to cause vitellogenin production in male fish at environmental concentrations of 1.0 ng L^{-1} (*4*). Vitellogenin is an egg yolk precursor protein expressed in female fish, but is dormant in male fish. Exposure to EDCs can cause the expression of vitellogenin in a dose dependent manner in male fish, therefore, the use of vitellogenin expression in male fish can be used as a marker of exposure to estrogenic EDCs. The hazards of chronic low concentration exposures to hormones in humans are unknown.

Environmental Sources of Hormones

Most hormones are produced and released into the environment by humans (e. g. human urine), livestock (e. g. animal manure), and wildlife (*7, 8*). Johnson et al. (*9*) reported that on average 1.6 μg day^{-1} of 17β-estradiol is excreted by human males, and 2.3–2.5 μg day^{-1} of 17β-estradiol is excreted by females. Johnson et al. (*9*) also reports that pregnant women can excrete up to 259 μg day^{-1} of 17β-estradiol. These human sources of hormones can enter the environment from sewage treatment plants (*10*).

Possible exposure to reproductive hormones (estrogens and androgens) may come from animal manures that are applied to agricultural fields. Estrogens and androgens are naturally produced and excreted by animals, and can also be administered as growth promoters. The potential for manure-borne hormone release into the environment is significant from concentrated animal feeding operations (CAFOs) where thousands of individual animals are housed on relatively small tracts of land. For instance, on the eastern shore of Maryland, it has been estimated that 200,000 metric tons yr^{-1} of broiler chicken manure is produced at numerous CAFOs. This amount of litter is estimated to contain 30 ng g^{-1} (30 ppb) of 17β-estradiol, which would be equivalent to 6 kg of this estrogen being applied to fields (*8*). In 1997 the entire U.S. poultry industry produced over ten billion kg of broiler litter, 90% of which was applied to pastures and hayfields. In a worst-case situation, this amount of manure could potentially contain 270 kg of 17β-estradiol, which has the capacity to contaminate 2.7×10^{14} L of water at a 1.0 ng L^{-1} concentration. A survey by Lorenzen et al. (*11*) found estrogenic activity levels of 5965 μg kg^{-1} (E2 equivalents per dry wt.) in manure from finishing pigs and androgenic activity levels of 1737 μg kg^{-1} (testosterone equivalents per dry wt.) in manure from pregnant dairy cows.

Following manure application in a field setting, 17β-estradiol was found in a nearby free flowing stream at concentrations of 5 ng L^{-1} (*8*). In another experiment where manure was applied to a field, the concentrations of 17β-estradiol in surface runoff reached 20-2530 ng L^{-1} (*12*). Renner (*13*) reported that the androgen (trenbolone acetate) hormone-adulterated runoff from cattle feedlots could significantly be decreasing fathead minnow egg production as well as affecting other aspects of endocrine function related to reproduction. Furthermore, Nichols et al. (*14*), Shore et al. (*15*), and Peterson et al. (*16*) have identified 17β-estradiol (highest concentration of 37.6 ng L^{-1}) in aquifers underlying areas where animal wastes have been applied.

Environmental Fate and Transport of Hormones in Soil-Water Systems

After steroidal hormones are released into the environment, they undergo several fate and transport processes, which have been reported by various researchers. They possess low volatility and water solubility, and are very hydrophobic. For example, 17β-estradiol has a water solubility reported to be 13 mg L^{-1} at 20°C, a vapor pressure of 2.3×10^{-10} mm Hg, and an octanol-water partition coefficient of 3.94 (*17*). These properties indicate that 17β-estradiol and testosterone would bind strongly to soil particles and not be dissolved or transported in soil water (*18, 19*).

Jacobsen et al. (*20*) conducted various laboratory studies to investigate the fate of 17β-estradiol and testosterone in different organic matrices (e. g. manure and biosolids) under aerobic conditions. Their result indicated that at the end of 6 d of incubation, 47% and 36% of testosterone in silt loam soil was mineralized to CO_2 in manured and unmanured treatments, respectively; however, no mineralization occurred in the sterilized soil. Different sterilization experiments indicated that there was no mineralization of testosterone and 17β-estradiol in sterile soil, which indicated that microbial processes play the predominant role in hormone degradation under aerobic conditions. Although some studies have identified the persistence of 17β-estradiol and testosterone in various soils, the fate and transport of these two hormones in the environment is still not fully understood, especially in sub-surface soil and water environments where limiting oxygen conditions often occur. Up to now, no such experiments have been done to study the fate of steroidal hormones under anaerobic conditions. Limited studies have been done on the persistence and sorption (*17, 24*) of estrogens and androgens in soil, biosolids, and sediment. 17β-Estradiol (75%) and testosterone (60%) were extensively mineralized to CO_2 at 1h in municipal treatment plant biosolids (*21*). Fifty percent of a 17β-estradiol dose dissipated from loam, sandy loam, and silty loam soils in under 0.5d, mostly converted to estrone and non-extractable residues (*22*), and similar dissipation results were observed for the oral contraceptive 17α-ethinyl estradiol except that only non-extractables were observed rather than extractable metabolites (*23*). Rapid sorption of 17β-estradiol (0.1 μg/mL) from the aqueous phase to sediment (3 g per 200 mL water) was observed (0.5h 4.0 – 9.4 μg/g/h; *17*), and similar reports of high affinity of 17β-estradiol to sediment have been reported elsewhere (*24*). However, little is known about the fate and transport of these hormones in the environment, therefore, our laboratory initiated a series of laboratory studies with 17β-estradiol and testosterone to investigate 1) their sorption, transport, and biodegradation, 2) develop models of these compounds in soil-water systems, and 3) describe their movement into soils potentially into surface and ground water resources.

Soil Sorption of Hormones

A series of batch experiments were conducted in the laboratory to determine the sorption of 17*β*-estradiol and testosterone in native soils and sand. Briefly, [4-^{14}C]-radiolabeled 17*β*-estradiol or testosterone (structures in Figure 1) was added to vials containing 8.0 ml of 0.01 M $CaCl_2$ to create solution concentrations of 0.138, 0.069, 0.015 and 0.0015 mg/L (17β-estradiol), or 0.738, 0.406, 0.0738, and 0.00738 mg/L (testosterone). Quantities of radioactivity ranged from 22,000 to 2,200,000 dpm/vial. These concentrations were chosen because they represent the range of concentrations reported in animal manures and biosolids applied to agricultural fields (*7, 11*). Portions of five different soils or sand (1.6 g) were added to the vials. The soils were Bearden-silty clay loam, Gardena-clay loam, Glyndon-sandy clay loam, LaDelle-silt loam, and Sioux-loam, which were collected in North Dakota, and represented a variety of native soil textures (Table I). These soil-water mixtures were agitated by rotation, and after 0.5, 1, 5, 24, 48, and 168 h, the supernatants (100 μL) were assayed for radioactivity after centrifugation at 1700 rpm (380 x g) for 20 min.

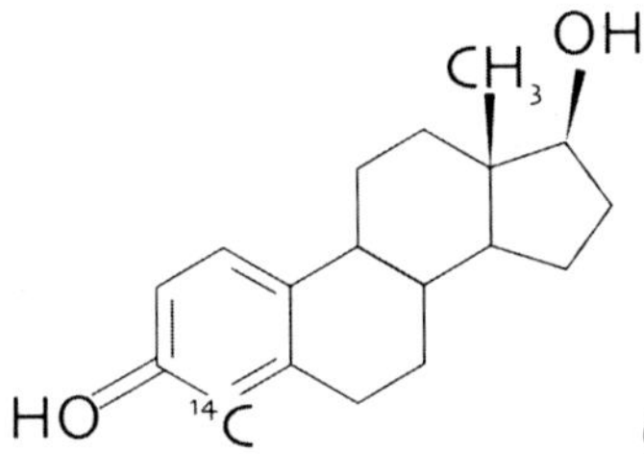

(A)

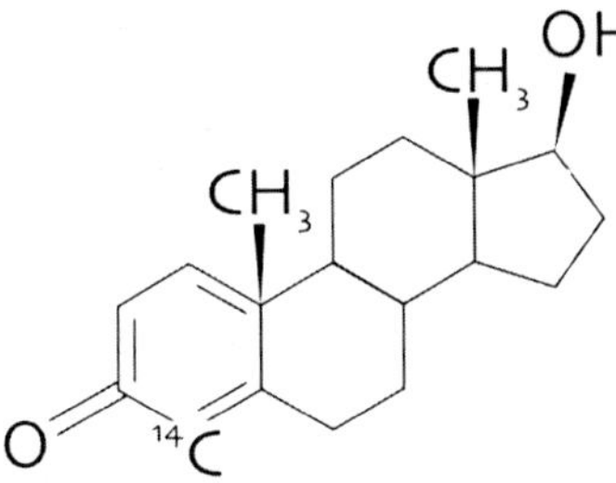

(B)

Figure 1. The structure of [4-^{14}C]17β-estradiol (A) and [4-^{14}C]testosterone (B) radiolabeled hormones used in the batch, soil miscible-displacement, and incubation experiments described in the present review to determine their environmental fate and transport.

Estradiol Results

17β-Estradiol sorption was very high for all soils measured (*18*), as indicated by high partition coefficients (Table II). Correlations of 17β-estradiol sorption to various soil fractions were determined for the initial aqueous concentrations of 0.0015 mg L^{-1} at 48 h. A general positive correlation between sorption and silt, clay and organic content was found (Table III). These correlations reflect a relationship between sorption and cation exchange capacity surface area, and the hydrophobic nature of the soils. A negative correlation existed between sorption and sand content.

Table I. Soil characteristics used for the sorption, miscible-displacement and incubation experiments described in this paper.

Soil Series	Taxonomic Description	Organic Matter Content (%)	Specific Surface Area[a] (m^2 g^{-1})	pH	Texture (Sand:Silt:Clay)
Bearden - silty clay loam	Fine-silty, mixed, superactive, frigid Aeric Calciaquoll	7.5	175	6.4	15:51:34
Gardena - clay loam	Coarse-silty, mixed, superactive, frigid Pachic Hapludoll	5.3	154	6.4	29:44:27
Glyndon - sandy clay loam	Coarse-silty, mixed, superactive, frigid Aeric Calciaquoll	3.3	123	7.9	52:27:21
LaDelle - silt loam	Fine-silty, mixed, superactive, frigid Cumulic Hapludoll	9.2	151	7.3	12:62:26
Sioux - loam	Sandy-skeletal, mixed, frigid Entic Hapludolls	7.5	106	7.8	46:34:21
Sand	-----	0.0	-----	n/a	100:0:0
Hecla-Hamar - loamy fine sand	Sandy, mixed, frigid Oxyaquic Hapludolls	2.2	-----	6.9	14:19:67

[a] Specific surface area indicates the surface area of a porous media contained in unit mass and is defined as: specific surface area = (surface area)/(mass of soil).

Testosterone Results

Soil sorption of testosterone increased for five hours, but then began to decrease thereafter (*19*). It was hypothesized that the observed desorption was caused by production of metabolites of higher polarity, which sorbed more weakly to the solid phase. As with 17β-estradiol, there were positive correlations of sorption to soil particle size, specific surface area, and organic matter content.

Table II. Organic-content normalized partition coefficients (Log_{oc}) and the calculated first-order degradation rate constants (μ) determined for 17β-estradiol and testosterone following miscible-displacement experiments in various soils. (Ref: *18, 19*)

	17β-estradiol		Testosterone	
	Log K_{OC}	μ (h^{-1})	Log K_{OC}	μ (h^{-1})
Bearden-Silt clay loam	6.02	1.494	3.30	0.404
Gardena-Clay loam	6.07	2.800	2.97	0.496
Glyndon-Sandy clay loam	7.18	0	4.56	0.600
LaDelle-Silt loam	5.85	0.480	3.03	0.493
Sioux-Loam	5.94	0.284	3.07	0.493

Fate and Transport of Hormones in Soil-Water Systems

The fate and transport of 17β-estradiol and testosterone was determined with soil types mentioned above (Table I). Each soil was packed into individual glass columns (8.4 x 15.2 cm, diameter x height) with stainless steel end caps. Each column was wetted with a weak salt solution (0.01M $CaCl_2$). A pulse of [^{14}C]17β-estradiol (0.65 μCi, 0.2mg) was applied to the surface of the soil column in 40 mL of 0.01 M $CaCl_2$, and eluted with the 0.01 M $CaCl_2$ solution for at least 7 to 12 relative pore volumes over 24h. The column effluent was fraction collected and analyzed for ^{14}C using liquid scintillation counting. Thin layer chromatography (TLC) analyses (5 x 20 cm silica gel, 1:1:2 tetrahydrofuran:ethyl acetate:hexane) were used to determine the presence of metabolites in the column effluent. At the end of the experiment, the distribution of resident ^{14}C in the column was determined by combustion analysis.

Table III. The linear regression parameters of sorbed concentrations of 17β-estradiol to soil fractions [=slope×(soil fraction)+intercept], and corresponding coefficient of determination (r^2) *(Ref: 18).*

Soil fraction	Slope (μg g^{-1})	Intercept (μg g^{-1})	r^2
%Sand	-0.000044	0.0076	0.90
%Silt	0.000058	0.0038	0.92
%Clay	0.000101	0.0037	0.46
%Organic Carbon	0.000287	0.0044	0.62

17β-Estradiol Results

There was a significant redistribution of 17β-estradiol-derived ^{14}C throughout the columns, although approximately 80% of the radioactivity was present in the top 3 cm of the column (Figure 2; *18*). Generally less than 1% of the applied dose eluted from the columns, and none was determined to be parent compound by TLC analyses. The effluent either contained a highly polar metabolite and/or sporadic detections of estriol. TLC analyses of the soil extracts from within each column (except Glyndon sandy-clay loam, for which no estrogen elution was observed) indicated that the majority of the sorbed ^{14}C was 17β-estradiol, sporadic detections of the more polar estriol, a second metabolite of very high polarity, and third metabolite which was identified as estrone. These results indicated that 17β-estradiol entered the soil columns, readily partitioned to the solid phase, and underwent rapid transformation to form three metabolites of different polarity. The lower polarity estrone and 17β-estradiol were adsorbed to the soil, while the higher polarity metabolites were more mobile in the column and more readily transported in the aqueous phase.

Testosterone Results

The majority (>80%) of the [^{14}C] recovered from the extraction from each soil column was detected in the top 3 cm of each column; however, 5 to 13% of the applied dose was also detected in the lower 3 cm of each column, depending on soil type (Figure 2; *19*). Therefore, testosterone demonstrated a high degree of redistribution throughout these soil columns. Unlike 17β-estradiol, testosterone as the dominant compound present in the column effluents, although the

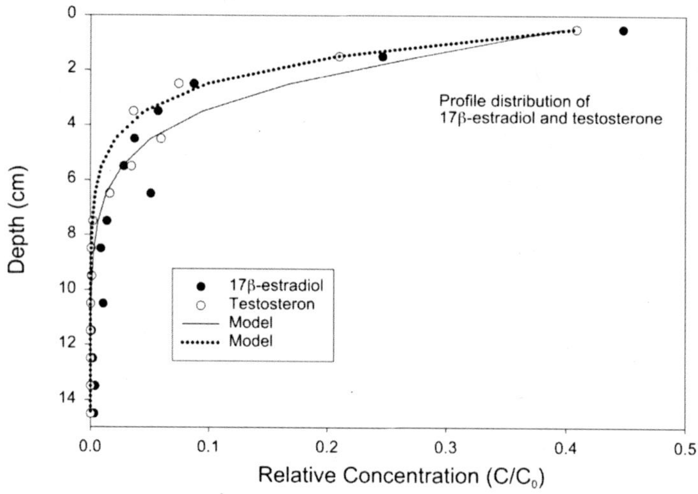

Figure 2. Depth profile of sorbed [^{14}C]17β-estradiol- or testosterone-derived radioactivity into Glyndon-sandy clay loam soil. The best-fit models of both hormones indicate that testosterone had a slightly greater ability to migrate through soil than did 17β-estradiol. Radioactivity at each depth was determined by combustion analysis (Ref: 26).

amount represented only a small portion of the applied dose. Sorption isotherms calculated for testosterone and then compared to other 17β-estradiol studies (*17, 18*) showed a lower solid phase sorption for testosterone than 17β-estradiol. The significance of these data is that they help explain field study results in which testosterone migrated deeper and possessed a higher mobility than 17β-estradiol. Complicating this picture, however, is the fact that the degradation rate constants (μ) are reported to be approximately four times as large for testosterone as for 17β-estradiol, i.e. $0.252 \pm 0.012\ h^{-1}$ (*21*) and $0.06 \pm 0.12\ h^{-1}$ (*25*), respectively. Therefore, the fate and transport behavior of testosterone compared to 17β-estradiol is that testosterone has the greater potential to migrate to groundwater, despite its higher inherent ability to be biologically transformed.

Incubation experiments

The persistence and fate of 17*β*-estradiol and testosterone were each investigated under the following four soil microcosms: (1) native soil under aerobic conditions, (2) native soil under anaerobic conditions, (3) sterilized soil under aerobic conditions, and (4) sterilized soil under anaerobic conditions. Aerobic conditions were maintained by continuous bathing with a stream of air, while a stream of helium maintained anaerobic conditions. The soil was a Hamar- mixed, frigid typic Endoaquolls (loamy fine sandy) with a bulk density of 1.54 g cm^{-3}, porosity of 0.42, 2.23% organic matter, and containing 14.0% clay; 19.0% silt; and 67% sand (Table I). Sterile soils were prepared by autoclaving for 40 min at 122°C. [4-^{14}C]17β-Estradiol or [4-^{14}C]testosterone in 0.01 M $CaCl_2$ was added to native and sterile soils and incubations were conducted for 5d at 25°C. Gaseous $^{14}CH_4$ and $^{14}CO_2$ were trapped in series with either Bray's solution (*27*) or 3N NaOH, respectively.

Water-soluble and acetone-soluble ^{14}C compounds were determined for each soil microcosm within minutes following the conclusion of the 5-day study period. Metabolite composition of the extracts was accomplished using TLC. Bound [^{14}C] was further fractionated as described by Kaplan and Kaplan (*28*) to determine the distribution among various organic matter fractions (e.g. humic acid, fulvic acid, and humin).

While only 6% of 17β-estradiol was mineralized to $^{14}CO_2$ in native soils under aerobic conditions in 120h, about 63% of the testosterone was mineralized at 5d (Table IV; *25*). The resistance of 17β-estradiol to mineralization when compared to testosterone was just as dramatic under anaerobic conditions at 120h, i.e. 0.9% and 46%, respectively. Furthermore, this mineralization of either 17*β*-estradiol or testosterone was biologically mediated, because it could be abolished in autoclaved soils regardless of oxidative state (Table IV). In native soils under anaerobic conditions, 2% of testosterone and no 17*β*-estradiol were degraded to $^{14}CH_4$ (Table IV). For testosterone, TLC analyses on the soil extracts indicated that >83% of the extractable testosterone was transformed to metabolites regardless of oxidative state for natural soils. However, when the soil was autoclaved, >87% of the extractable testosterone was parent compound. These values were >84% and >73%, respectively, for the 17β-estradiol experiments.

The first-order mineralization rate constants (k) of 17β-estradiol and testosterone also reflected the role of biological activity and oxygen status. The k values also indicated that testosterone was more readily mineralized than 17β-estradiol. The k values were determined by fitting the mineralization process equation, $\frac{\partial(^{14}CO_2)}{\partial t} = k\ [^{14}C](t)$, to the time(t)-series of $^{14}CO_2$ concentrations trapped in the NaOH solution. The k values for testosterone under aerobic and anaerobic conditions in natural soil were 0.012 h^{-1} and 0.004 h^{-1}, respectively. For 17β-estradiol, the aerobic and anaerobic k values for natural soil were 0.0006 h^{-1} and 0.0001 h^{-1}, respectively. The importance of biological activity for hormone mineralization was indicated by the lower k rates under anaerobic conditions, and by the fact that neither hormone was converted to CO_2 in autoclaved soil.

These findings were similar to those reported by Colucci et al. (*22*), who conducted a series of incubation experiments to investigate the persistence of hormones in soils under aerobic condition, and found that 8.2% of initial 17*β*-estradiol was mineralized to CO_2 at 30ºC at 72h. Jacobsen et al. (20) also reported that under aerobic conditions 47% and 36% of testosterone (in silt loam soil with 0.076 g g^{-1} moisture at 30°C) was mineralized to CO_2 in the manured and unmanured treatment, respectively, following 6 d incubations. Layton et al. (*21*) found that 70–80% of 17β-estradiol and 55-65% of testosterone was mineralized to CO_2 in biosolids under aerobic conditions, and the same research demonstrated that microbial populations clearly can adapt in different waste systems, since much lower mineralization was observed for 17β-estradiol when incubated with industrial waste (4% mineralization). Table IV shows that 50–73% of [^{14}C]-spiked17β-estradiol and its metabolites, and 19–54% of [^{14}C]-spiked testosterone and/or its metabolites were bound to humic substances. These findings indicated that assessing fate and risks of estrogenic and androgenic hormones in the environment can be overestimated, because reproductive hormones may be either sorbed or incorporated into humic substances in the soil, which may thereby reduce their bioavailability and toxicity.

Fate and Transport Process Modeling

A conceptual model of hormone fate and transport was beginning to emerge based on laboratory observations (*18, 19, 25*). Aqueous hormone inputs into soil could lead to one of two hormone fates, (a) aqueous-phase degradation to metabolites, and/or ultimate mineralization to CO_2, or (b) soil solid-phase sorption, which was initially reversible (water and/or acetone extractable), but ultimately becomes irreversible (non-extractable). This conceptual model could then be converted to multiple first order expressions, which simultaneously described the transformation, kinetic sorption, transport, and mineralization of each hormone. The mathematical model was tested against the behavior of each hormone in a native, undisturbed soil. The usefulness of the models was shown in being able to reasonably simulate a particular field condition.

Undisturbed core samples of Hamar soil series were collected using pre-cleaned stainless steel cylinders (15 x 30 cm). The end-capped core was pumped with 0.01 M $CaCl_2$ solution, and either [4-^{14}C]17β-estradiol or [4-^{14}C]testosterone was applied. Six column volumes were eluted, elution fractions

Table IV. Fractionation of [^{14}C]17β-estradiol and [^{14}C]testosterone, and their degradation-derived products in natural and autoclaved soils under aerobic and anaerobic conditions after 5-d incubation (Ref: *25*).

	Fraction of applied dose recovered			
	Aerobic		Anaerobic	
	Natural	Autoclaved	Natural	Autoclaved
	-----%Dose--------		------%Dose---------	
17β-Estradiol				
Trapped ^{14}C				
$^{14}CO_2$	6	0.2	0.9	0
$^{14}CH_4$	0	0	0	0
Extractable ^{14}C				
H_2O	2	2	2	1
Acetone	10	26	17	23
Subtotal	12	28	19	24
Non-extractable				
Humic acids	37	31	37	24
Fulvic acids	17	13	22	15
Humus	19	23	11	11
Subtotal	73	67	70	50
Total	**91**	**95**	**90**	**74**
Testosterone				
Trapped ^{14}C				
$^{14}CO_2$	63	0.2	46	0
$^{14}CH_4$	0	0	2	0
Extractable ^{14}C				
H_2O	0.4	0.2	0	0.5
Acetone	3	25	16	37
Subtotal	3.4	25.2	16	37.5
Non-extractable				
Humic acids	3	11	5	9
Fulvic acids	9	2	0	3
Humus	7	41	20	37
Subtotal	19	54	25	49
Total	**85**	**79**	**89**	**87**

were collected, and CO_2 was trapped in 3N NaOH solution by a column design which purged the effluent with helium. Post-study degradation of 17β-estradiol or testosterone was abolished by addition of antimicrobial mercuric chloride to both column eluants and to the column itself at the conclusion of the experiment. In the case of [^{14}C]17β-estradiol application, it was known that the only species expected in the eluant of the Hamar undisturbed column would be estrone and an unidentified polar metabolite (*18*), and in fact, these two metabolites were the only constituents of the effluent (*25*). The elution behavior of the radiolabeled metabolites was then successfully modeled using an advective dispersive model to describe the simultaneous transport and multiple first-order kinetic processes of reversible/irreversible sorption/desorption and transformation of the parent hormones and their metabolites. The kinetic transformation and sorption parameter estimates from the column studies agreed well with the independently determined values from the batch sorption studies. The modeled effluent concentration predictions of the hormone metabolites agreed well with the measured values. Also, confidence that the correct fate and transport processes were discerned and modeled was supported by the narrow confidence interval of each parameter. In addition, a forward prediction of estradiol-derived radioactivity in the sorbed fractions also produced an excellent description of observed data in both the reversibly (soil extracts) and irreversibly (non-extractable) sorbed compartments, thus exemplifying the confidence in the model and the quantification of the process parameters.

Conclusions and future studies

Although the comprehensive model proposed in this study was successful in describing the batch and column studies, there are still many uncertainties (e.g. unsaturated water flow, heterogeneous soil properties, background hormone concentration, facilitated transport, etc.) as these results are applied to a complex field scenario. Further laboratory studies are needed in order to address these and other issues so that a better understanding of hormone detections in the environment is obtained. As a result, the development of a more robust model will be beneficial for more accurate predictions and quantification of potential exposures. Field studies consistently detect hormones in both surface water (Baronti et al., 2000; Kolpin et al., 2002; Tabata et al., 2001), groundwater (Peterson et al., 2000), and soil (Finlay-Moore et al., 2000; Herman and Mills, 2003). However, despite the present state of research on hormones that show the potential for hormone transport into groundwater to be extremely low, a mechanism must still exist in the field, which can explain the consistent detection of these hormones in the environment, e.g. macropore flow or colloidal facilitated transport. Therefore, studies are currently underway in our laboratory to extend our laboratory understanding of hormone fate and transport to the field. A swine facility has been secured where the waste is being utilized as a fertilizer in adjacent fields. The predominant soil at the farm is a Hecla-Hamar loamy fine sand (sandy, mixed, frigid oxyaquic hapludolls) developed from glacial outwash and is 67% sand. The research at this farm is intended to produce information about land-applied, manure-borne hormones from various manure types. Lysimeters (60 cm depth) and wells (250 cm depth) have been installed in these fields both upgradient and downgradient of groundwater movement. Plans are in place to monitor 17β-estradiol and testosterone movement into these shallow groundwater systems during multiple growing seasons. SPE cleanup and LC/MS analysis methods have been developed to monitor hormone presence in ground water at the low ppt level. Data will analyze hormone migration into ground water versus drainage (rainfall) over time. Possible hypotheses of hormone transport will be tested including preferential flow, hormone-conjugate transport followed by reversion to parent compound, and movement of historical burdens of these hormones.

The use of trade, firm, or corporation names in this publication is for the information and convenience of the reader. Such use does not constitute an official endorsement or approval by the U.S. Department of Agriculture or the Agricultural Research Service of any product or service to the exclusion of others that may be suitable.

The technical assistance of C. Pfaff, B. Magelky, and G. Harrington is greatly appreciated. This research was funded in part by a National Science Foundation (NSF) Grant Nos. 0244169. and 0730492. Any opinions, findings, and conclusions or recommendations expressed in this material are those of the authors and do not necessarily reflect the views of the NSF.

References

1. U. S. Congress. 1999. 106th Congress. Bill H.R. 1712. To amend the Federal Water Pollution Control Act to authorize an estrogenic substances screening program. To amend the Federal Water Pollution Control Act to authorize an estrogenic substances screening program.
2. U. S. Congress. 2005. 109th Congress lst session. Bill S. 1500. Long title: A bill to authorize the National Institute of Environmental Health Sciences to develop multidisciplinary research centers regarding women's health and disease prevention and to conduct and coordinate a research program on hormone disruption, and for other purposes. Short title: Environmental Health Research Act of 2005.
3. Kolpin, D.W.; Furlong, E. T.; Meyer, M. T.; Thurman, E. M.; Zaugg, S. D.; Barber, L. B.; Buxton, H. T. Pharmaceuticals, hormones, and other organic wastewater contaminants in U.S. streams, 1999-2000: A national reconnaissance. *Environ. Sci. Technol.* **2002**, *36*, 1202 -1211.
4. Purdom, C. E.; Hardiman, P. A.; Bye, V. J.; Eno, N. C.; Tyler, C. R.; Sumpter, J. P. Estrogenic effects of effluents from sewage-treatment works. *Chem.Ecol.* **1994**, *8*, 275-285.
5. Routledge, E. J.; Sheahan, D.; Desbrow, C.; Brighty, G. C.; Waldock, M.; Sumpter, J. P. Identification of estrogenic chemicals in STW effluent: In vivo responses in trout and roach. *Environ. Sci. Technol.* **1998**, *32*, 1559-1565.
6. Baronti, C.; Curini, R.; D'Ascenzo, G.; Di Corcia, A.; Gentili, A.; Samperi, R. Monitoring natural and synthetic estrogens at activated sludge sewage treatment plants in a receiving river water. *Environ. Sci. Technol.* **2000**, *34,* 5059-5066.
7. Shore, L. S.; Shemesh, M. Naturally produced steroid hormones and their release into the environment. *Pure Appl. Chem.* **2003**, *75*, 1859-1871.

8. Shore, L. S.; Shemesh, M.; Cohen, R. The role of estradiol and estrone in chicken manure silage in hyperestrogenism in cattle. *Aust. Vet. J.* **1998**, *65*, 68-68.
9. Johnson, A. C.; Belfroid, A.; Di Corcia, A. C. Estimating steroid oestrogen inputs into activated sludge treatment works and observations on their removal from the effluent. *Sci. Total Environ.* **2000**, *256*, 163-173.
10. Desbrow, C.; Routledge, E. J.; Brighty, G. C.; Sumpter, J. P.; Waldock, M. Identification of estrogenic chemicals in STW effluent 1. Chemical fractionation and in vitro biological screening. *Environ. Sci. Technol.* **1998**, *32*, 1549-1558.
11. Lorenzen, A.; Hendel, J. G.; Conn, K. L. S.; Kwabiah, A. B.; Lazarovitz, G.; Massé, D.; McAllister, T. A.; Topp, E. Survey of hormone activities in municipal biosolids and animal manures. *Environ. Toxicol.* **2004**, *19*, 216-225.
12. Finlay-Moore, O.; Hartel, P. G.; Cabrera, M. L. 17 *β* -estradiol and testosterone in soil and runoff from grasslands amended with broiler litter. *J. Environ. Qual.* **2000**, *29,* 1604-1611.
13. Renner, R. Do cattle growth hormones pose an environmental risk? *Environ. Sci. Technol. Magazine* **2002**, *36*, 194A–197A.
14. Nichols, D.J.; Daniel, T. C.; Edwards, D. R.; Moore, P. A.; Pote, D. H. Use of grass filter strips to reduce 17β-estradiol in runoff from fescue-applied poultry litter. *J. Soil Water Conserv.* **1998**, *53,* 74-77.
15. Shore L. S.; Correll, D. L.; Chakraborty, P. K. In *Animal waste and the land-water interface;* Steele, K., Ed.; Relationship of fertilization with chicken manure and concentration of estrogens in small streams. Lewis Publications, Boca Raton, FL, 1995; pp. 155-162.
16. Peterson, E. W.; Davis, R. K.; Orndorff, H. A. 17 *β* -estradiol as an indicator of animal water contamination in mantled karst aquifers. *J. Environ. Qual.* **2000**, *29,* 826-834.
17. Lai, K.M.; Johnson, K. L.; Scrimshaw, M. D.; Lester, J. N. Binding of waterborne steroid estrogens to solid phases in river and estuarine systems. *Environ. Sci. Technol.* **2000**. *34*, 3890-3894.
18. Casey, F. X. M.; Larsen, G. L.; Hakk, H.; Šimůnek, J. Fate and transport of 17β-estradiol in soil-water systems. *Environ. Sci. Technol.* **2003**, *37*, 2400-2409.
19. Casey; F. X. M.; Hakk, H.; Šimůnek, J.; Larsen, G. L. Fate and transport of testosterone in agricultural soils. *Environ. Sci. Technol.* **2004**, *38*, 790-798.
20. Jacobsen, A.; Lorenzen, A.; Chapman, R.; Topp, E. Persistence of testosterone and 17*β*-estradiol in soils receiving swine manure or municipal biosolids. *J. Environ. Qual.* **2005**, *34*, 861-871.
21. Layton, A. C.; Gregory, B. W.; Seward, J. R.; Schultz, T. W.; Sayler, G. S. Mineralization of steroidal hormones by biosolids in wastewater treatment systems in Tennessee U.S.A. *Environ. Sci. Technol.* **2000**, *34*, 3925-3931.
22. Colucci, M. S.; Bork, H.; Topp, E. Persistence of estrogenic hormones in agricultural soils: I. 17*β*-estradiol and estrone. *J. Environ. Qual.* **2001**, *30*, 2070-2076.
23. Colucci M. S.; Topp E. Persistence of estrogenic hormones in agricultural soil: II. 17a-Ethynylestradiol. *J. Environ. Qual.* **2001**, *30*, 2077-2080.

24. Jurgens, M.D.; Johnson, A.C.; Williams, R.J. Fate and behavior of steroid oestrogens in rivers: Scoping study R&D technical report No. P161. Environment Agency:UK. **1999**.
25. Fan, Z.; Casey, F. X. M.; Hakk, H.; Larsen, G. L. Persistence and fate of 17β-estradiol and testosterone in agricultural soils. *Chemosphere* **2007**, *67*, 886-895.
26. Casey, F. X.; Larsen, G. L.; Hakk, H.; Fan, Z. The fate of manure-borne, land-applied hormones. 2006 ASA-CSSA-SSSA Meeting. Indianapolis, IN, Nov. 12-16, 2006.
27. Bray, G.A. A simple efficient liquid scintillator for counting aqueous solutions in a liquid scintillation counter. *Analyt. Biochem.* **1960,** *1*, 279-285.
28. Kaplan, D. L.; Kaplan, A. M. Thermophilic biotransformations of 2,4,6-trinitrotoluene under simulated composting conditions. *Appl. Environ. Microbiol.* **1982**, *35*, 949-954.
29. Fan, Z.; Casey, F. X. M.; Hakk, H.; Larsen, G. L. Discerning the fate and transport of testosterone in undisturbed soil: Model development and experimental evaluation. *J. Environ. Qual.* **2007**, *36*, 864-873.
30. Tabata, A.; Kashiwa, S.; Ohnishi, Y.; Ishikawa, H.; Miyamoto, N.; Itoh, M. Estrogenic influence of estradiol-17 β ; p-nonylphenol and bisphenol on Japanese Medaka (*Oryzias iatipes*) at detected environmental concentrations. *Water Sci. Technol.* **2001**, *43*, 109-116.
31. Herman, J. S.; Mills, A. L. Biological and hydrogeological interactions affect the persistence of 17 β -estradiol in an agricultural watershed. *Geobiology* **2003**, *1*, 141-151.

Chapter 4

Fate of Chlorate Salts Excreted from Animals

D. J. Smith[1], C. E. Oliver[2], H. Hakk[1], R. C. Anderson[3], and G. L. Larsen[1]

[1]USDA ARS, Biosciences Research Laboratory, 1605 Albrecht Blvd, Fargo, ND 58105-5674
[2]Department of Animal and Range Sciences, North Dakota State University, Fargo, North Dakota 58105
[3]USDA ARS, Food and Feed Safety Research, 2881 F&B Road, College Station, Texas 77845

Colonization of food animals with human pathogens such as *Escherichia* coli O157:H7 and *Salmonella* has been a bane to the food animal industry for years. In the United States, *E. coli* O157:H7 is considered an adulterant, and meat animal products contaminated with the pathogen must be disposed at great cost to the producer. A new pre-harvest food safety technology, based upon the oral administration of chlorate salts to food animals, selectively eliminates gram-negative pathogens such as *E. coli* O157:H7 and *Salmonella* species in economically important livestock species. Residue trials have demonstrated that chlorate salts are rapidly and extensively eliminated as parent chlorate in the excreta of cattle, swine, and broilers. Under proposed dosing regimens, environmental burdens of excreted chlorate could be significant, especially in areas of intensive food animal production. Chlorate is highly water soluble and its transport through soil columns was not restricted by a variety of soils composed of mixtures of loams, clays, or sands. In soil column studies, chlorate was sparingly converted to chloride ion by a few soil types, but in batch studies utilizing a mixture of soil, urine, and feces, chlorate was rapidly reduced to chloride under both aerobic and anaerobic conditions. Rapid reduction under in vitro conditions implies a short environmental half-life of chlorate for most animal production waste-management systems.

Human Pathogens in Food Animals

Recent estimates have placed the rate of *Salmonella* and *Escherichia. coli* O157:H7 infections within the United Sates at greater than 1.4 million cases per year (*1, 2*) resulting in over 450 deaths. Outbreaks and sporadic episodes of both pathogens have been perceived to be associated with food animals or with food animal production systems, even though raw produce and unpasteurized juices are important sources of food-borne pathogens (*3*). For consumers in the United States, the biggest risk factors associated with sporadic salmonellosis episodes were international travel (odds ratio of 62.3), having a pet lizard (odds ratio 3.0), having a bird in the home (odds ratio of 2.7), and eating undercooked eggs (risk factor 2.1) (*4*). The largest risk factors for contracting sporadic *E. coli* related illness were associated with drinking untreated surface waters, recent association with a farm, and eating undercooked hamburger (*5*), whereas outbreaks of *E. coli* illnesses are also associated with a variety of sources, such as meat, unpasteurized milk and juice, produce, contaminated municipal water sources, recreational waters, contact with infected persons, and contact with cattle (*5*).

Although several sources of, and risk factors for, *Salmonella* and pathogenic *E. coli* strains are encountered by today's consumers, a perception exists that consumption of food animal products conveys the greatest risks for illness. Misconceptions are common. For example, in a review article written by a physician, for physicians, it was stated that "Recent estimates have suggested that 50% of all feed lot (sic) cattle are colonized with *E. coli* O157:H7 and up to 90% of lots of hamburger are contaminated" (*6*). Although ruminant animals do serve as reservoirs of *E. coli* O157:H7, the overall prevalence of the pathogens in feedlot animals is not 50% (*7*) and according to the USDA FSIS (*8*), less than 1%, not "up to 90%", of hamburger lots tested positive for *E. coli* during the time period when the cited article was published (Figure 1).

Misconceptions notwithstanding, food-animal production systems are major reservoirs of both Salmonella and shigella toxin-producing E. coli strains (*9, 10*). Salmonella presence in poultry and contamination of poultry products by Salmonella species remains a significant problem, because contamination of ground poultry products has actually increasing in recent years (*11*). The US FDA considers E. coli O157:H7 an "adulterant" and food animal products contaminated with the pathogen requires the immediate destruction of contaminated carcasses or product lots. Because in some large processing establishments a "lot" of ground beef may be 10,000 pounds, the routine presence of *E. coli* O157:H7 in products could soon bankrupt a company. In response to the issue of food-borne pathogen transmission, animal producer groups in the United States have invested well over 20 million dollars for food safety research in recent years. Indeed, some progress is being made; the United States Department of Agriculture announced in 2005 that the "Healthy People 2010" goals for reducing the incidence of *E. coli* O157 infections had been met, 6 years ahead of schedule (*12*). The FSIS announced in 2007 that the numbers of *E. coli* outbreaks had increased in 2006, mainly due to large incidences involving contaminated lettuce and spinach.

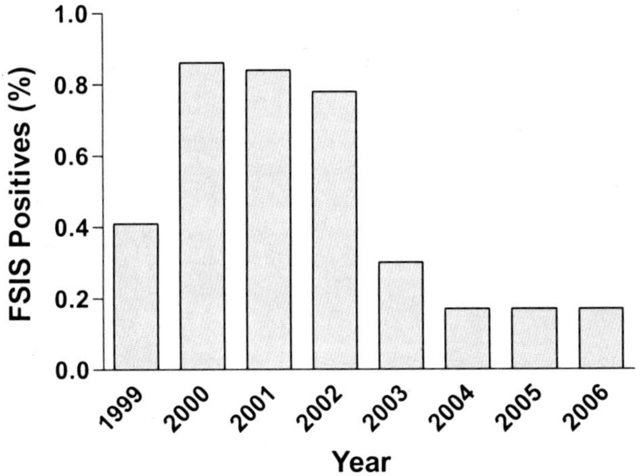

Figure 1. Prevalence (% positives) of E. coli O157:H7 in production lots of ground beef inspected by the US Food Safety and Inspection service from 1999 to 2006 (8). In 2006, 20 lots of ground beef tested positive for the pathogen out of a total of 11,779 lots tested.

Chlorate as a Pre-Harvest Food Safety Tool

Within the Agricultural Research Service (ARS), the research arm of the US Department of Agriculture, a concerted effort has been underway to develop pre-harvest food safety tools that will decrease the incidence of pathogens in live animals. Reducing the incidence or the prevalence of pathogens prior to animal harvest ("on farm" interventions) has been targeted because it has great potential to significantly reduce human exposures to pathogens such as *E. coli* and *Salmonella* via environmental or foodborne exposures (*13, 14, 15*).

A promising pre-harvest food safety technology has recently been developed at the USDA ARS laboratory in College Station, TX. The technology is based on the use of chlorate salts to selectively kill Gram-negative pathogens that express respiratory nitrate-reductases (*16*). Examples of such pathogens are both *E. coli* O157:H7 and *Salmonella*. Hackenthal et al. 1964, as cited by Stouthammer (*17*), have hypothesized that chlorate (ClO_3^-) exerts its lethal effect on the pathogens after it is chemically reduced to chlorite (ClO_2^-) by respiratory nitrate reductase (Figure 2); chlorite production by micro-organisms was reported as early as 1925 (*18*). Selectivity occurs because the respiratory nitrate reductase is an intracellular enzyme, and because respiratory nitrate reductases are not commonly expressed in enteric bacteria.

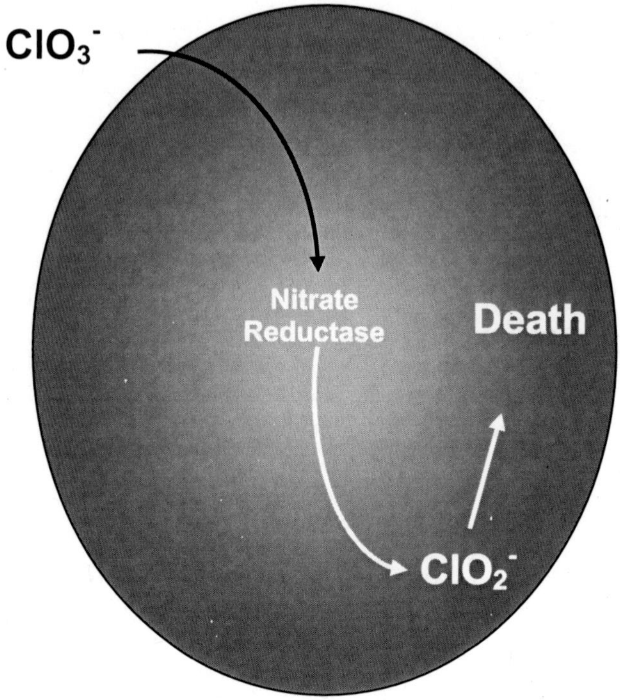

Figure 2. Idealized schematic of the mechanism through which chlorate is hypothesized to have lethal effects in nitrate-respiring bacteria. Intracellular respiratory nitrate is capable of reducing chlorate to chlorite. Chlorite is highly toxic to bacteria and kills those bacteria in which it is produced.

Chlorate salts have been shown to be effective at reducing the numbers of Gram-negative pathogens in live animals for the most economically important food animal species in the United States. Table 1 summarizes published studies demonstrating the effectiveness of chlorate in broilers, cattle, laying hens, sheep, swine, and turkeys. Although a number of dosing regimens have been used, experimental chlorate products (ECP) have typically been delivered to animals in either drinking water formulations or as proprietary feed additive formulations. Doses have ranged fairly widely from 14 mg/kg body weight in swine (*19*) to over 4 g/kg in sheep (*20*). It is expected that should a commercial product be approved for use, the targeted dose for most species will be less than 100 mg/kg body weight.

Table 1. Activity of Chlorate Salts against Gram-negative Pathogens

Species	*Chlorate Dose*[1]	*Vehicle*	*Exposure Period*	*Activity Against: O157*	*Salmonella*	*Ref.*
Swine	14.1 mg/kg	Water	gavage, 2x		✓	19
	21.2 mg/kg	Water	gavage, 3x	✓		51
	42.5 mg/kg	Water	gavage, 3x	✓		51
	30-40 mg/kg	Water	24 h		✓	52
	30-40 mg/kg	Water	36 h		✓	52
	30-40 mg/kg	Water	36 h		✓	52
	60-80 mg/kg	Water	36 h		✓	52
	80 mg/kg	Water	5 d		✓	53
	~50 mg/kg[2]	Feed	28 d		✓	54
Cattle	1.25 mM in vitro	NA	24 h	✓		55
	5.0 mM in vitro	NA	24 h	✓		55
	525 mg/kg	Water	24 h	✓		56
	147 mg/kg	Water	12 h	✓		57
	100 mg/kg feed, 147 mg/kg water	Feed+Water	24 hr	✓		57
	100 mg/kg feed, 144 mg/kg water	Feed+Water	5 d	✓		57
	498 mg/kg feed, 127 mg/kg water	Feed+Water	24 hr	✓		57
	100 mg/kg	Feed	24 hr	✓		57
	100 mg/kg per day	Feed	5 d	✓		57
	497 mg/kg feed	Feed	24 hr	✓		57
Sheep	1100 mg/kg	Feed	Single	✓		20
	2200 mg/kg	Feed	Single	✓		20
	4400 mg/kg	Feed	Single	✓		20
	400 mg/kg	Water	24 hr		✓	58
Turkeys[3]	137 mg/kg	Water	38 h		✓	59
	222 mg/kg	Water	39 h		✓	59
	490.2 mg/kg	Water	40 h		✓	59
	816 mg/kg	Water	41 h		✓	59
	490.2 mg/kg	Water	14 h			59
	490.2 mg/kg	Water	26 h			59
	490.2 mg/kg	Water	38 h			59
Broilers[4]	337 mg/kg	Water	48 h		✓	60
Broilers[5]	148 mg/kg	Water	10 h		✓	61
	232 mg/kg	Water	10 h		✓	61
	710 mg/kg	Water	10 h		✓	61
	24 mg/kg	Water	10 h		✓	61
	138 mg/kg	Water	24 h		✓	61
	422 mg/kg	Water	48 h		✓	61
Hens[6]	282 mg/kg/d	Water	9 d		✓	62

[1]Approximate dose, mg/kg body weight
[2]Estimated intake assuming 0.6 kg/d feed intake (800 ppm dietary chlorate) for 15 kg hogs
[3]Assume a body wt of 16 kg
[4]Assume a body wt of 2.5 kg
[5]Assume body wts of 2 kg
[6]Assume a water intake of 11 mL/h and a body weight of 1.5 kg

Chlorate Metabolism and Disposition in Food Animals

Because chlorate salts have broad activity against Gram-negative pathogens across species and because of the potential impact that chlorate could have on the production of pathogen free or pathogen "lite" food animals, further development of chlorate as a pre-harvest food safety tool has been in process. After the demonstration of a technical effect in live animals, one of the first requirements in the development of a new feed or water additive is to determine whether tissue residues will preclude its use for food safety reasons. To this end, metabolism and residue studies of chlorate have been conducted at the USDA Biosciences Laboratory in Fargo, North Dakota. These studies have demonstrated that tissue residues of chlorate will probably not limit its development as a feed or water additive for use in cattle (*21*), swine (*22*), or broilers (*23*). Regardless of species, residues of chlorate in edible tissues are typically less than 100 ppb when moderate doses (40 mg/kg) were used with a 24-hour withdrawal period. For most species and tissues, residues fell under provisional Safe Tissue Concentrations (STC) provided by the US FDA Center for Veterinary Medicine even when high doses were used. In each of these studies, which used ^{36}Cl-labeled chlorate, the only metabolite of chlorate detected in edible tissues or in excreta was chloride ion. Because chloride is a naturally occurring nutrient, its formation and presence in tissues of chlorate treated animals is considered to be of no risk to consumers of chlorate treated animals.

The finding that chloride was the sole metabolite of chlorate was in conflict with reports published in the early 1980's (*24, 25*) indicating that chlorite, a potentially toxic intermediate of chlorate, was formed from chlorate in rats and that it was stable enough to be excreted in appreciable quantities. Because chlorite is biologically active, the US-FDA established provisional STC for chlorite as well as chlorate (Dubek et al., personal communication). No chlorite, however, has been detected in excreta or tissues of non-ruminant animals such as swine or poultry (*22; 23*) or in ruminant animals (*21, 26*). Further, when metabolism and disposition studies with chlorate were conducted in rats, Hakk et al (*27*) were unable to replicate the work of Abdel-Rahman (*24, 25*) by finding chlorite in urine or tissues of rats. Using synthetic [^{36}Cl]chlorite standards, Hakk et al. (*27*) demonstrated that the analytical methods used by Abdel Rahman et al (*25*) produced artifacts leading to the erroneous conclusions that chlorite was a stable metabolic end-product of chlorate metabolism in non-ruminants. Indeed, Hakk et al. (*27*) demonstrated that chlorite is not stable in rat or bovine urine. In addition, Oliver et al. (*28*) demonstrated that chlorite has a half-life of only 4.5 min in live ruminal fluid. Thus, residue and metabolism studies have clearly indicated that chloride is the only chlorate metabolite having food-safety or environmental relevance.

Table 2. Elimination of Chlorate from Cattle, Swine, Broilers and Rats (% of total [^{36}Cl]chlorate dose)

Species	*Dose*	*Time of Collection (hours)*								*Total*	*Ref.*
		0-12	*12-24*	*24-30*	*24-36*	*36-48*	*48-54*	*48-56*	*48-72*		
	(mg/kg bw)										
Cattle	21	1.1	2.2		2.8	0.7				6.8	21
	42	3.4	5.9		2.8	1.2				13.3	21
	63	9.1	7.2		5.9	3.2				25.3	21
	63	6.7	1.8		8.1	7.8		10.5		34.9	26
	131	12.4	6.5		12.3	6.0		8.3		45.5	26
Swine	20	50.8	26.4	3.9						81.1	22
	40	62.7	12.5	8.4						83.6	22
	60	55.1	25.5	2.9						83.5	22
Broilers[1]	164	14.2	7.4		14.1	6.7	0.2			42.6	23
	292	7.7	9.6		16.2	6.9	0.1			40.5	23
	407	6.1	10.8		16.8	2.4	0.1			36.2	23
Rats	3	72.1	75.6			76.3			76.5	76.5	27

[1]Data are estimated for the 0-12, 12-24, 24-36, and 36-48 hour time periods; the percentage excreted as parent chlorate was only measured the 48-54 hour excreta. The percentages of the total radioactive residue that was excreted as parent chlorate were used to estimate the total amount of chlorate excreted at the earlier time points.

These metabolism and disposition studies have also shown that parent chlorate is extensively absorbed, and depending upon species, substantial percentages of parent chlorate are excreted in urine. Table 2 shows that 7 to 84 percent of the total amount of dosed chlorate was eliminated intact from broilers, non-ruminants, and ruminants. The proportion of parent chlorate excreted was highly dependent upon the species tested with non-ruminant animals excreting large amounts of unchanged chlorate in the urine. In cattle intra-ruminally dosed with [^{36}Cl]chlorate, absorption of radioactivity was rapid with a half-life of 0.5 to 1 h (*29*), but the majority of the radioactivity in serum was in the form of chloride ion. Other studies have shown that the rumen has a great capacity for the reduction of chlorate to chloride (*28*), but whether the reduction was chemically mediated or through bacterial action remains unknown. From the standpoint of chlorate's efficacy against gastrointestinal pathogens, its absorption from the gastrointestinal lumen, or its biological or chemical reduction in the gastrointestinal tract, removes active ingredient (chlorate) from its intended site of action and likely reduces efficacy at distal portions of the gastrointestinal tract. Clearly there is a need for chlorate formulations that will allow the efficient delivery of the anion to the lower gastrointestinal tracts of both ruminant and non-ruminant animals.

Environmental Fate of Chlorate Salts

Rapid absorption and excretion of parent chlorate by target animals has the potential to introduce significant quantities of chlorate into the local environments of feedlots, poultry houses, and(or) swine operations. For example, if a pork producer used 40 mg/kg of chlorate in pre-market hogs during the 24-72 hour period prior to market, a typical market hog (118 kg) would receive approximately 4.7 g of chlorate; of this, approximately 4.0 grams would be excreted in the urine. In 2003, about 80% of hogs marketed in the United States were produced by operations that annually marketed 10,000 or more animals, with about 40% of all pork produced by operations of 500,000 or more hogs marketed per year (*30*). For a pork operation of 20,000 animals, the annual amount of chlorate excreted could exceed 94 kg and for a facility that handles 400,000 animals annually, as much as 1,880 kg of chlorate could be excreted.

Although chlorate is reduced to chloride to a much greater extent in cattle than it is in other species, significant quantities of parent chlorate could enter waste streams in cattle production facilities that use chlorate salts. For example, a 590 kg market steer receiving 40 mg/kg of chlorate and excreting 13% of the oral dose as intact chlorate would eliminate over 3 grams of chlorate within 48-hours of treatment. In a feedlot that handled 100,000 animals per year, the cumulative burden of chlorate would be about 300 kg of parent chlorate per year. If the amount of intact chlorate increased to 30% of the dose, the corresponding burden for a feedlot of 100,000 animals would be 708 kg. With these figures in mind, and with the knowledge that intensive food-animal production systems raise large numbers of animals on a fairly small amount of land, the fate of chlorate after elimination from the animal is of interest.

Although chlorate has been used as an herbicide for over a century (*31*) very little information is available in the open literature on the transport and fate of chlorate in soils. As early as 1928, Åslander (*32*) demonstrated that chlorate was readily transported through soils, but stated that chlorates decomposed "rather slowly", being measured in weeks rather than days or hours. Decomposition of chlorate increased the concentrations of chloride in their test systems, presumably through bacterial action (*32*). One widely cited paper (*33*) indicates that under "favorable moisture conditions", which were undefined by the author, chlorate "completely" decomposed over a week's time at "summer temperatures". The *Pesticide Manual* (*34*) indicates that chlorate survives in soil for "0.5 to 5 years" depending upon a number of variables such as application rate, soil organic matter, moisture, and temperature. More recent studies (*35, 36*) have shown that chlorate salts used to induce the flowering of the Asian Longan trees survived from 1 to > 8 weeks depending upon a number of variables, including prior chlorate treatment. Chlorate disappearance was most rapid on soils without prior exposure to chlorate.

Industrial use of chlorine dioxide for the bleaching of pulp wood generates appreciable quantities of waste waters containing chlorate with concentrations as high as 115 mg/L (*37*). To remediate these waters, effluent treatment lagoons have been developed that utilize bacteria for the reduction of chlorate to chloride ion. Chlorate reduction in such test systems has been measured at rates as high as 3.2 mmol L^{-1} h^{-1} (267 mg L^{-1} h^{-1}); using test bioreactors, the complete removal of chlorate was accomplished with hydraulic retention times of 6 h (*38*). Thus, the environmental stability of chlorate varies widely from an estimate of 5 years to as little as 6 hours, depending upon the environment in which chlorate is found.

For animal agriculture, the excretion of chlorate has environmental implications in a number of areas. First, the sodium salt of chlorate has a solubility of 790 mg/mL in ice-cold water and 2.3 g per mL at 100° C (*39*). Chlorate's high water solubility renders it susceptible to leaching and/or being present in run-off water after rain events. Chlorate does not bind to soil and is rapidly transported through soils. Figure 3 shows the mean transport of radioactivity dosed as [^{36}Cl]chlorate through soil columns (8.5 x 15 cm) composed of loam, silty-loam, and clay-loam soil types. Radiochlorine was not bound to soil particles and there was quantitative recovery of radioactivity from each soil. The potential for leaching is significant for areas in which well water is the major source of water.

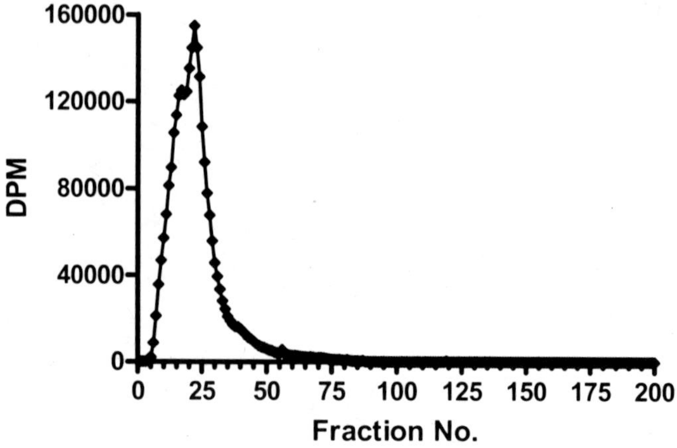

Figure 3. Mean transport of [^{36}Cl]chlorate salts through soil columns packed with loam (n=2), silty-loam (n=2), and clay-loam (n=2) soils. Soil columns (8.5 x 15 cm) were packed with 1 kg of soil and were hydrated with a 0.01 M $CaCl_2$ solution. Each column was loaded with 7 mg of sodium [^{36}Cl]chlorate of 98.1% radiochemical purity (1.9% Na[^{36}Cl]). Columns were eluted with 0.01 M $CaCl_2$ and eluate was collected into fractions of approximately 25 mL each. Radioactivity in an aliquot of each fraction was determined by liquid scintillation counting.

Although the solubility of chlorate is high, it is susceptible to either chemical or biological reduction during the transport process through the soil column. For example, Figure 4 shows the composition of radioactivity applied to soil columns as a function of fraction number collected from the column. Data presented in Figure 4 were compiled from the analysis of various fractions eluted from soil columns containing either loam, silty-loam, or clay-loam soils that had been loaded with 7 mg of [^{36}Cl]chlorate (98.1% radiochemical purity). The data clearly show that as the elution volume increases, the percentage of radiochlorine present as chlorate decreases. Although these data indicate that the majority of chlorate is eluted unchanged, the capacity of chlorate transformation in a rapidly flowing soil system was demonstrated.

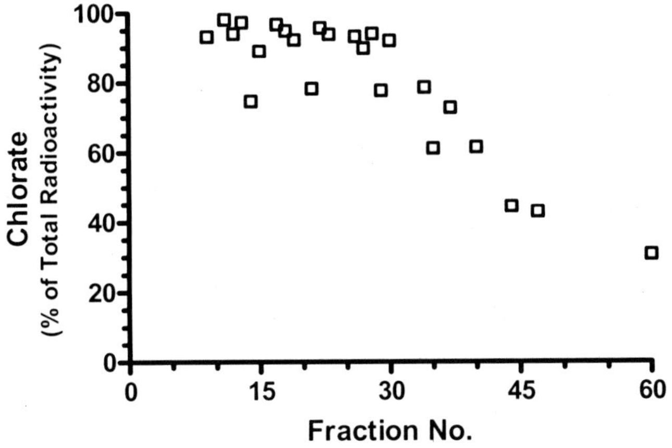

Figure 4. Variation in chemical composition of radiochlorine applied to soil columns as [^{36}Cl]chlorate (98.1% radiochemical purity) and eluted from loam (n=2), silty-loam (n=2), and clay-loam (n=2) soils columns using 0.01 M $CaCl_2$. Data are expressed as the percentage radioactivity that eluted as parent chlorate as a function of the fraction number collected from soil columns (see Figure 3). As fraction number increased, the percentage of the soil radioactivity converted to chloride ion increased indicating that either biological or chemical reduction of chlorate was occurring, even when chlorate was rapidly transported through the soil.

In addition to the concern that chlorate could leach into ground water, there is also concern that chlorate salts excreted from animals could have secondary effects after manures are applied to fields. Chlorate salts have been use for over a century as weed killers and defoliants (*31, 34*), but chlorate has biological activities in plants that extend beyond herbicidal activities. For example, chlorate salts applied to the root or foliage of the tropical Longan tree (*40, 41*) induces flowering, enabling the tree to bear fruit year around rather than on an annual basis. Additionally, recent work has shown that application of 8 mM chlorate (851 mg/L sodium chlorate equivalents) to orchid plants increases the number of floral buds, and increased the time to flowering, with the subsequent result of reducing flower size (*42*). Therefore, application of manure containing appreciable quantities of chlorate could adversely affect growth of plants.

Concentrations of chlorate excreted into urine of treated animals have varied widely, depending upon dose and time of urine collection after dosing. In swine, urinary concentrations of parent chlorate have exceeded 2000 mg/L (*22*) and in cattle dosed with large amounts of chlorate, urinary concentrations of parent chlorate have exceeded 5,000 mg/L (*26*). Because feedlot soils and animal waste systems are substantially different than agricultural soils and are substantially different than waste streams from pulp beach bleaching plants, a series of experiments were set up to measure the longevity of chlorate in beef cattle wastes. To this end, mixtures of 1 g of soil; 1 g of dried, ground (2 mm) cattle feces, and 1 mL of cattle urine containing, 0, 50, 100, or 200 ppm of

[^{36}Cl]chorate, were combined with autoclaved or live buffered cattle feces extracts containing approximately 1 x 10^8 colony forming units of *E. coli* O157:H7 and *Salmonella enterica* serotype Typhimurium. Vials were incubated under aerobic or sealed anaerobic conditions at 5, 20, or 30 °C for 0, 0.5, 1, 3, 7, 14, or 28 days. At sample harvest, vials were diluted with sterile phosphate-buffered saline, centrifuged, and the supernate harvested. Supernate was assayed for *E. coli* and *Salmonella* (results not reported) and for chlorate. Soluble silver [^{36}Cl]chlorate salts were analyzed after precipitation of insoluble silver [^{36}Cl]chloride salts with silver nitrate (*28*).

Figure 5 shows the degradation of chlorate by *Salmonella enterica* and *E. coli* O157:H7, and by a mixture of *Salmonella enterica, E. coli* O157:H7 and fecal bacteria. Regardless of whether anaerobic or aerobic conditions were present, chlorate was rapidly reduced to chloride ion at 20 and 30 °C with chlorate half-lives being less than 25 hours. At 5 °C, chlorate half lives ranged from 3 to 15 days. Under the experimental conditions used in this study, chlorate was not completely reduced to chloride in any of the incubations, reaching an asymptote at 18-20% of the original starting material. An explanation for the cessation of chlorate reduction is not readily apparent, but similar result of chlorate reduction by mixed cultures from activated sludge occurred when the electron donor, acetate, had been completely consumed (*43, 44*). Conversely, chlorate reduction was complete with the addition of excess electron donor (*43, 44*). Similar results for the degradation of chlorate occurred when chlorate was incubated with a "fertile soil" –soil on which a sheep feedlot previously existed– that was either supplemented with dextrose or not supplemented. Under the conditions of dextrose supplementation, chlorate (400 ppm) was completely reduced, but when dextrose was not added, 5 to 27.5% of the chlorate remained (*45*). For the results presented in Figure 5, pure bacteria or mixed bacterial cultures were employed, but nutrients available for growth and metabolism by the bacteria were limited to those available in 1 mL of cattle urine, 1 mL of phosphate buffered saline cattle feces extract, and 1 g of dried, ground manure. In the incubations in which *E. coli* and *Salmonella* had to compete with fecal bacteria, no growth of either *E. coli* or *Salmonella* occurred (data not shown), but in incubations in which only *E. coli* and *Salmonella* were added, bacterial growth was about 1 to 1.5 log units over a 12 to 24 hour time period (data not shown), thereafter bacterial numbers declined. Thus, bacteria were rapidly growing and it is likely that the reduction of chlorate ceased due to a limiting nutrient.

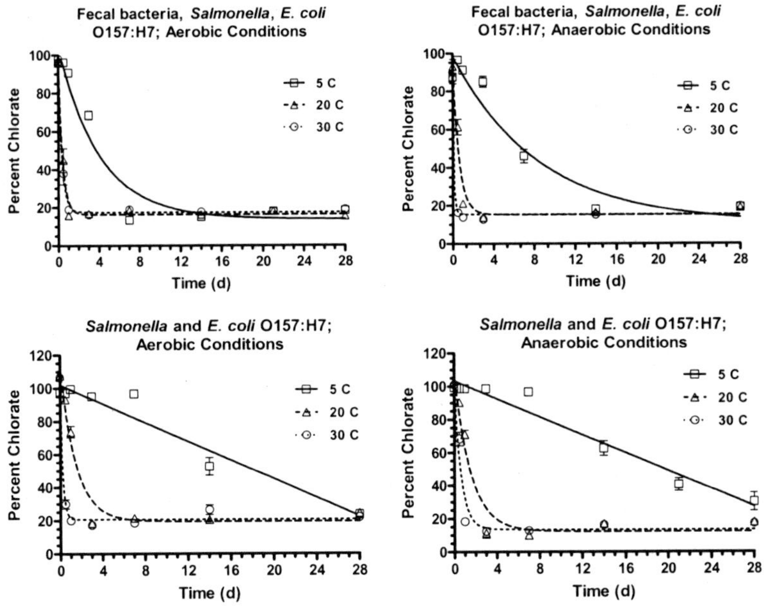

Figure 5. Chlorate degradation in a urine, feces, and soil mix by Salmonella, E. coli O157:H7, and fecal bacteria from cattle (top panel) or by a mix of Salmonella and E. coli O157:H7 (bottom panel) under aerobic (left panels) and anaerobic (right panels) conditions at 5, 20, and 30 °C.

Aerobic reduction of chlorate was rapid and extensive at 20 and 30 °C (Figure 5), regardless of the presence of fecal bacteria. Although the redox potentials of the aerobic and anaerobic systems were not measured, it can be assumed that in the anaerobic system, the redox potential favored chlorate reduction. In the aerobic system, however, the reduction of chlorate is counterintuitive, since one would expect that other electron acceptors would be more readily available. Shrout and Parkin (*46*), however, conducted a series of studies in which perchlorate was reduced to chloride by a mixed lactate-enriched culture with redox potentials as great as +180 mV and in cultures with dissolved oxygen concentrations as great as 4.8 mg/L. Thus, while it may be surprising that aerobic degradation of chlorate occurred, it is not unprecedented.

Because most modern animal production facilities have either lagoon or composting capabilities, and because (per)chlorates are rapidly degraded to chloride by bacteria in environments ranging from those described as pristine to industrial lagoons (*47, 48, 49, 50*), the likelihood that significant quantities of chlorate would survive degradation that would occur in the feedlot and a lagoon system is low. In support of this idea are the studies of Nelson (*45*) who showed that in soybean fields treated with 272 kg/acre of chlorate and fertilized with sheep manure chlorate reduction was complete. In contrast, similarly treated fields fertilized with sodium nitrate or ammonium sulfate, always had measurable quantities of chlorate. Because Nelson used a heavy chlorate application rate (273 kg per acre), measured chlorate 4 months after its application, and had microbial loads much less than would be expected in lagoon systems, it is reasonable to expect that in commercial settings chlorate would not survive for an extended time in a lagoon. Nevertheless, further research specifically investigating the fate of chlorate in lagoon systems will test the veracity of this hypothesis.

The use of trade, firm, or corporation names in this publication is for the information and convenience of the reader. Such use does not constitute an official endorsement or approval by the U.S. Department of Agriculture or the Agricultural Research Service of any product or service to the exclusion of others that may be suitable.

The technical assistance of D. Ellig, J. M. Giddings, J. Holthusen, B. Magelky, C. Pfaff, and C. Thomas is greatly appreciated. This research was funded in part by beef and veal producers and importers through their $1-per-head checkoff and was produced for the Cattlemen's Beef Board and state beef councils by the National Cattlemen's Beef Association.

References

1. Voetsch, A. C.; Van Gilder, T. J.; Angulo, F. J.; Farley, M. M.; Shallow, S.; Marcus, R.; Cieslak, P. R.; Deneen, V. C.; Tauxe, R. V. FoodNet estimate of the burden of illness caused by nontyphoidal *Salmonella* infections in the United States. *Clin. Infect. Dis.*, **2004**, *38(Suppl. 3)*, S127-134.
2. Mead, P. S.; Slutsker, L.; Dietz, V. McCaig, L. F.; Bresee, J. S.; Shapiro, C.; Griffin, P. M.; Tauxe, R. V. Food-related illness and death in the United States. *Emerg. Infect. Dis.*, **1999**, *5*, 607-625.
3. Burnett, S. L.; Beuchat, L. R. Food-borne pathogens. Human pathogens associated with raw produce and unpasteurized juices, and difficulties in decontamination. *J. Ind. Microbiol. Biotechnol.*, **2001**, *27*, 104-110.
4. Marcus, R.; Varma, J. K.; Medus, C.; Boothe, E. J.; Anderson, B. J.; Crume, T.; Fullerton, K. E.; Moore, M. R.; White, P. L.; Lyszkowicz, E.; Voetsch, A. C.; Angulo, F. J. Re-assessment of risk factors for sporadic *Salmonella* serotype Enteritidis infections: a case-control study in five FoodNet Sites, 2002-2003. *Epidemiol. Infect.* **2006**, *135*, 84-92.
5. Voetsch, A. C.; Kennedy, M. H.; Keene, W. E.; Smith, K. E.; Rabatsky-Ehr, T.; Zansky, S.; Thomas, S. M.; Mohle-Boetani, J.; Sparling, P. H.; McGavern, M. B.; Mead, P. S. Risk factors for sporadic Shiga toxin producing *Escherichia coli* O157 infections in FoodNet sites, 1999-2000. *Epidemiol. Infect.*, **2006**, Dec 6; 1-8, doi:10.1017/S0950268806007564.
6. Oldfield, E. C. III. Emerging foodborne pathogens: Keeping your patients and your families safe. *Rev. Gastrointest. Disord.*, **2001**, *1*, 177-186.
7. Elder, R. O.; Keen, J. E.; Siragusa, G. R.; Barkocy-Gallagher, G. A.; Koohmaraie, M.; Laegreid, W. W. Correlation of enterohemorrhagic *Escherichia coli* O157 prevalence in feces, hides, and carcasses of beef cattle during processing. *PNAS*, **2000**, *97*, 2999-3003.
8. USDA Food Safety Inspection Service. Microbiological results of ground beef products analyzed for *Escherichia coli* O157:H7, summarized by calendar year. *www.fsis.usda.gov*, **2007**, http://www.fsis.usda.gov/science/Ecoli_0157_Summary_-Tables/index.asp

9. Patrick, M. E.; Adcock, P. M.; Gomez, T. M.; Altekruse, S. F.; Holland, B. H.; Tauxe, R. V.; Swerdlow, D. L. *Salmonella* enteritidis infections, United States, 1985-1999. *Emerg. Infect. Dis.*, **2004**, *10*, 1-7.
10. Beutin, L. Emerging enterohaemorrhagic *Escherichia coli*, causes and effects of the rise of a human pathogen. *J. Vet. Med. B.*, **2006**, *53*, 299-305.
11. USDA Food Safety Inspection Service. Microbiology. Progress report on *Salmonella* testing of raw meat and poultry products (1998-2005). *www.fsis.usda.gov*, **2006**, http://www.fsis.usda.gov/science/Progress_Report_Salmonella_Testing/index.asp
12. USDA Food Safety Inspection Service. Foodborne illnesses continue downward trend: 2010 health goals for *E. coli* 0157 reached. *www.fsis.usda.gov.*, **2005**, http://www.fsis.usda.gov/News_&_Events/NR_041405_02/index.asp
13. Callaway, T. R.; Anderson, R. C.; Edrington, T. S.; Elder, R. O.; Genovese, K. J.; Bischoff, K. M.; Poole, T. L.; Yung, Y. S.; Harvey, R. B.; Nisbet, D. J. Preslaughter intervention strategies to reduce food-borne pathogens in food animals. *J. Anim. Sci.*, **2003**, *81*, E17-E23.
14. Vugia, D.; Hadler, J.; Chaves, S.; Blythe, D.; Smith, K.; Morse, D.; Cieslak, P.; Jones, T.; Conquist, P.; Goldman, D.; Guzewich, J.; Angulo, F.; Griffin, P.; Tauxe, R. Preliminary FoodNet data on the incidence of foodborne illness, selected sites, United States, 2002. *MMWR*, **2003**, *52(15)*, 340-343.
15. Loneragan, G. H.; Brashears, M. M. Pre-harvest interventions to reduce carriage of *E. coli* O157 by harvest-ready feedlot cattle. *Meat Sci.*, **2005**, *71*, 72-78.
16. Anderson, R. C.; Nisbet, D. J.; Stanker, L. H. Use of chlorate ion or preparations thereof for reduction of food borne pathogens. *United States Patent*, **2004**, no. 6,761,911 B2.
17. Stouthammer, A. H. Nitrate reduction in *Aerobacter aerogenes* I. Isolation and properties of mutant strains blocked in nitrate assimilation and resistant against chlorate. *Arch. Mikrobiol.*, **1967**, *56*, 68-75.
18. Quastel, J. H.; Stephenson, M.; Whetham, M. D. XLVII. Some reactions of resting bacteria in relation to anaerobic growth. *Biochem. J.*, **1925**, *19*, 304-317.
19. Anderson, R. C.; Buckley, S. A.; Callaway, T. R.; Genovese, K. J.; Kubena, L. F.; Harvey, R. B.; Nisbet, D. J. Effect of sodium chlorate on *Salmonella* Typhimurium concentrations in the weaned pig gut. *J. Food Prot.*, **2001**, *64*, 255-258.
20. Edrington, T. S.; Callaway, T. R.; Anderson, R. C.; Genovese, K. J.; Jung, Y. S.; McReynolds, J. L.; Bischoff, K. M.; Nisbet, D. J. Reduction of *E. coli* O157:H7 populations in sheep by supplementation of an experimental sodium chlorate product. *Small Ruminant Res.*, **2003**, *49*, 172-181.
21. Smith, D. J.; Oliver, C. E.; Caton, J. S.; Anderson, R. C. Effect of sodium [^{36}Cl]chlorate dose on total radioactive residues and residues of parent chlorate in beef cattle. *J. Agric. Food Chem.*, **2005**, *53*, 7352-7360.
22. Smith, D. J.; Anderson, R. C.; Huwe, J. K. Effect of sodium [^{36}Cl]chlorate dose on total radioactive residues and residues of parent chlorate in growing swine. *J. Agric. Food Chem.* **2006**, *54*, 8648-8653.

23. Smith, D. J.; Byrd, J. A.; Anderson, R. C. Total radioactive residues and residues of [^{36}Cl]chlorate in market size broilers. *J. Agric. Food Chem.*, **2007**, *55*, 5898-5903.
24. Abdel-Rahman, M. S.; Couri, D.; Bull, R. J. The kinetics of chlorite and chlorate in the rat. *J. Am. Coll. Toxicol.*, **1984,** *3*, 261-267.
25. Abdel-Rahman, M. S.; Couri, D.; Jones, J. D. Chlorine dioxide metabolism in rat. *J. Environ. Pathol. Toxicol.*, **1980**, *3*, 421-430.
26. Smith, D. J.; Anderson R. C.; Ellig, D. A.; Larsen, G. Tissue distribution, elimination, and metabolism of dietary sodium [^{36}Cl]chlorate in beef cattle. *J. Agric. Food Chem.*, **2005**, *53*, 4272-4280.
27. Hakk, H.; Smith, D. J.; Shappell, N. W. Metabolism and distribution of sodium [^{36}Cl]chlorate in rats. *J. Agric. Food Chem.*, **2007**, *55*, 2034-2042.
28. Oliver, C. E.; Bauer, M. L.; Caton, J. S.; Anderson, R. C.; Smith, D. J. The in vitro reduction of sodium [^{36}Cl]-chlorate in bovine ruminal fluid. *J. Anim. Sci.*, **2007**, (in press).
29. Oliver, C. E.; Craigmill, A. L.; Caton, J. S.; Anderson, R. C.; Smith, D. J. Pharmacokinetics of ruminally dosed sodium [^{36}Cl]chlorate in beef cattle. *J. Vet. Pharmacol. Ther.*, **2007**, *30*, 358-365.
30. National Pork Board. Pork Checkoff-Quick Facts-Stats. http://www.pork.org/newsandinformation/quickfacts/stats8.aspx. Accessed April 10, 2007.
31. Anonymous. Destruction of prickly pear. *Qld. Agric. J.*, **1901**, *9*, 460-462.
32. Åslander, A. Experiments on the eradication of Canada Thistle, *Cirsium arvense*, with chlorates and other herbicides. *J. Agric. Res.*, **1928**, *36*, 915-934.
33. Harper, H. J. The use of sodium chlorate in the control of Johnson grass. *J. Am. Soc. Agron.*, **1930**, *22*, 417-422.
34. Tomlin, C. *The Pesticide Manual, 10th Edition*; The British Crop Protection Council and the Royal Society of Chemistry: Surrey and Cambridge, **1994**; pp. 916-917.
35. Sutigoolabud, P.; Senoo, K.; Ongprasert, S.; Hisamatsu, M. Suppression of chlorate degradation in longan plantation soil after multiple applications of chlorate and accelerated degradation by sugar amendment. *Soil Sci. Plant Nutr.*, **2004**, *50*, 293-296.
36. Sutigoolabud, P.; Senoo, K.; Ongprasert, S.; Mizuno, T.; Tanaka, A.; Obata, H.; Hisamatsu, M. Decontamination of chlorate in longan plantation soils by bio-stimulation with sugar amendment. *Soil Sci. Plant. Nutr.*, **2004**, *50*, 249-256.
37. Rosemarin, A.; Lehtinen, K-J.; Notini, M.; Mattsson, J. Effects of pulp mill chlorate on Baltic sea algae. *Environ. Pollut.*, **1994**, *85*, 3-13.
38. Kroon, A.G.M.; van Ginkel, C. G. Biological reduction of chlorate in a gas-lift reactor using hydrogen as an energy source. *J. Environ. Qual.*, **2004**, *33*, 2026–2029.
39. *CRC Handbook of Chemistry and Physics*, 63rd ed.; Weast, R. C., Astle, M. J., Eds.; CRC Press: Boca Raton, FL, 1982; p B147.
40. Manochai, P.; Sruamsiri, P.; Wiriya-alongkorn, W.; Naphrom, D.; Hegele, M.; Bangerth, F. Year around off season flower induction in longan

(*Dimocarpus longan*, Lour.) trees by $KClO_3$ applications: potentials and problems. *Sci. Hortic.*, **2005**, *104*, 379-390.

41. Hegele, M.; Bangereth, F.; Napthrom, D.; Sruamsiri, P.; Manochai, P. Control of flower induction in tropical/subtropical fruit trees by phytohormones using the example of longan and mango. *Acta. Hortic.*, **2006**, *727*, 217-276.
42. Li, G. S.; Duan, J.; Chen, Z. L.; Zeng, S. J.; Jiang, Y. M.; Joyce, D. C. KClO3 applications affect *Phalaenopsis* orchid flowering. *Sci. Hortic.*, **2006**, *110*, 362-365.
43. van Ginkel, G. C.; Plugge, C. M.; Stroo, C. A. Reduction of chlorate with various energy substrates and inocula under anaerobic conditions. *Chemosphere*. **1995**, *31*, 4057-4066.
44. Malmqvist, Å.; Welander, T.; Gunnarsson, L. Anaerobic growth of microorganisms with chlorate as an electron acceptor. *Appl. Environm. Microbiol.*, **1991**, *57*, 2229-2232.
45. Nelson, R. T. Studies of microbial activity, chlorate reduction, and chlorate toxicity in soils treated with sodium chlorate. *J. Agric. Res.*, **1944**, 221-237.
46. Shrout, J. D.; Parkin, G. F. Influence of electron donor, oxygen, and redox potential on bacterial perchlorate reduction. *Water Res.*, **2006**, *40*, 1191-1199.
47. Coates, J. D.; Michaelidou, U.; Bruce, R.; O'Connor, S. M.; Crespi, J.N.; and Achenbach, L.A. Ubiquity and Diversity of Dissimilatory, (Per) chlorate-Reducing Bacteria, *Applied and Environmental Microbiology*, **1999**, *65*, 5234-5241.
48. Wu, J.; Unz, R. F.; Zhang, H.; Logan, B. E. Persistence of perchlorate and the relative numbers of perchlorate- and chlorate-respiring microorganisms in natural waters, soils, and wastewater, *Bioremed. J.*, **2001**, *5*, 119-130.
49. Simon, R.; Weber, E. J. Reduction of perchlorate in river sediment. *Environ. Toxicol. Chem.*, **2006**, *25*, 899-903.
50. Tan, K.; Anderson, T. A.; Jackson, W. A. Degradation kinetics of perchlorate in sediments and soils. *Water Air Soil Pollut.*, **2004**, 245-259.
51. Anderson, R. C.; Callaway, T. R.; Buckley, S. A.; Anderson, T. J.; Genovese, K. J.; Sheffield, C. L.; Nisbet, D. J. Effect of oral sodium chlorate administration on *Escherichia coli* O157:H7 in the gut of experimentally infected pigs. *Int. J. Food Microbiol.*, **2001**, 125-130.
52. Anderson, R. C.; Hume, M. E.; Genovese, K. J.; Callaway, T. R.; Jung, Y. S.; Edrington, T. S.; Poole, T. L.; Harvey, R. B.; Bischoff, K. M.; Nisbet, D. J. Effect of drinking-water administration of experimental chlorate ion preparations on *Salmonella enterica* serovar Typhimurium colonization in weaned and finished pigs. *Vet. Res. Commun.*, **2004**, *28*, 179-189.
53. Patchanee, P.; Bahnson, P. B.; Crenshaw, T. D. Chlorate and disinfectant modify *Salmonella enterica* shedding in weaned pigs. Proceedings, 6th International Symposium on the Epidemiology and Control of Foodborne Pathogens in Pork. Rohnert Park, CA; 2005; 139-141.
54. Burkey, T. E.; Dritz, S. S.; Nietfeld, J. C.; Johnson, B. J.; Minton, J. E. Effect of dietary mannanoligosaccharide and sodium chlorate on the growth performance, acute-phase response, and bacterial shedding of weaned pigs

challenged with *Salmonella enterica* serotype Typhimurium. *J. Anim. Sci.*, **2004**, *82*, 397-404.

55. Anderson, R. C.; Buckley, S. A.; Kubena, L. F.; Stanker, L. H., Harvey, R. B.; Nisbet, D. J. Bactericidal effect of sodium chlorate on *Escherichia coli* O157:H7 and *Salmonella* Typhimurium DT104 in rumen contents in vitro. *J. Food Prot.*, **2000**, *63*, 1038-1042.
56. Callaway, T. R.; Anderson, R. C.; Genovese, K. J.; Poole, T. L.; Anderson, T. J.; Byrd, J. A.; Kubena, L. F.; Nisbet, D. J. Sodium chlorate supplementation reduces *E. coli* O157:H7 populations in cattle. *J. Anim. Sci.*, **2002**, *80*, 1683-1689.
57. Anderson, R. C.; Carr, M. A.; Miller, R. A.; King, D. A.; Carstens, G. E.; Genovese, K. J.; Callaway, T. R.; Edrington, T. S.; Jung, Y. S.; McReynolds, J. L.; Hume, M. E.; Beier, R. C.; Elder, R. O.; Nisbet, D. J. Effect of experimental chlorate preparations as feed and water supplements on *Escherichia coli* colonization and contamination of beef cattle and carcasses. *Food Microbiol.*, **2005**, *22*, 439-447.
58. Callaway, T. R.; Edrington, T. S.; Anderson, R. C.; Genovese, K. J.; Poole, T. L.; Elder, R. O.; Byrd, J. A.; Bischoff, K. M.; Nisbet, D. J. *Escherichia coli* O157:H7 populations in sheep can be reduced by chlorate supplementation. *J. Food Prot.*, **2003**, *66*, 194-199.
59. Moore, R. W.; Byrd, J. A.; Knape, K. D.; Anderson, R. C.; Callaway, T. R.; Edrington, T.; Kubena, L. F., Nisbet, D. J. The effect of an experimental chlorate product on *Salmonella* recovery of turkeys when administered prior to feed and water withdrawal. *Poultry. Sci.*, **2006**, *85*, 2101-2105.
60. Jung, Y. S.; Anderson, R. C.; Byrd, J. A.; Edrington, T. S.; Moore, R. W.; Callaway, T. R.; McReynolds, J.; Nisbet, D. J. Reduction of *Salmonella* Typhimurium in experimentally challenged broilers by nitrate adaptation and chlorate supplementation in drinking water. *J. Food Prot.*, **2003**, *66*, 660-663.
61. Byrd, J. A.; Anderson, R. C.; Callaway, T. R.; Moore, R. W.; Knape, K. D.; Kubena, L. F.; Ziprin, R. L.; Nisbet, D. J. Effect of experimental chlorate product administration in the drinking water on *Salmonella* Typhimurium contamination of broilers. *Poultry Sci.*, **2003**, *82*, 1403-1406.
62. McReynolds, J.; Kubena, L.; Byrd, J.; Anderson, R.; Ricke, S.; Nisbet, D. Evaluation of *Salmonella enteritidis* in molting hens after administration of an experiemental chlorate product (for nine days) in the drinking water and feeding an alfalfa molt diet. *Poultry Sci.*, **2005**, *84*, 1186-1190.

Chapter 5

Influence of MnO_2 on the transformation of oxy- and chlortetracycline in pond water

Kennedy F. Rubert IV[1], Curtis J. Hedman[2], Tan Guo[2], Joel A. Pedersen[1,3,*]

[1]Molecular and Environmental Toxicology Center, University of Wisconsin, Madison, WI 53706
[2]Wisconsin State Laboratory of Hygiene, Madison, WI 53718
[3]Department of Soil Science, University of Wisconsin, Madison, WI, 53706

The antibiotics oxytetracycline (OT) and chlortetracycline (CT) underwent non-photolytic abiotic transformation in a natural surface water (pH 8) in the absence and presence of suspended manganese dioxide (MnO_2), a solid-phase oxidant naturally occurring in soils and sediments. In the absence of MnO_2, the CT concentration declined by >60% after 20 d, while OT did not degrade appreciably. Adding MnO_2 increased CT transformation and resulted in the formation of epi-*iso*-chlortetracycline and the appearance of a hydroxylated product with a chemical structure differing substantially from the parent compound. Approximately 40% of the CT was converted to *iso*-chlortetracycline and epi-*iso*-chlortetracycline regardless of the presence or absence of MnO_2. In the presence of MnO_2, 91% of the OT and 79% of the CT were transformed after 20 d. The initial rate of CT degradation exceeded that of OT. The dominant OT transformation product, *N*-desmethyl-oxytetracycline, accounted for up to 10% of the OT loss. The major transformation products are expected to possess reduced biological activity.

Introduction

Tetracycline antibiotics are widely used to treat human and animal diseases and to promote livestock growth; they are occasionally employed to treat disease in plants and fish. The widespread use of tetracycline antibiotics and their detection in a variety of environmental compartments has raised concern that they may promote the spread of antibiotic resistance in microorganisms (*1*). The occurrence of these antibiotics has been reported in soils, surface waters and sediments. Reported surface water concentrations range from 0.1 to 10 $\mu g \cdot L^{-1}$ for oxytetracycline (OT) and from 0.1 to 12.1 $\mu g \cdot L^{-1}$ for chlortetracycline (CT) (*2-4*). Reported sediment concentrations range from 6.9 to 285,000 $\mu g \cdot kg^{-1}$ for OT and 1.1 to 30.8 $\mu g \cdot kg^{-1}$ for CT (*4-7*). Concentrations reported for CT in soils fertilized with contaminated manure have been as high as 39 $\mu g \cdot kg^{-1}$ (*8*).

In addition to concern about their ability to promote selection of antibiotic resistant microorganisms, tetracyclines can be taken up by some food plants, alter microbial community structure and cause toxicity to non-target organisms. Chlortetracycline can be taken up by onion, cabbage and maize from soil fertilized with animal manure (*9*), and form glutathione conjugates during detoxification in maize (*10*). O'Reilly and Smith (*11*) reported that OT began to alter microbial metabolic activity in sediments at 20,000 $\mu g \cdot kg^{-1}$; minimum effect concentrations are expected to depend on soil and sediment properties (*e.g.*, content of organic carbon, clay minerals, iron oxides). Fe(III) reduction was inhibited in bacterial communities extracted from soil by OT concentrations ranging from 13,800 to 267,000 $\mu g \cdot kg^{-1}$ and CT concentrations in the range of 1200 to 106,000 $\mu g \cdot kg^{-1}$ (*12*). In one study, sediment OT concentrations of 50,000 $\mu g \cdot kg^{-1}$ were capable of altering microbial respiration (*13*). Mixtures of tetracyclines (viz. OT, CT, tetracycline and doxycycline) with additive concentrations >100 $\mu g \cdot L^{-1}$ caused toxicity to phytoplankton in aquatic microcosms (*14*). While the toxicological endpoints reported to date occurred at OT and CT concentrations exceeding most of those reported in the environment, these toxicological studies examined only a small subset of environmental conditions and did not account for the effects of tetracycline antibiotic degradation products.

The fate of tetracycline antibiotics in soils and sediments is governed by sorption, complexation with metal cations, and transformation. Tetracycline antibiotics sorb strongly to soils in a manner that increases with decreasing pH (*14, 15*), decreasing nominal particle size (*16*) and increasing cation exchange capacity (*14, 16, 17*). Tetracycline antibiotic sorption to soils is influenced by the clay minerals (*18-21*), organic matter (*22-25*), Fe oxides (*26*) and Al oxides (*27*), and each of these components increases the number of cation exchange sites. Ion bridging, complexation, H-bonding and hydrophobic partitioning are thought to control the sorption of tetracyclines to various constituents in soils and sediments (*17*) in a manner that depends on the pH-dependent speciation of the antibiotic (*19*).

Formation of complexes with polyvalent cations such as Fe (*28, 29*), Al (*29, 30*), Zn (*31, 32*), Ca (*31, 33-35*), Mg (*31, 33-36*), Mn (*37*), Mo, W (*29*), Zr (*38*) and Cu (*39, 40*) have been reported in the laboratory. Cation bridging by Fe^{3+}

and Al^{3+} enhances sorption of OT to organic matter (*24*). Metal complexation and sorption can decrease the bioavailability of these antibiotics by altering their molecular conformation (*41*) and rendering them inaccessible to organisms (*42*), but these processes do not remove the antibiotics from the environment. Complexation with metal cations may even inhibit degradation of OT in alkaline media (*43*).

Biotic and abiotic processes can contribute to the transformation of tetracycline antibiotics in aquatic and terrestrial environments. Tetracycline antibiotics can be degraded by sediment and sludge microorganisms (*44*). Tetracycline antibiotics are susceptible to hydrolysis; elevated temperatures (*e.g.*, 43 °C) and high pH (*e.g.*, 10) accelerate this process (*21, 41*). Photolysis is expected to be important in surface waters and at the soil surface (*45*). A number of tetracycline photolysis products have been reported (*46-50*). Titanium and iron oxides can catalyze tetracycline photodegradation (*49-51*). However, the photic zone in soils is typically restricted to the top ~0.5 mm (*52*). In absence of light, cations such as Mn^{2+} (pH 8-9.5) and Cu^{2+} (pH 4-6) can catalyze degradation of tetracycline antibiotics in aqueous media (*53*). Oxides of $Mn^{III/IV}$, ubiquitous in soil and sediment environments, can also mediate the degradation of tetracycline antibiotics (*54*).

Investigations of tetracycline antibiotic transformation in pond water microcosms (*55*), soil interstitial water (*56*), and hog lagoons (*57*) revealed formation of a variety of products, most of which were present at levels substantially less than those of the parent compounds. The highest level of any transformation product found to date was *iso*-chlortetracycline (iCT) occurring at concentrations up to 15,000 $\mu g \cdot L^{-1}$ in manure samples from hog waste lagoons (*57*). Identification of tetracycline transformation products and assessment of their environmental risks warrants investigation. The parent compounds are rapidly degraded under a range of environmental conditions producing a variety of transformation products (*58, 59*). Investigations of numerous organic contaminants (*e.g.*, phenylurea herbicides, organochlorine insecticides) has demonstrated that degradation products can be toxicologically relevant (*60-63*). The same may be true of some classes of antibiotics. The need to identify additional tetracycline antibiotic transformation products likely to be formed in the environment served as impetus for the present study. Identification of transoformation products is necessary for subsequent investigation of their toxicity and persistence.

The manganese oxide mineral birnessite (δ-MnO_2) commonly occurs in soils and sediments (*64-70*). Mn oxides are generally more reactive than other minerals due to their high reduction potentials (E_H^0 = +1.23 V for MnO_2; *71*). Aqueous suspensions of synthetic MnO_2 have been shown to mediate the rapid transformation of several tetracycline antibiotics (*54, 72*). The rate of OT transformation by MnO_2 was highly pH-dependent, and OT and CT were transformed by MnO_2 at dramatically different rates (*54*). Although the abundance of $Mn^{III/IV}$ oxides in soils and sediments is typically less than that of phyllosilicates, tectosilicates, Al and Fe oxides, they can significantly impact the chemical properties of soils because of their prevalence as surface coatings and high reactivity (*73*). In addition, Mn oxides are the most common minerals yet reported to facilitate transformation of tetracycline antibiotics in absence of sunlight (*54*, *72*). (Interaction of OT with hematite and goethite surfaces does not appear to result in appreciable degradation (*26, 74*)). Most Mn oxides are strongly negatively charged and adsorb cations under pH conditions typical of soil and sediment porewaters (*64, 75*). Manganese oxide minerals can facilitate oxidation (*76*), polymerization (*77, 78*) and dealkylation (*79*) of organic compounds bearing functional groups in common with the tetracyclines (i.e., *N*-alkyl and phenolic moieties). Native $Mn^{III/IV}$ oxides have been implicated in the oxidation, polymerization and irreversible sorption (bound residue formation) of a variety of aromatic amines in whole soils (*76, 80*); birnessite-mediated bound residue formation of phenols in soils has also been reported (*81*).

The objectives of this study were to (1) assess the non-photolytic abiotic degradation of OT and CT in a natural surface water in the presence and absence of MnO_2, (2) identify major transformation products in these reactions and (3) evaluate the potential biological activity of identified products using published structure-activity relationships (*41*).

Materials and Methods

Reagents and Preparation

Chlortetracycline hydrochloride and OT were purchased from Acrōs. Reagents and solvents (HPLC grade or better) were purchased from Fisher Scientific. All solutions were prepared using double deionized water (ddH_2O; Barnstead NANOpure Ultrapure Water System, Dubuque, IA). Laboratory ware was prepared by soaking for > 12 h in 10% HNO_3 followed by rinsing with ddH_2O. Glassware was silanized to minimize sorption of the antibiotic, and photolysis was prevented by conducting experiments in a darkroom and storing solutions in amber glassware. MnO_2 (apparent oxidation state of +3.94 (*82*)) was prepared following Murray (*83*) with modifications (*54*), stored in aqueous suspension at 4°C and used within one week of preparation.

Tetracycline Antibiotic Transformation in Pond Water

Surface water was collected from an earthen settling pond receiving a continuous effluent from a coldwater fish hatchery. The water was passed through a 0.22-μm filter (Millipore type GV) to remove suspended particles and the majority of bacteria, and analyzed for common cations and anions, pH, and dissolved organic carbon (Table I). Tetracycline antibiotic degradation was examined under four conditions (treatments) at two antibiotic concentrations: (1) ddH_2O (100 μM), (2) pond water (10 and 100 μM), (3) pond water amended with 490 μM MnO_2 (10 and 100 μM), and (4) pond water amended with 490 μM $MnCl_2$ (100 μM). Incubations were conducted in 10-mL fluorinated ethylene-propylene (FEP) tubes (reactors) at 10°C in the dark under continuous shaking. Reactors were sacrificed at each time point investigated. Figure 2 presents a schematic of the sampling and analysis of reactors. MnO_2-mediated reactions were quenched by removing the oxide from suspension by filtration (0.2-µm filter) or by dissolution with oxalic acid (4 mg·mL^{-1}) (*54*). The presence of culturable microorganisms in the reactors was assessed by plating 100-µL aliquots from the last reactor sacrificed for each treatment onto Peptone-yeast or Difco Antibiotic Medium 3 (Becton Dickinson and Company, Franklin Lakes, NJ) media solidified with 0.15% agar and counting colonies after 24 and 96 h incubation at 28°C.

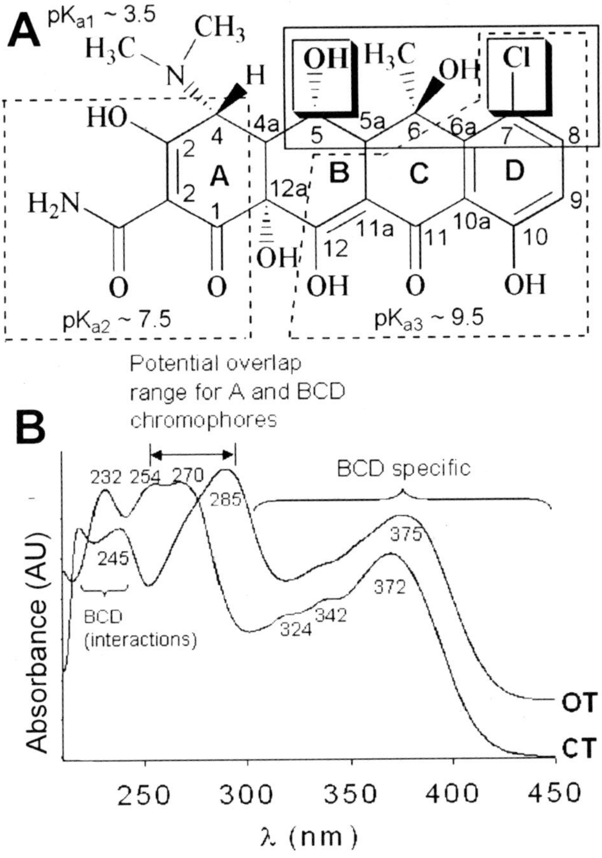

Figure 1. (A) Oxytetracycline (C_5 = -OH; C_7 = -H) and chlortetracycline (C_5 = -H; C_7 = -Cl) possess structural moieties requisite for ribosome binding and structural moieties that may undergo alteration without loss of biological activity (shown in boxes) (41). Alteration of moieties within the solid box does not generally reduce antibacterial activity (41). (B) The tetracycline ring structure possesses two distinct UV absorbing chromophores (the A and BCD ring structures enclosed in dashed boxes). The UV absorbance of the A and BCD chromophores overlaps between λ = 254 and 285 nm. The BCD chromophore has a specific absorbance band from λ = 320 to 400 nm. Changes in the UV absorbance spectra of OT and CT reflect changes to A and BCD chromophores.

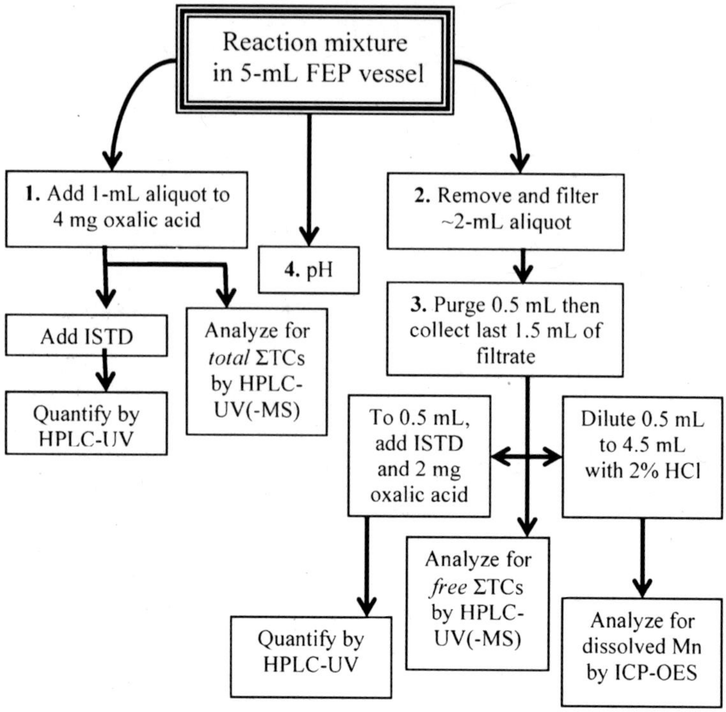

Figure 2. Sample processing and analysis scheme. Steps 1-4 were performed in order, for each reaction at each time point. ΣTCs refers to the sum of OT or CT and their UV chromophore-bearing transformation products. Samples treated with oxalic acid provided a measurement of total antibiotic and products (ΣTCs) while those filtered provided a measurement of free ΣTCs (not bound to particles or in aggregates >0.2 μm). ISTD = internal standard.

LC Separation

Reaction mixtures were separated on a Phenomenex Luna $C_{18(2)}$ column (150 mm × 3.00 mm) using an Agilent 1100 Series HPLC system (Agilent Technologies, Palo Alto, CA) following a method similar to that of Lindsey *et al.* (*84*). Sample injections were of 20 μL. Analytes were eluted at a flow rate of 0.36 $mL{\cdot}min^{-1}$ using a dynamic gradient of 0.3% aqueous formic acid (pH 2.6) and methanol at 30°C (0-5 min: 95% aqueous, 5-20 min: 95-0% aqueous, 20-25 min: 0-5% aqueous), followed by a 5-min post-column re-equilibration time. When the objective of an analysis was to quantify only OT or CT, we employed the more rapid isocratic separation method described previously (*54*).

Table I. Pond Water Composition

pH = 7.8			
[Alkalinity][a] = 3.2 mM			
DOC (<2 μm)[b] = < 10 $mg\cdot L^{-1}$			
Cation[c]	*Conc. ($mg\cdot L^{-1}$)*	*Anion*[d]	*Conc. ($mg\cdot L^{-1}$)*
Ca^{2+}	22.04	Cl^-	36.97
Mg^{2+}	37.71	SO_4^{2-}	19.50
Na^+	15.60	NO_3^-	23.58
K^+	1.34	NO_2^-	<0.01
Mn	0.001	F^-	0.16
Zn^{2+}	0.008	PO_4^{3-}	<0.02
B	<0.02	Br^-	<0.01
Fe	<0.001		
Cu	<0.005		
Al	<0.05		

[a][Alkalinity] = $[Na^+] + [K^+] + 2[Ca^{2+}] + 2[Mg^{2+}] - [Cl^-] - 2[SO_4^{2-}] - [NO_3^-]$ (*85*)

[b]Dissolved organic carbon (DOC) using Carlo-Erba elemental analyzer (Soil and Plant Analysis Laboratory) following 0.2 μm filtration.

[c]ICP-OES (UW Soil and Plant Analysis Laboratory)

[d]Ion chromatography (UW Soil and Plant Analysis Laboratory).

Ultraviolet Absorbance Data Acquisition

Losses of OT and CT and evolution of chromophore-bearing products were monitored by ultraviolet (UV) absorbance spectroscopy (λ = 210 – 400 nm). UV absorbance spectra were acquired using an Agilent 1100 Series photodiode array detector (4-mm slit width, 2 nm increments), and were extracted from total wavelength chromatograms with baseline subtraction. Data collection and analysis were performed using Analyst (version 1.4.1, Applied Biosystems, Foster City, CA, USA) or ChemStation (Rev. A.09.01) software. Quantitative analysis of CT and iCT was performed using tetracycline as an internal standard, and quantitation of OT, *N*-desmethyl-oxytetracycline (dmOT) and 4-epi-oxytetracycline (eOT) was accomplished using CT as an internal standard. *N*-desmethyl-oxytetracycline and epi-*iso*-chlortetracycline (eiCT) concentrations were estimated using standard curves prepared for OT and iCT, respectively, because these transformation products contain the same chromophores as the parent compound. Analyte UV absorbance was monitored over wavelength ranges including at least one absorbance maximum for each compound. Standard curves had linear regression coefficients (R^2) ≥ 0.99 within the calibration intervals [OT and CT, 1-10 or 10-100 μM (n = 4); iCT, 5-100 μM (n = 5)]. The potential for interference by co-eluting products was assessed by comparison of UV absorbance spectra for analytes separated from the reaction mixture with the UV absorbance spectra for standards (OT, CT and iCT); in the cases of iCT and eiCT a non-absorbing wavelength was concurrently monitored (360 nm) to verify that compounds absorbing at higher wavelengths did not co-elute with either of these compounds.

Mass Spectrometry

Positive electrospray ionization (ESI) mass spectrometry (MS) was performed using a Sciex API 4000 (Foster City, CA, USA) triple quadrupole mass spectrometer operating in single quadrupole (Q1) scanning mode or triple quadrupole (Q3) mode. For Q1 scans, the ESI needle (capillary) potential was 3000 V, curtain gas pressure was 20 psi, nebulizer gas pressure was 35 psi, drying gas pressure was 30 psi, declustering potential was 51.0 V, and entrance potential was 10.0 V. Collision-induced dissociation was performed in 10-eV steps up from 10 to 60 eV. The instrument acquired spectra of unit mass resolution with a scan step size of 0.1 amu from 10 – 2000 *m/z*. A centroid filter was applied to mass range scans following extraction from the total ion chromatogram (TIC), and the *m/z* ranges were truncated in cases where no peaks appeared above or below a certain value.

Inductively Coupled Plasma – Optical Emission Spectroscopy (ICP-OES)

Manganese passing through a 0.2-μm filter was measured by ICP-OES (Perkin-Elmer, Waltham, MA, USA) with quantitation on the 259.4 nm emission band and confirmation on the 257.6 nm emission band. Samples from the filtered reactions were stored at −80°C; prior to analysis samples were thawed and diluted 10-fold with 2% HCl in polystyrene conical tubes (Falcon®). External standardization with $MnCl_2$ diluted in the same manner as the samples provided a linear calibration curve from 0.1 to 1000 μM ($n = 6$, $R^2 = 1.000$) that was sufficiently stable over the course of the analyses.

Results and Discussion

Quenching of MnO_2-mediated Reactions

Two methods were used to quench reactions mediated by MnO_2: (1) filtration to remove reactive MnO_2 and (2) oxalic acid addition to dissolve MnO_2 and complex metal cations. Analysis of samples quenched by the filtration method yields concentrations of non-sorbed antibiotic and products; analysis of oxalic acid-quenched samples yields total concentrations of antibiotic and products. To ensure consistency of results, these treatments were also used in the processing of samples lacking MnO_2 amendment. Oxalic acid was also added to the filtered samples to lower pH and suppress conversion of CT to iCT (*41*). We previously demonstrated these two methods produced equivalent recoveries of OT from laboratory water buffered to pH 5.6 (*54*). In the present study, concentrations of both antibiotics measured in ddH_2O differed by <5% for the two methods. In subsequent discussion, we refer to *total* antibiotics or products (as determined by oxalic acid treatment without filtration) unless specified otherwise.

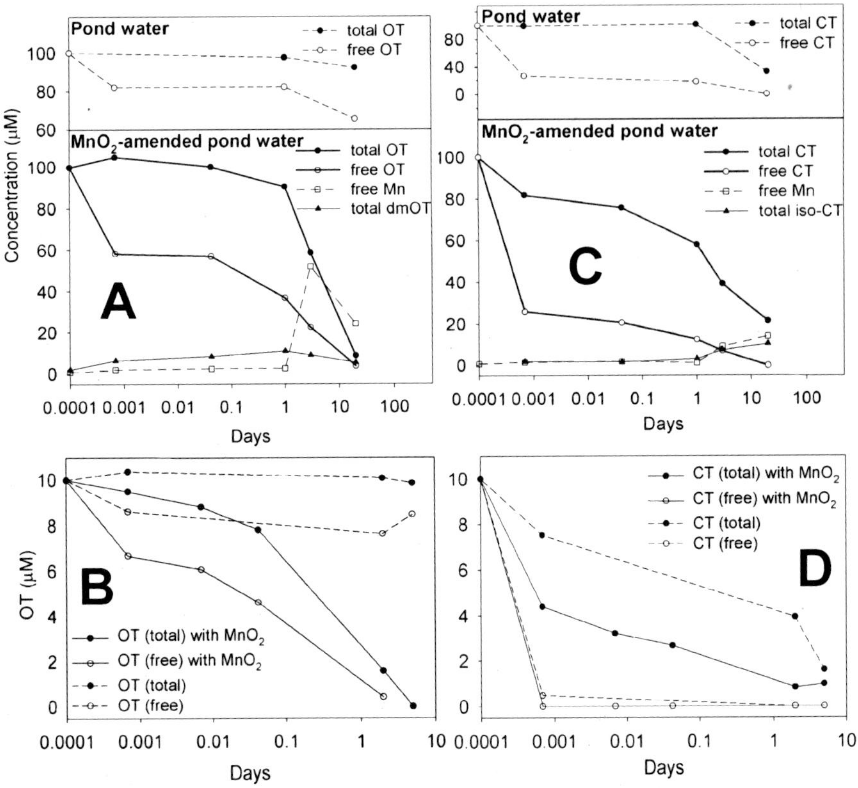

Figure 3. OT (A, B) and CT (C, D) were transformed in pond water (A, C) and in MnO_2-amended pond water (B, D) at pH 7.5 – 8.0 and 10 °C in absence of light. Antibiotic transformation in MnO_2-amended reactions was accompanied by dissolution of the oxide and the formation of unique products (see Figure 4). Oxalic acid was added to dissolve MnO_2 and provide an estimate of total antibiotic remaining; and filtration was used to physically remove MnO_2 particles > 0.2 μm and provide an estimate of the non-sorbed (free) antibiotic. Although recoveries by filtration and oxalic acid addition were comparable in ddH_2O, filtered reactions exhibited consistently lower OT or CT recoveries in pond water, suggesting dissolved organic material and/or cations may facilitate antibiotic retention by the filter.

Degradation of Tetracycline Antibiotics in Pond Water

Oxytetracycline degraded slowly in the filtered pond water in absence of light. In experiments with initial OT concentrations, $[OT]_0$ = 10 μM, >95% remained after 5 d; when $[OT]_0$ = 100 μM , >90% remained after 20 d (Figure

3). Chlortetracycline degraded to a larger extent in the pond water than did OT. At an initial CT concentration, $[CT]_0$ = 10 μM, 84% of the CT was degraded after 5 d; at $[CT]_0$ = 100 μM, 67% of the CT transformed after 20 d.

Oxytetracycline recovery from pond water by the filtration method was only 75-85% of that by the oxalic acid method in the surface water, even though the antibiotic concentration was >100-fold lower than its aqueous solubility at pH 8 (*41*) and no visible precipitates formed. Similar results were found for CT. The differences in recovery may have been due to formation of complexes between OT, cations and organic matter (*23, 24, 41*) or inorganic colloids > 0.22 μm that were retained by the filter.

Products from Tetracycline Antibiotic Transformation in Pond Water

The extent of OT degradation was small over 20 d in the surface water. One transformation product and two manufacturing contaminants were identified (Figure 4A and Table II). We identified the OT transformation product 4-epi-oxytetracycline (eOT) by LC-UV-MS in the surface water after 20-d incubation (~7% of the OT transformed to eOT). This transformation product exhibits 6-25% of the antibacterial activity of OT against a variety of bacteria (*86*). Although eOT quantitation at low levels was complicated by incomplete chromatographic separation from OT and another product, much of the OT loss appeared to be accounted for by epimerization. The two putative manufacturing contaminants identified were 2-acetyl-2-decarboxamido-oxytetracycline (ADOT) and *N*-desmethyl-oxytetracycline (dmOT) (*56*). ADOT retains ~3% of the antibacterial activity of OT (*87*). The amount of ADOT remained at ~1.5% of the amount of OT (by peak area) over the course of the incubations in pond water and ddH_2O. The amounts of dmOT in ddH_2O and pond water ranged from 0-3% at time points < 1 d; poor chromatographic separation from low levels of eOT prevented estimating the amount of dmOT present at 20 d. In control incubations in ddH_2O, we observed the same products as in pond water, as well as eight new ones. Formation of these products may have been suppressed by the pond water matrix.

Chlortetracycline was rapidly transformed in ddH_2O (pH = 6.8 - 7.4) and the natural pond water to yield two major products: iCT and eiCT (Figure 4B and Table II). The antibacterial potency of iCT is only 0.09% that of CT (as assessed using aerobic sludge bacteria; *86*); the biological activity of eiCT is expected to be even less because epimers of tetracyclines characteristically show less biological activity (*86*). After 20 d, the sum of CT, iCT and eiCT accounted for ~88% of the 100 μM CT initially present in ddH_2O (33 μM CT, 2 μM eiCT and 52 μM iCT) and ~74% in the pond water (33 μM CT, 9 μM eiCT, and 32 μM iCT). Chlortetracycline disappeared more rapidly at 10-fold lower concentration (10 μM), declining by 84% after 5 d, and the loss was accompanied by formation of iCT and eiCT. The differences in CT loss and product formation between pond water and ddH_2O were likely attributable to differences in the water matrices. One additional product, identified by LC-UV-MS as 5,11a-dihydroxy-*iso*-chlortetracycline (dhiCT, referred to as M510 in (*57*)) was identified in the ddH_2O control after 20 d; dhiCT had a retention time similar to CT and may have been present as a contaminant in the original CT preparation but masked by the larger CT peak initially present. Our results suggest that constituents in the pond water matrix, possibly Ca^{2+}, Mg^{2+} (*59*) and Mn^{2+} cations (*53*) may influence the extent of CT transformation. At a concentration <25% that of tetracycline, Mn^{2+} facilitated transformation of that antibiotic, suggesting Mn^{2+} may act in part as a catalyst (*53*).

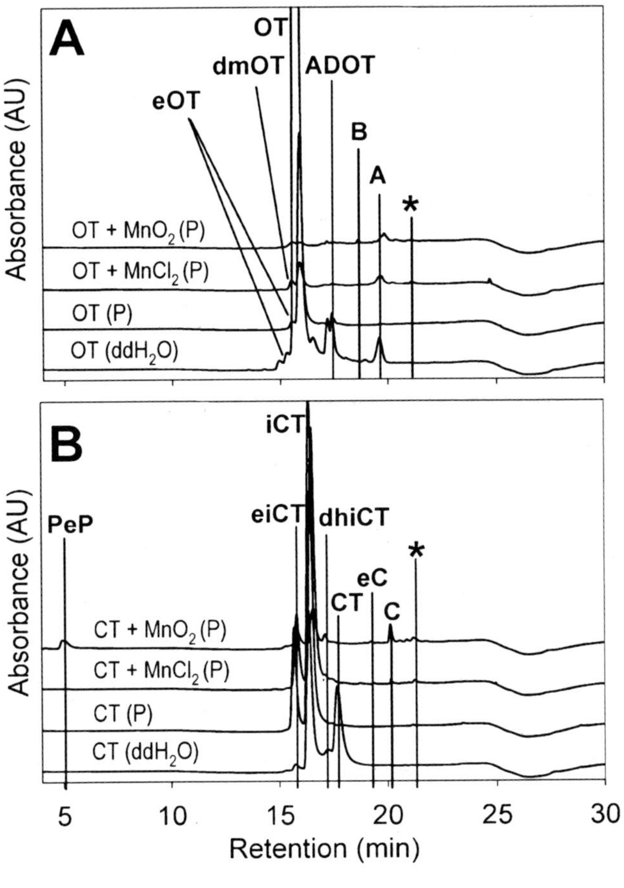

Figure 4. Transformation products and impurities observed after 20-d incubations of (A) OT or (B) CT at 10°C (pH 7.5 – 9.0, dark) in deionized water (ddH$_2$O) or pond water (P) with or without added $MnCl_2$ (49 μM) or MnO_2 (490 μM) as determined by HPLC-UV following sample filtration (0.2 μm) and in absence of oxalic acid. Peaks: (A) eOT (4-epi-oxytetracycline), dmOT (N-desmethyl-oxytetracycline), OT (oxytetracycline), ADOT (2-acetyl-2-decarboxamido-oxytetracycline), A and B; (B) PeP (polyhydroxylated product; Figure 5), eiCT (epi-iso-chlortetracycline), iCT (iso-chlortetracycline, CT (chlortetracycline), dhiCT (5,11a-dihydroxy-iso-chlortetracycline), eC And C (putative epimeric products). Asterisks (*) designate peaks corresponding to products which were not discussed in this text.

Table II. OT and CT Transformation Products and Impurities

Transformation products and impurities of OT	*Transformation products and impurities of CT*
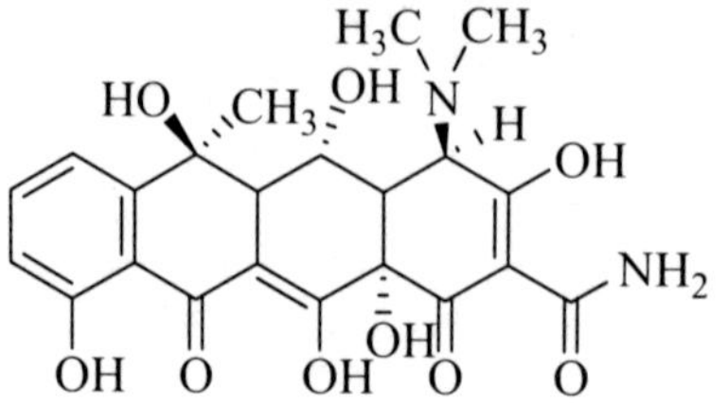 4-epi-oxytetracycline (eOT) *m/z*: **461.4** 444.1 283.1 201.2 λ_{peak}: 220 256 278 322 365 nm	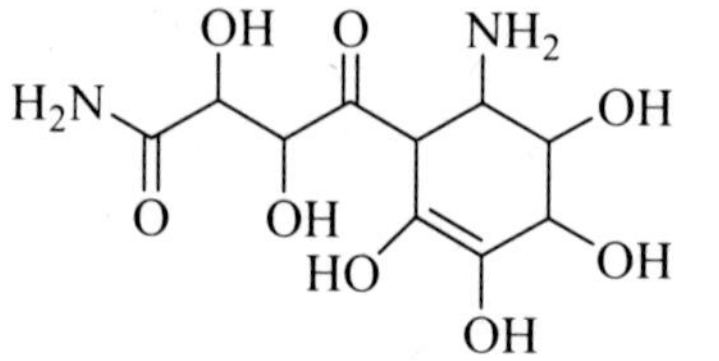 polyhydroxy product (PeP) *m/z*: **293.4** 160.3 132.2 114.3 86.1 λ_{peak}: 270 nm
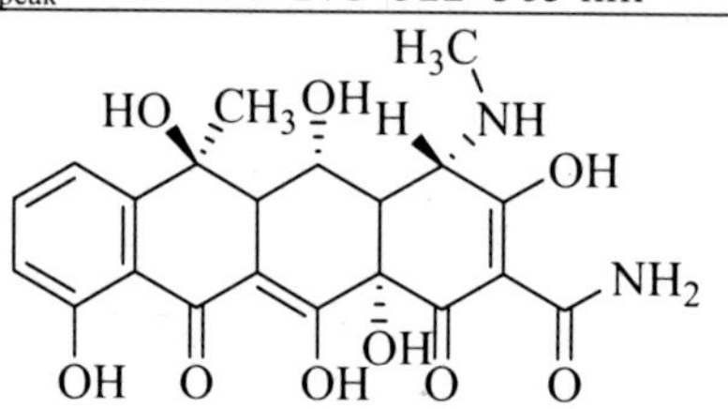 *N*-desmethyl-oxytetracycline (dmOT) *m/z*: **447.5** 430.3 412.1 283.2 200.9 λ_{peak}: 222 274 320 360 nm	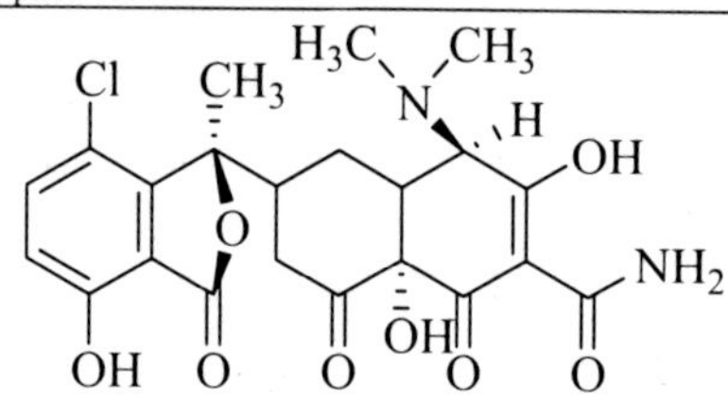 epi-*iso*-chlortetracycline (eiCT) *m/z*: **479.2** 462.5 λ_{peak}: 220 246 272 316 nm
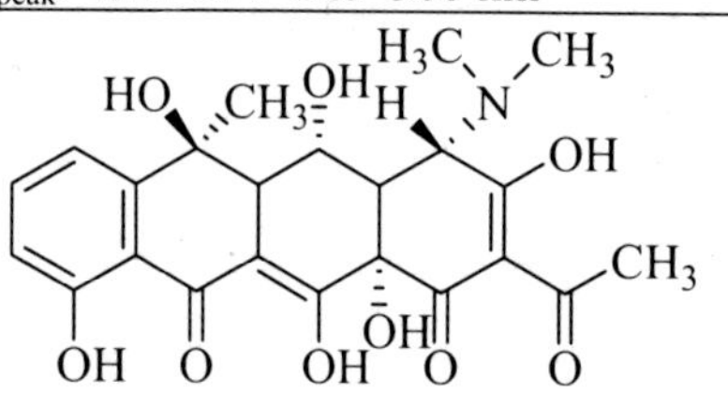 2-acetyl-2-decarboxamido-oxytetracycline (ADOT)* *m/z*: **461.4** 443.2 380.4 283.1 201.2 λ_{peak}: 220 278 360 nm	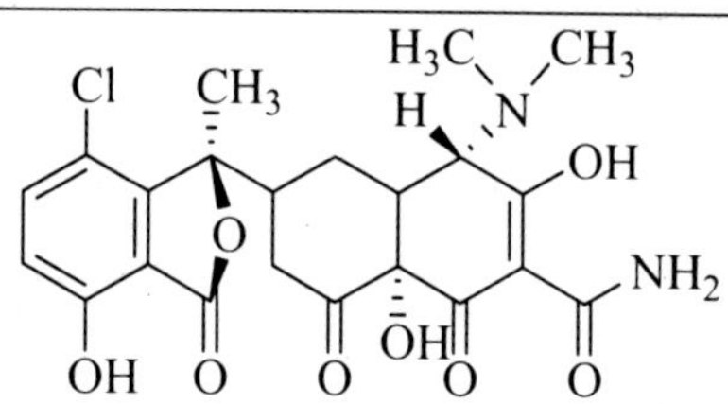 *iso*-chlortetracycline (iCT) *m/z*: **479.2** 462.5 λ_{peak}: 220 246 274 316 nm

Table II (Continued). OT and CT Transformation Products and Impurities

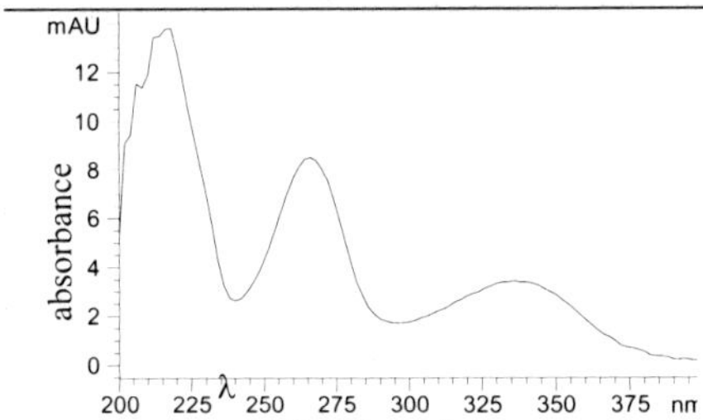 Product A λ_{peak}: 215 265 335 nm	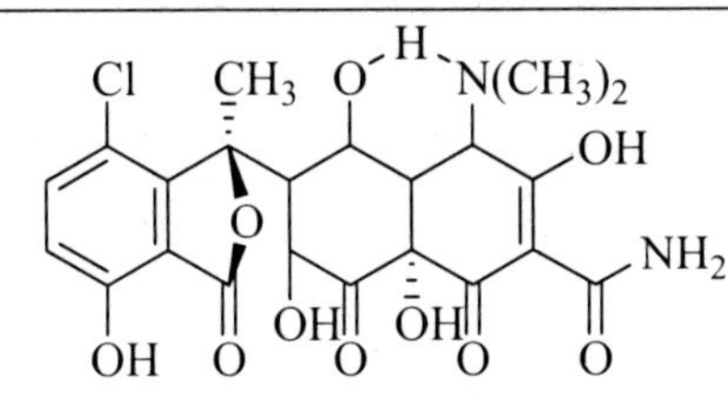 5,11a-dihydroxy-*iso*-chlortetracycline (dhiCT) *m/z*: **510.7** 478.8 452.4 λ_{peak}: 218 250 268 316 nm
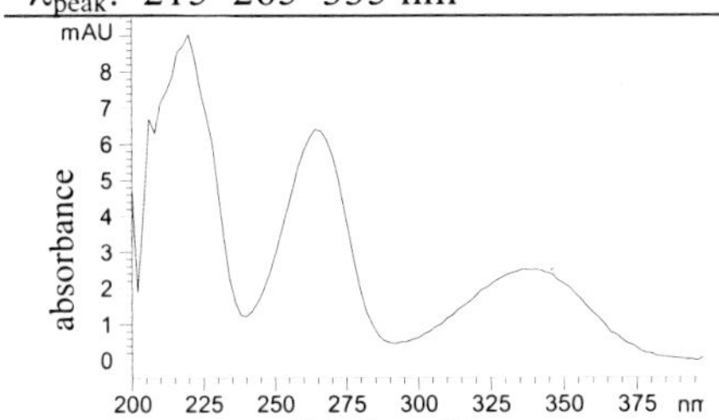 Product B λ_{peak}: 220 265 340 nm	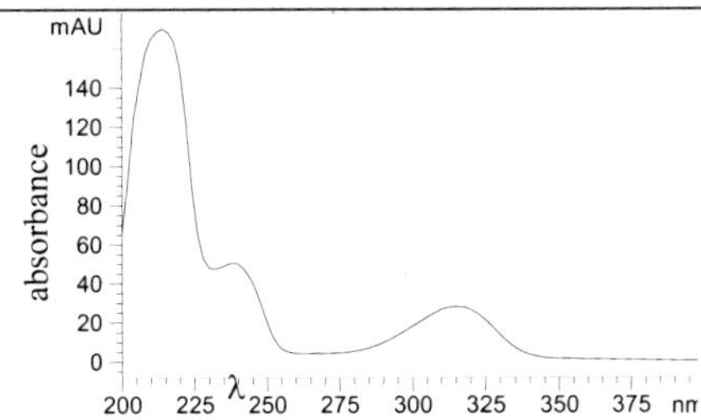 Product C λ_{peak}: 215 240 315 nm

*, impurity in OT or CT

NOTE: λ_{peak} refers to maxima of major UV absorbance peaks.

NOTE: Figure 5 details the structural assignment of PeP.

Influence of MnO_2 on the Rates Tetracycline Antibiotic Transformation in Pond Water

MnO_2 was capable of sorbing OT in pond water (Figure 3). Although OT recoveries ($[OT]_0$ = 100 μM) in MnO_2-free pond water were 15-25% lower than ddH_2O controls, addition of MnO_2 (490 μM) to the pond water roughly doubled the amount of OT (to ~40%) irrecoverable by filtration at a given time point, indicating substantial sorption of OT to the Mn oxide. This result is consistent with our previous findings that surface sorption is an integral step in the mechanism of MnO_2-mediated transformation of tetracycline antibiotics (*54*). More rapid initial degradation of CT ($[CT]_0$ = 10 and 100 μM) by MnO_2 and larger interference from the MnO_2-free pond water matrix prevented us from assessing the degree of sorption of CT to MnO_2.

The addition of 490 μM MnO_2 accelerated OT and CT transformation in the pond water (Figure 3). The average oxidation state of the MnO_2 was 3.94 (*82*). Manganese species with lower oxidation states may have also contributed to transformation of OT and CT, possibly due to reactions with Mn^{2+} (*53*) or to reactions with Mn^{III}(s) formed by oxidation of Mn^{2+} by dissolved O_2 (*88*). Chen and Huang (*53*) hypothesized that complexation of Mn^{2+} by tetracycline facilitated oxidation of the Mn^{2+} to Mn^{III} by molecular O_2 at pH 9, and that the Mn^{III}-bound tetracycline was in turn oxidized by the Mn^{III}. To investigate this possibility, we added an amount of $MnCl_2$ (49 μM) approximately equal to the maximum dissolved concentration of Mn in the MnO_2-amended pond water and examined the transformation of OT and CT. After 20 d exposure, 64% of the initial 100 μM OT and 78% of the CT disappeared. For the 100 μM antibiotic solutions, these losses were larger than observed in pond water alone (8% and 67% for OT and CT), but less than for MnO_2-amended pond water (91% and 79% for OT and CT). The larger antibiotic losses in the presence of MnO_2 relative to $MnCl_2$ alone suggested that Mn^{IV} (or less likely, Mn^{III}) was in part responsible for OT and CT transformation.

Birnessite showed higher reactivity toward OT in pond water than in laboratory water at pH 7.8. The estimated pseudo-first-order loss rates for 10 and 100 μM OT in pond water amended with 490 μM MnO_2 and at 10°C were 4.2 ± 0.7 and 13 ± 1 $\mu M \cdot d^{-1}$. The transformation rates for 10 and 100 μM OT by 490 μM MnO_2 in laboratory water buffered to pH 7.8 and 25°C were 0.003 and 0.01 $\mu M \cdot d^{-1}$ (computed from results in (*54*)). Accounting for the effect of temperature on the rate of OT transformation, we estimate that MnO_2 was $\sim 10^3$-fold more reactive toward OT in the pond water relative to laboratory water buffered to pH 7.8. Chlortetracycline declined much more rapidly than OT in the presence of MnO_2, with 56% and 18% losses over the first minute for initial antibiotic concentrations of 10 and 100 μM. The CT losses observed in the presence of the oxide were larger than those in absence of the oxide, and could not be approximated by a pseudo-first-order model over any time interval.

Birnessite-mediated transformation of OT and CT was accompanied by dissolution of the oxide (Figure 3) and, in absence of a buffer, an increase in solution pH. In absence of antibiotic, $[Mn]_{aq}$ remained < 0.1 μM in the pond water. Reaction of MnO_2 with 10 μM OT or CT produced no measurable Mn (aq) over 5 d. When antibiotic concentrations were increased to 100 μM, OT exposure resulted in liberation of ~52 μM Mn (aq), and CT produced ~9 μM Mn (aq) after 3 d. The apparent free Mn:antibiotic ratios computed for day 3 were approximately 1:1 for OT and 1:7 for CT. Failure to observe Mn dissolution during early stages of our incubations may have been due to resorption of Mn^{2+} to the oxide surface or conproportionation reactions involving conversion of Mn^{IV} from the oxide and soluble Mn^{II} to insoluble Mn^{III}, both of which are favorable in our pH range of interest (*88*).

Products of Tetracycline Antibiotic Transformation in the Presence of MnO_2 in Pond Water

Twenty-day incubations of OT with MnO_2 resulted in the appearance of three chromophore-bearing products, one of which was uniquely attributed to MnO_2 and all of which appeared to be present at much lower levels than the parent compound. The apparent concentration of the manufacturing contaminant ADOT decreased markedly over the course of the incubation in the presence of MnO_2 and to a lesser extent in the presence of $MnCl_2$. The most abundant UV peak by area, A, appeared to be a poorly separated mixture of compounds. This peak had similar intensity in samples from reactions conducted in the presence of $MnCl_2$ and MnO_2. A compound eluting 1.2 min prior to peak A (viz. peak B) formed only in the presence of MnO_2.

We previously identified dmOT (formerly designated OT' (*54*)) as a product of MnO_2-mediated transformation of OT in laboratory water between pH 3 and 9.2. This product, formed by demethylation of the dimethylamino moiety at C-4 of OT, formed transiently in the MnO_2-mediated OT transformation process. In pond water, dmOT accumulated to roughly 10% of the parent compound initially present after 1 d with nearly half in the bound form, and declined slowly thereafter (Figure 3). Formation of dmOT (~8 μM) from 100 μM OT after 20 d in $MnCl_2$ (49 μM)-amended pond water suggests that $MnCl_2$ treatment can induce formation of this product, although whether dmOT formation was due exclusively to less oxidized Mn species (viz. Mn^{2+}, Mn^{III}) expected in these solutions remains uncertain. Lower pH (≤5) favored more rapid formation of dmOT in laboratory water, but this product preferentially accumulated at higher pH (≥8) where degradation was evidently suppressed (data not shown). The amount of dmOT in solution appeared to stabilize and eventually decline at pH values between 3 and 8, suggesting that dmOT underwent subsequent transformations. The structural similarity between dmOT and OT suggests that dmOT may be susceptible to many of the transformations available to OT. The dimethylamino moiety of OT plays a role in ribosome binding (*41*). Alteration at this site is expected to result in diminished antibacterial activity (*41*).

Losses of CT after 20-d exposure to MnO_2 in pond water were accompanied by the appearance of five major products (Figure 4B and Table II). The CT transformation products included the previously described dhiCT, iCT and eiCT (*vide supra*), two products formed in the presence of $MnCl_2$, and one formed in the presence of MnO_2. LC-UV-MS^2 analysis of the product formed exclusively in the presence of MnO_2 revealed the formation of a polyhydroxylated product (PeP) (Table II and Figure 5) which lacked Cl. This product also formed from OT ($[OT]_0$ = 1000 μM, data not shown) in the presence of MnO_2, suggesting that both antibiotics share a common terminal degradation pathway in the presence of MnO_2. Product C (Figure 4 and Table II) formed from CT in the presence of $MnCl_2$ and MnO_2 and had a UV spectrum suggesting its structure differed considerably from CT. A compound eluting 0.9 min before product C (eC) and formed in the presence of $MnCl_2$ and MnO_2 had a UV absorbance spectrum which was nearly identical to C, suggesting it may have been an epimer of that compound.

The relative proportions of iCT and eiCT forming from CT appeared to depend on the solution matrix. In ddH_2O, eiCT accounted for ~3% of the CT transformed, whereas in pond water eiCT accounted for ~13% of the CT transformed, and in pond water amended with MnO_2 it accounted for ~6%. The eiCT:iCT ratios after 20 d for these three treatments were 0.04, 0.28 and 0.44, respectively. These different proportions may reflect epimerization at C-4 occurring independently of the transformation of CT to iCT. Epimerization at C-

4 occurs relatively readily by proton exchange (*41*). Recoveries of eiCT and iCT from samples treated by filtration or acidification were the same within the margin of analytical error within a particular treatment group (MnO_2 or pond water matrix only), suggesting that iCT and eiCT do not exhibit the same propensity for sorption or complexation as does CT.

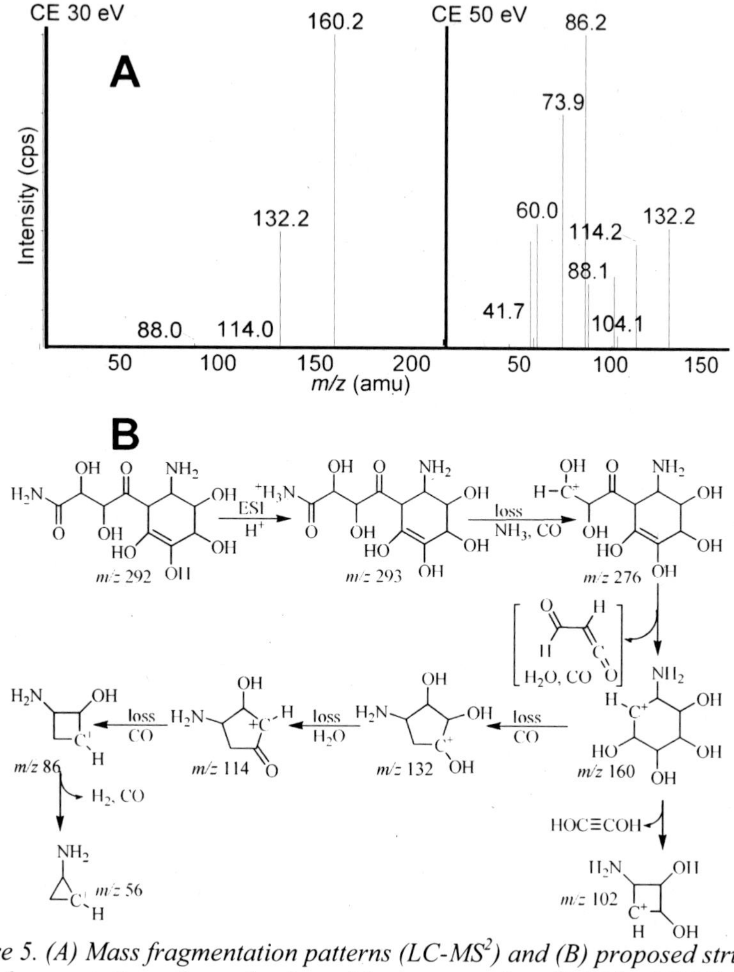

Figure 5. (A) Mass fragmentation patterns (LC-MS²) and (B) proposed structure and fragmentation scheme for the polyhydroxy product (PeP) by LC-MS-TOF and LC-MS². This product formed in pond water as a result of MnO₂-mediated transformation of CT (initially 100 and 1000 μM) and OT (initially 1000 μM).

Environmental Implications

We demonstrated that suspended manganese oxide accelerates the abiotic, non-photolytic transformations of OT and CT in surface water. Chlortetracycline is expected to be rapidly converted to *iso*-CT and epi-*iso*-CT in slightly alkaline waters under low light conditions (as simulated in the present study). In contrast,

OT degraded slowly under the same conditions. While OT and CT transformation was observed in $MnCl_2$-treated pond water, the levels of added Mn^{2+} substantially exceeded those expected in natural waters; therefore, significant transformation of tetracycline antibiotics by Mn^{2+} (aq) in most environmental settings appears unlikely. Large differences in OT and CT recovery determined by filtration or acidification of pond water in absence of added $MnCl_2$ or MnO_2 suggests the imporatnace of tetracycline antibiotic association with aquatic colloids in natural waters. Birnessite appeared more reactive toward OT in pond water relative to laboratory water buffered to the same pH. The ability of other specific constituents of the pond water matrix to enhance tetracycline antibiotic transformation in the presence of MnO_2 warrants further investigation.

The products formed by abiotic, non-photolytic transformations of OT or CT in pond water of similar composition to that used in this investigation are unlikely to pose substantial risks to the environment at the concentrations detected to date. Structure-activity relationships for tetracycline antibiotics suggest that major alteration of the parent compounds will markedly reduce their antibacterial activity (*41*). Oxytetracycline and CT were investigated at concentrations in excess of those reported in the environment, and the transformation products bearing characteristics similar to the parent compounds had apparent concentrations much less than the parent compounds. Products appearing in total ion chromatograms (i.e., in LC-MS^2 analysis) but lacking UV chromophores appeared at low abundance relative to the parent compound, and were not investigated as they were expected to lack the antibacterial activity associated with the parent compound. The two notable exceptions in our study were iCT and eiCT, the levels of which accounted respectively for 32% and 9% of the parent compound after 20 d in pond water. Although the biological activities of iCT and eiCT are more than 1000× less than that of the parent compound, these compounds have been reported at relatively high concentrations (1910 and 15,800 $\mu g.L^{-1}$) in hog waste lagoons, and iCT showed higher stability than the parent compound (*57*).

Acknowledgements

This research was funded by the USDA Cooperative State Research, Education and Extension Service project WIS04902 and the University of Wisconsin Sea Grant Institute supported by grants from the National Sea Grant College Program, National Oceanic and Atmospheric Administration, U.S. Department of Commerce, and the State of Wisconsin, Federal grant NA1GRG2257, project R/AQ-40.

References

1. Boxall, A. B. A.; Fogg, L. A.; Kay, P.; Blackwell, P. A.; Pemberton, E. J.; Croxford, A. *Toxicol. Lett.* **2003**, *142,* 207-218.

2. Kolpin, D. W.; Furlong, E. T.; Meyer, M. T.; Thurman, E. M.; Zaugg, S. D.; Barber, L. B.; Buxton, H. T. *Environ. Sci. Technol.* **2002**, *36,* 1202-1211.
3. Dietze, J. E.; Scribner, E. A.; Meyer, M. T.; Kolpin, D. W. *Int. J. Environ. Anal. Chem.* **2005**, *85*, 1141-1152.
4. Koeypudsa, W.; Yakupitiyage, A.; Tangtrongpiros, J. *Aquac. Res.* **2005**, *36,* 570-577.
5. Kim, S.; Carlson, K. *Environ. Sci. Technol.* **2007**, *41*, 50-57.
6. Samuelsen, O.B.; Torsvik, V.; Ervik, A. *Sci. Total Environ.* **1992**, *114*, 25-36.
7. Hamscher, G.; Priess, B.; Nau, H. *Archiv für Lebensmittelhygiene*. **2006,** *57,* 97-101.
8. Hamscher, G.; Pawelzick, H. T.; Höper, H.; Nau, H. *Environ. Toxicol. Chem.* **2005**, *24*, 861-868.
9. Kumar, K.; Gupta, S. C.; Baidoo, S. K.; Chander, Y.; Rosen, C. J. *J. Environ. Qual.* **2005,** *34,* 2082-2085.
10. Farkas, H. M.; Berry, J. O.; Aga, D. S. *Environ. Sci. Technol.* **2007,** *41,* 1450-1456.
11. O'Reilly, A.; Smith, P. *Aquaculture* **2001,** *196,* 13-26.
12. Thiele-Bruhn, S. *Environ. Toxicol. Chem.* **2005,** *24,* 869-876.
13. Vaclavik, E.; Halling-Sørensen, B.; Ingerslev, F. *Chemosphere* **2004,** *56,* 667-676.
14. Sassman, S. A.; Lee, L. S. *Environ. Sci. Technol.* **2005**, *39*, 7452-7459.
15. MacKay, A. A.; Figueroa, R. A. Mechanisms of antibiotic sorption to agricultural soils. *Abstr. ACS Natl. Meet., Am. Chem. Soc., Div. Environ. Chem.* Aug 22-26, **2004**, 1426-1430.
16. Jones, A. D.; Bruland, G. L.; Agrawal, S. G.; Vasudevan, D. *Environ. Toxicol. Chem.* **2005**, *24*, 761-770.
17. Tolls, J. *Environ. Sci. Technol.* **2001,** *35*, 3397-3406.
18. Pils, J. R. V.; Laird, D. A. *Environ. Sci. Technol.* **2007**, *41*, 1928-1933.
19. Figueroa, R. A.; Leonard, A.; Mackay, A. A. *Environ. Sci. Technol.* **2004**, *38*, 476-483.
20. Kulshrestha, P.; Giese, R. F.; Aga, D. S. *Environ. Sci. Technol.* **2004**, *38,* 4097-4105.
21. Doi, A. M.; Stoskopf, M. K. *J. Aquat. Anim. Health.* **2000**, *12*, 246-253.
22. Kim, S.; Eichhorn, P.; Jensen, J. N.; Weber, A. S.; Aga, D. S. *Environ. Sci. Technol.* **2005**, *39*, 5816-5823.
23. Gu, C.; Karthikeyan, K. G.; Sibley, S. D.; Pedersen, J. A. *Chemosphere* **2007**, *66*, 1494-1501.
24. MacKay, A. A.; Canterbury, B. *J. Environ. Qual.* **2005**, *34*, 1964-1971.
25. Alegakis, A. K.; Tzatzarakis, M. N.; Tsatsakis, A. M.; Vlachonikolis, I. G.; Liakou, V. *J. Environ. Sci. Health, Part B*. **2000**, *35*, 559-569.
26. Figueroa, R. A.; MacKay, A. A. *Environ. Sci. Technol.* **2005**, *39*, 6664-6671.
27. Gu, C.; Karthikeyan, K. G. *Environ. Sci. Technol.* **2005**, *39*, 2660-2667.
28. Hassan, S. S. M.; Amer, M. M.; Ahmed, S. A. *Mikrochim. Acta* **1985**, *3*, 165-175.
29. Ding and Fu, 1993, Ding, X.; Fu, C. *Talanta* **1993**, *40*, 641-644.

30. Zhang, X.; Liu, Y.; Wang, L.; Lin, C. *Journal of China University of Science and Technology* **2001**, *31*, 213-217.
31. Novák-Pékli, M.; Mesbah, M. E.; Pethő, G. *J. Pharm. Biomed. Anal.* **1996**, *14*, 1025-1029.
32. Brion, M.; Lambs, L.; Berthon, G. *Agents Actions* **1985**, *17*, 229-242.
33. Lambs, L.; Brion, M.; Berthon, G. *Agents Actions* **1984**, *14,* 743-750.
34. Lambs, L.; Venturini, M.; Révérend, B. D.; Kozlowski, H.; Berthon, G. *J. Inorg. Biochem.* **1988**, *33*, 193-210.
35. Lambs, L.; Révérend, B. D.; Kozlowski, H.; Berthon, G. *Inorg. Chem.* **1988**, *27*, 3001-3012.
36. Tongaree, S.; Goldberg, A.M.; Flanagan, D.R.; Proust, R.I. *Pharm. Dev. Technol.* **2000**, *5*, 189-1999.
37. Tantuvay, L.; Khan, F. *Bull. Electrochem.* **2004**, *20*, 327-331.
38. Croubels, S. ; Baeyens, W. ; Peteghem, C. V. *Analytica Chimica Acta.* **1995**, *303*, 11-16.
39. Brion, M.; Lambs, L.; Berthon, G. *Inorg. Chim. Acta* **1986**, *123*, 61-68.
40. Jezowska-Bojczuk, M.; Lambs, L.; Kozlowski, H.; Berthon, G. *Inorg. Chem.* **1993**, *32*, 428-437.
41. Mitscher, L. A. *The Chemistry of the Tetracycline Antibiotics;* Marcel Dekker: New York, NY, 1978.
42. Thiele-Bruhn, S. *J. Plant Nutr. Soil Sci.* **2003**, *166*, 145-167.
43. Clive, D. L. J. Chemistry of tetracyclines. *Quarterly Reviews* **1968**, *22*, 435.
44. Ingerslev, F.; Torang, L.; Loke, M. L.; Halling-Sørensen, B., Nyholm, N. *Chemosphere* **2001,** *44,* 865-872.
45. Werner, J. J.; Arnold, W. A.; McNeill, K. *Environ. Sci. Technol.* **2006,** *40,* 7236-7241.
46. Hlavka, J. J.; Bitha, P. *Tetrahedron Lett.* **1966**, 3843-3846.
47. Hasan, T.; Allen, M.; Cooperman, B. S. *J. Org. Chem.* **1985**, *50*, 1755-1757.
48. Davies, A. K.; McKellar, J. F.; Phillips, G. O.; Reid, A. G. *J. Chem. Soc. Perkin Trans. II* **1979,** 369-375.
49. diPaola, A.; Addamo, M.; Garcia-Lopez, V. A. E.; Loddo, V.; Marci, G.; Palmisano, L. *Fresenius Environ. Bull.* **2004,** *13*, 1275-1280.
50. Oka, H.; Ikai, Y.; Kawamura, N.; Yamada, M.; Harada, K.; Ito, S.; Suzuki, M. *J. Agric. Food Chem.* **1989,** *37*, 226-231.
51. Bautitz, I. R.; Nogueira, R. F. P. Degradation of tetracycline by photo-Fenton process - Solar irradiation and matrix effects. *J. Photochem. Photobiol., A* **2007,** *187*, 33-39.
52. Herbert, V.R.; Miller, G.C. *J. Agric. Food Chem.* **1990,** *38*, 913-918.
53. Chen, W. R.; Huang, C. H. *Abstr. ACS Natl. Meet., Am. Chem. Soc., Div. Environ. Chem.* **2006,** 231.
54. Rubert, K. F.; Pedersen, J. A. *Environ. Sci. Technol.* **2006,** *40,* 7216-7221.
55. Sanderson, H.; Ingerslev, F.; Brain, R. A.; Halling-Sørensen, B.; Bestari, J. K.; Wilson, C. J.; Johnson, D. J.; Solomon, K. R. *Chemosphere* **2005,** *60,* 619-629.
56. Halling-Sørensen, B.; Lykkeberg, A.; Ingerslev, F.; Blackwell, P.; Tjørnelund, J. *Chemosphere.* **2003,** *50,* 1331-1342.

57. Eichhorn, P.; Aga, D. S. Identification of a photooxygenation product of chlortetracycline in hog lagoons using LC/ESI-ion trap-MS and LC/ESI-time-of-flight-MS. *Anal. Chem.* **2004,** *76,* 6002-6011.
58. Pouliquen, H.; Delpepee, R.; Larhantec-Verdier, M.; Morvan, M. L.; Le Bris, H. *Aquaculture* **2007,** *262,* 23-28.
59. Søeborg, T.; Ingerslev, F.; Halling-Sørensen, B. *Chemosphere* **2004,** *57,* 1515-1524.
60. Bonnet, J. L.; Bonnemoy, F.; Dusser, M.; Bohatier, J. *Environ. Toxicol.* **2007,** *22,* 78-91.
61. Tixier, C.; Sancelme, M.; Ait-Aissa, S.; Widehem, P.; Bonnemoy, F.; Cuer, A.; Truffaut, N.; Veschambre, H. *Chemosphere* **2002,** *46,* 519-526.
62. Bondy, G.; Armstrong, C.; Coady, L.; Doucet, J.; Robertson, P.; Feeley, M.; Barker, M. *Food Chem. Toxicol.* **2003,** *41,* 291-301.
63. Kralj, A. B.; Franko, M.; Trebse. P. *Chemosphere* **2007,** *67,* 99-107.
64. Essington, M.E. *Soil and Water Chemistry;* CRC Press: New York, NY, 2004.
65. Post, J. E. *Proc. Natl. Acad. Sci.* **1999,** *96,* 3447–3454.
66. Cornell, R.M.; Giovanoli, R. *Clays Clay Miner.* **1988,** *36,* 249-257.
67. Taylor, R.M.; McKenzie, R.M.; Norrish, K. *Aust. J. Soil Res.* **1964,** *2,* 235-248.
68. Chukhrov, F.V.; Gorschkov, A.I. *Tans. R. Soc. Edinburgh: Earth Sci.* **1981,** *72,* 195-200.
69. Koljonen, T.; Lahermo, P.; Carlson, L. *Bull. Geol. Soc. Finl.* **1976,** *48,* 111-135.
70. Carlson, L.; Koljonen, T.; Lahermo, P.; Rosenberg, R. J. *Bull. Geol. Soc. Finl.* **1977,** *49,* 159-173.
71. Chang, R. *Chemistry*; McGraw-Hill: New York, NY, 1998; p 766.
72. Bialk, H. M.; Pedersen, J. A. Oxidative coupling of human and veterinary pharmaceuticals with monomeric humic precursors. In *Humic Substances and Soil and Water Environment*; Martin-Neto, L., Milori, D. M. B. P., daSilva, W. T. L., Eds.; International Humic Substances Society: São Paulo, Brazil, 2004; pp 582-584.
73. Kämpf, N.; Scheinost, A. C.; Schulze, D. G. Oxide Minerals. In *Handbook of Soil Science*; Sumner, M. E., Ed.; Taylor and Francis: Boca Raton, FL, 2000, pp. F125-F168.
74. Rubert, K. F. ; Gao, J. ; Pedersen, J. A. Oxytetracycline transformation on mineral surfaces. 2004. *Abstr. Soc. Environ. Toxicol. Chem.* 2004
75. McKenzie, R. M. Manganese Oxides and Hydroxides. In *Minerals in Soil Environments,* 2nd ed.; Dixon, J. B.; Dixon, S. B., Eds.; Soil Science Society of America: Madison, WI, 1989, *pp*. 439-465.
76. Li, H.; Lee, L. S.; Schulze, D. G.; Guest, C. A. *Environ. Sci. Technol.* **2003**, *37*, 2686-2693.
77. Jokic, A.; Frenkel, A. I.; Vairavamurthy, M. A.; Huang, P. M. *Geophys. Res. Lett.* **2001,** *28*, 3899-3902.
78. Jokic, A.; Wang, M. C.; Liu, C.; Frenkel, A. I.; Huang, P. M. *Org. Geochem.* **2004,** *35*, 747-762.
79. Wang, D. J.; Shin, J. Y.; Cheney, M. A.; Sposito, G.; Spiro, T. G. *Environ. Sci. Technol.* **1999,** *33*, 3160-3165.

80. Li, H.; Lee, L. S.; Jafvert, C. T.; Graveel, J. G. *Environ. Sci. Technol.* **2000,** 34, 3674-3680.
81. Jung, J.-W.; Lee, S.; Ryu, H.; Kang, K.-H.; Nam. K. *J. Environ. Sci. Health A* **2008**, *43*, 255-261.
82. Gao, J. Ph.D. thesis, University of Wisconsin, Madison, WI, 2007.
83. Murray, J. W. *J. Colloid Interf. Sci.* **1974,** *46,* 357-371.
84. Lindsey, M. E.; Meyer, M.; Thurman, E. M. *Anal. Chem.* **2001,** *73,* 4640-4646.
85. Stumm, W.; Morgan, J. J. *Aquatic chemistry : Chemical Equilibria and Rates in Natural Waters*, 3rd Ed.; Wiley: New York, NY, 1996.
86. Halling-Sørensen, B.; Sengelov, G.; Tjornelund, J. *Arch. Environ. Contam. Toxicol.* **2002,** *42,* 263-271.
87. Lykkeberg, A. K.; Sengelov, G.; Cornett, C.; Tjørnelund, J.; Hansen, S. H.; Halling-Sørensen, B. *J. Pharm. Biomed. Anal.* **2004,** *34,* 559-567.
88. *Standard Potentials in Aqueous Solution*; Bard, A. J.; Parsons, R.; Jordan, J., Eds.; Marcel-Dekker: New York, NY, 1985.

Chapter 6

Comparative Biotransformation of Fluoroquinolone Antibiotics in Matrices of Agricultural Relevance

Heinz-Georg Wetzstein[1], Josef Schneider[2], and Wolfgang Karl[3]

[1]Bayer HealthCare AG, Animal Health, 51368 Leverkusen
[2]Bayer CropScience AG, Metabolism and Environmental Fate, 40789 Monheim am Rhein
[3]Bayer Industry Services GmbH, 51368 Leverkusen, Germany

Fluoroquinolones are synthetic antibiotics widely used to treat bacterial infections in humans and animals. Drug residues enter the environment via urine and feces. In the only soil degradation study yet published, ≤0.6% of the ^{14}C label applied with [2-^{14}C]sarafloxacin had been recovered as $^{14}CO_2$ after 80 days. This prompted our work on the fate of enrofloxacin in agricultural matrices. Pure cultures of basidiomycetous fungi thriving on native wheat straw could release up to 50% of the ^{14}C label, initially added with [4-^{14}C]enrofloxacin, as $^{14}CO_2$ within 8 weeks. However, in samples of pre-rotted wheat straw, cattle dung pats, a manure hill, and a soil only about 0.5 to 1% of $^{14}CO_2$ was produced over one year, while production rates of $^{14}CO_2$ from [piperazine-2,3-^{14}C]enrofloxacin were on the order of 7 to 40%. Such rates indicated a limited utility of recording $^{14}CO_2$ formation. Close to quantitative biotransformation in two soils, a plant-derived compost, and cattle dung ($t_{1/2}$ = 83 to 113 days) could only be demonstrated by monitoring the fate of parent drug based on chemical analysis. Most of the metabolites identified had lost antibacterial activity.

Fluoroquinolone Antibacterial Drugs

Fluoroquinolones (FQs) are synthetic antibiotics widely used to treat severe bacterial infections in humans and animals (*1*). Though the drug class is said to comprise more than 10,000 compounds, at present, only about half a dozen are utilized in each field, human and veterinary medicine. Annual FQ consumption in 30 industrialized countries has been estimated in 1997 at 800 t on the human and 50 t on the veterinary side (*2*). FQs inhibit bacterial type II topoisomerases, i.e., DNA gyrase and topoisomerase IV. These enzymes catalyze negative supercoiling of DNA or facilitate the separation of replicated, then interlinked daughter chromosomes, respectively. Hence, they are crucial for maintaining the physiological structure of DNA, and effective enzyme inhibition causes bacterial death (*1*, *3*). The work presented herein focuses on veterinary use of FQs, for which the primary indications are respiratory, gastro-intestinal tract, and skin infections in cattle, swine, dogs, cats, and additional species (*4*).

Intact FQs and metabolites enter the environment via urine and feces (*5*). On the human side, drug residues have been detected in wastewater streams (*6*, *7*). On the veterinary side, residues will be present in the waste of treated animals which, after a period of storage, may eventually be spread as fertilizer on agricultural fields and pastures. However, the mobility and bioavailability of FQs are greatly restricted because of strong binding to feces (see refs. 6, 14, 20, 26 in [*8*]) and soils (K_d = 260 to 5600 [*9*, *10*]), which is facilitated by a positive and a negative charge present at pH 7 at the distal N of the amine substituents and the carboxyl group, respectively (Figure 1). As tight binding may also reduce the rate of biodegradation, concerns have been expressed about FQ biodegradability in general, an accumulation of drug residues in agricultural soils with concomi-tant build up of selective pressure for bacterial FQ resistance, as well as poten-tial selection and spread of resistant clones via the food chain, possibly com-promising the utility of FQs for the treatment of infections in humans (*11-17*).

Enrofloxacin (ENR), the first veterinary FQ (Figure 1), is composed of a quinolone core carrying: a cyclopropyl substituent at N-1, an essential carboxylic group at C-3, a fluorine atom at C-6 (the F-aromatic bond has not yet been observed in a natural product [*18*]), and an amine substituent, ethylpiperazine (pip), at C-7. Considerable structural diversity is encountered among the veterinary FQs, as illustrated by sara-, dano-, marbo- and orbifloxacin (Figure 1). In our work, ENR and three additional FQs have been evaluated, including ciprofloxacin (CIP), pradofloxacin (PRA) - a new veterinary FQ under deve-lopment (*19*), and moxifloxacin (MFX; Figure 1). The latter two differ in the substituent at position C-8, a cyano and a methoxy group, respectively, but have in common a bicyclic type of amine, a specific pyrrolidinopiperidine (pyr; [*19*]). CIP and MFX are applied in human medicine (*1*, *20*) and served herein as structurally analogous controls. Of these four drugs, ten ^{14}C-labeled variants could be utilized: (i) Those of ENR comprised of [2-^{14}C]-, [4-^{14}C]-, [pip-2,3-^{14}C]-, [^{14}COOH]ENR, as well as its [^{14}COOH]-decyclopropyl-1-H congener. (ii) Of CIP, [4-^{14}C]- and [pip-2,3-^{14}C]-labeled molecules were used. The set of probes was completed by (iii) [2-^{14}C]- and [pyr-7-^{14}C]PRA as well as (iv) [3-^{14}C]MFX. Specific activities and purities were similar to those reported before (*8*, *21*), and all compounds had been purified immediately before usage.

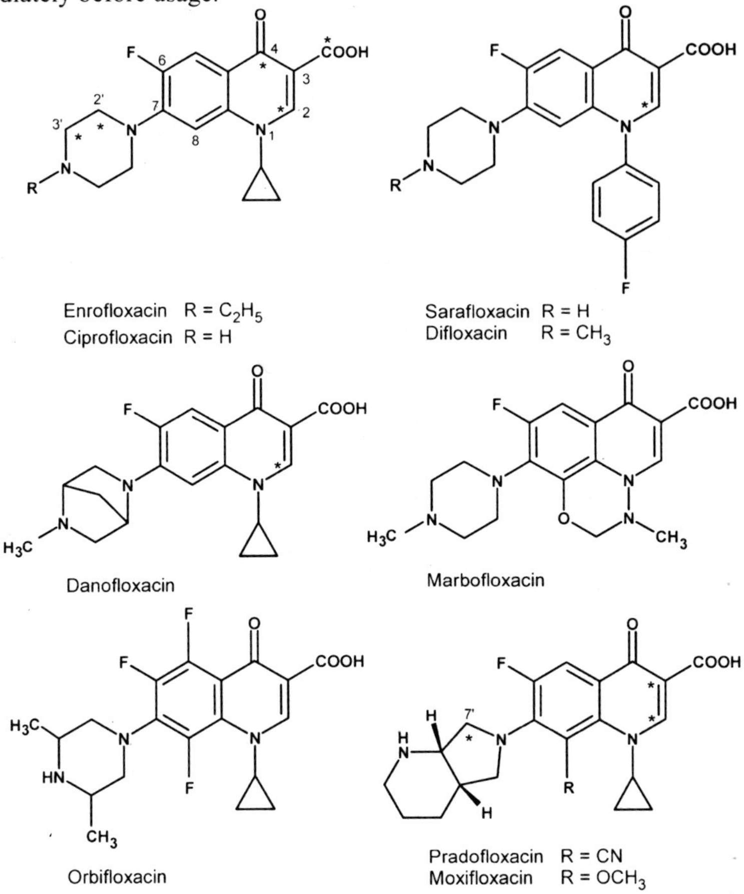

Figure 1. Chemical structures of veterinary fluoroquinolones and two reference compounds used in human medicine, cipro- and moxifloxacin. Alternative positions of the ^{14}C label are indicated (*). (Adapted with permission from reference 19. Copyright 2005 ASM Press.)

Assessing Biotransformation by Recording $^{14}CO_2$ Formation

The only report yet published on assessing biodegradation of a FQ in soils appeared in 1993 (*22*): [2-^{14}C]sarafloxacin had been added to three soils, but only ≤0.6% of the applied label could be recovered as $^{14}CO_2$ after 80 days (*23*). Despite this carefully executed analysis (compare with [*24*]), it had to be left undecided whether such rates of $^{14}CO_2$ formation indicated FQ biodegradation or rather long-term persistence, taking into consideration the radiochemical purity of sarafloxacin of 97%. As internal studies at Bayer AG had provided similar results for ENR, we chose an alternative experimental approach, initially defining optimal *in vitro* conditions for pure cultures of soil fungi, which may represent 50 to 80% of the soil biomass. Other groups studied bacteria (*25, 26*) and additional, quite different fungal taxa (*26-28*). We then embarked on investigating the fate of ENR in matrices of agricultural relevance such as wheat straw, cattle manure, and plant-derived humus before returning to agricultural soils, as will be described below.

Pure Cultures of Basidiomycetous Fungi on Wheat Straw

Ground native wheat straw, the first model matrix, was supplemented by drop-wise addition of an aqueous solution of [4-^{14}C]ENR, resulting in a final concentration of 10 μg ENR per g of straw. After inoculation with a fungal species, cultures were incubated at 25°C, in the dark, and under continuous aeration (*29*). Already in the second week, $^{14}CO_2$ was released at the maximum rate observed, 19% of the initially added ^{14}C per week, reaching a total of >50% $^{14}CO_2$ after 8 weeks. *Gloeophyllum striatum*, a species naturally causing brown-rot decay of wood by employing a Fenton-type, hydroxyl radical-based mechanism (*21, 30, 31*), was found to be most active. Such unexpectedly high rates of $^{14}CO_2$ formation suggested that biodegradability of ENR was no principal problem, if the drug was readily available, just as when bound to straw. The test system employed (type I) was similar to the FDA reference (*24*) and worked satisfactorily. However, its technical complexity limited the experimental capacity to ≤30 microcosms at a time.

Bioavailability – Just One of the Parameters Affecting Degradation of FQs

Our further approach was guided by the following observations (*29*): If an aqueous solution of [4-^{14}C]ENR had been added drop-wise to an agricultural soil, release rates of 0.09% of $^{14}CO_2$ were observed after 8 weeks (Figure 2), even much lower than the rates reported for sarafloxacin. Furthermore, combining the catalytic power of an active fungal wheat straw culture with the soil environment confirmed the low rates of $^{14}CO_2$ formation: [4-^{14}C]ENR had been added to either native or gamma-sterilized soil (a loam soil specified in [*29*]) at a final concentration of 10 μg/g (dry weight), and each type of soil was placed on top of a straw culture of *G. striatum*. With gamma-sterilized soil, a massive penetration of the soil by mycelium could be observed. After an adaptation phase of 4 weeks, the rate of $^{14}CO_2$ release reached 0.9% of the applied radioactivity per week (~5% of the rate of straw cultures), and such rates were stably maintained for several weeks (Figure 2). Therefore, at least a fraction of the soil-bound ENR (*9*) had remained accessible to the fungal hyphae.

However, when native loam soil containing [4-^{14}C]ENR was placed on top of active fungal cultures, after an attempted initial burst of $^{14}CO_2$ formation (Figure 2), only the low rates typical of soil were observed again (*29*). Obviously, even a massive inoculum of a fungus competent at FQ degradation could not compete with the native microbial community, though it must be remembered that *G. striatum* may not be typical for the soil ecosystem. The sparse aerial mycelium, which initially had developed, disappeared within two weeks. So, in addition to drug binding, either a competitive exclusion-type phenomenon, associated with the native microbial population, or the chemical environment in such soil, or both had prevented a high-potential degrader from releasing $^{14}CO_2$ at a rate, which would have been facilitated by the degree of ENR bioavailability observed with gamma-sterilized soil. It should be noted that the mode of ENR application to soil samples differed from agricultural practice: residues will be entering soils tightly bound to an organic matrix, i.e., animal waste.

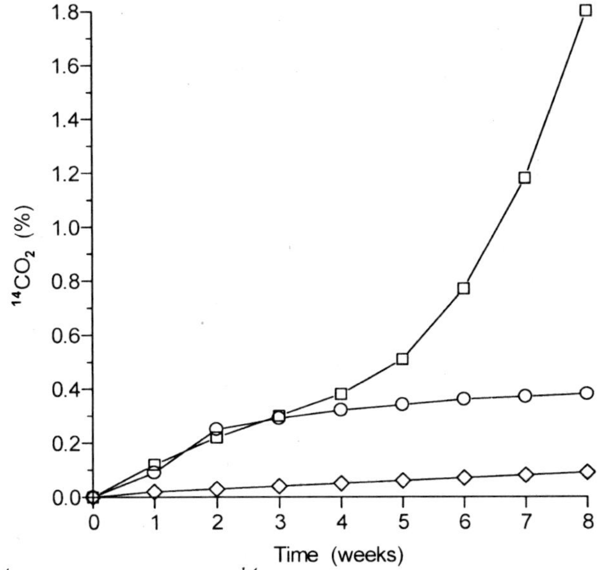

Figure 2. $^{14}CO_2$ release from [4-^{14}C]enrofloxacin bound to native agricultural soil (◊), and after placing such soil on top of wheat straw cultures of G. striatum (○). Alternatively, the soil had been gamma-irradiated previous to the addition of enrofloxacin and then placed on straw cultures of G. striatum (□). (Adapted with permission from reference 29. Copyright 1996 ASM Press.)

Test System II and its Application in Recording $^{14}CO_2$ Formation

A culture vessel more convenient than test system I (*24*) was made available to us by J. P. E. Anderson, who had characterized the consequences of storing unsupplemented soils under conditions used in degradation experiments, namely, drastic declines in the biomass and biodegradation activity occurring after approx. 70 days of incubation (*32*, *33*); this proved to be relevant for interpreting our soil experiments described below. Test system II comprised of a sterile 100-ml Duran® glass bottle with a terephthalate screw-cap (Schott, Mainz, Germany) and a perbuan cap liner, centrally holding a 10-ml test tube with 3 ml of 1 M NaOH as $^{14}CO_2$-absorbing solution. Above the NaOH, a paraffin-impregnated cotton wool plug was placed for absorbing volatile metabolites, though from FQs, no such metabolites have ever been detected.

Two bore holes of 5 mm in the test tube wall allowed for a free access of $^{14}CO_2$ to the NaOH solution.

Samples of 1 g of an agricultural matrix were supplemented by drop-wise addition of an aqueous FQ stock solution, providing the ^{14}C-label at ≥3.7 kBq (≥2 x 10^5 dpm), and incubated at room temperature in the dark to prevent photo-degradation effects. At each sampling time, the NaOH solution was replaced and the radioactivity determined by liquid scintillation counting (LSC). During the time required for this exchange (approx. 20 min), the atmosphere in the vessel was replaced by ambient air. The water content of each matrix (specified below) was monitored by weighing and re-adjusted, as up to 0.1 ml of water was lost per month from the matrix to the NaOH solution, and long-term incubation for up to 2 years was planned. In Figures 3 through 9, arrow heads mark the addition of 1 ml of sterile water, and numerals placed near the end of selected graphs indicate the total recovery of ^{14}C-label, which comprised: $^{14}CO_2$ produced, $^{14}CO_2$ collected upon combustion of the solid matrix at 900°C in a Harvey Oxidizer 500, and ^{14}C label recovered by rinsing of the culture vessels. Values shown represent the mean ± SD of at least three independent cultures.

The ecological significance of low rates of $^{14}CO_2$ formation remained uncertain, even when up to 7% of the ^{14}C label applied with [pip-2,3-^{14}C]ENR (the alternative amine label position) was released as $^{14}CO_2$ over one year, when applied to two German soils characterized below (Figure 3). The concentration of ENR was 33 μg/g of soil. In another screening experiment, rates of $^{14}CO_2$ formation from [4-^{14}C]ENR and [pip-2,3-^{14}C]ENR on the order of 0.06 to 0.39% and 0.8 to 11.6% of the ^{14}C applied, respectively, were determined after 8 weeks with 33 soils collected in Germany, France, Ireland, Australia, Philippines, United States, Honduras, Ecuador, Chile, and Ivory Coast, with soil pHs ranging from 3.4 to 7.8. Three grams of soil had been suspended in 10 ml of water and supplemented with 100 μg of ENR (*34*). Many agriculturally managed soils contain 1-3% of organic carbon (*35-37*). In an attempt to minimize retardation of the biodegradation process, potentially due to drug binding to the inorganic soil compartment, we then focused on organic matrices.

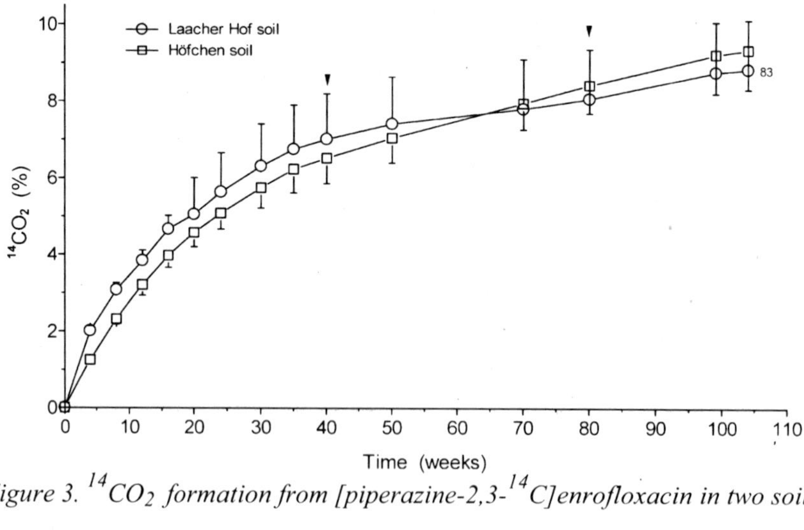

Figure 3. $^{14}CO_2$ formation from [piperazine-2,3-^{14}C]enrofloxacin in two soils.

Wheat Straw Pre-rotted in Agricultural Soil

Pre-rotted straw was expected to provide the authentic microbiota involved in the degradation of straw left behind on fields. So native wheat straw was disintegrated with a blender, packaged at quantities of 20 g in medical gauze, autoclaved, and wrapped in a mosquito net-type polyamide tissue, Monodur PA 750 N (Verseidag-Techfab GmbH, Geldern-Walbeck, Germany), with an open surface area of 54%; the width of the squared meshes was 750 ± 75 μm. In mid April 1997, such packages were placed within the upper 10 cm of a fallow field located at 51°04'16"N, 6°53'59"E, which had not been treated with a herbicide, pesticide or fungicide for several years. After 4 to 52 weeks of soil exposure at the moderate middle European climate, several packages were recovered each sampling time. The loamy sand soil was largely excluded by sampling from the core of the packages. Aliquots of 1 g of pre-rotted straw were supplemented with 3 ml of water containing a ^{14}C-labeled FQ; if just 1 ml of water had been applied, the rate of $^{14}CO_2$ formation was declining from week 26 onwards (note also in this context the stimulatory effect of water added after 90 to 100 weeks in Figures 4 and 5). The final drug concentration was on the order of 66 to 110 μg/g (dry weight), as the water content of pre-rotted straw ranged from 50 to 70% during the year.

The $^{14}CO_2$ release kinetics obtained indicated that 0.5 to 2% of the initially added ^{14}C was released as $^{14}CO_2$ from [4-^{14}C]ENR, [4-^{14}C]CIP*, [2-^{14}C]ENR*, [^{14}COOH]ENR, [2-^{14}C]PRA, and [3-^{14}C]MFX after one year; [4-^{14}C]ENR has been selected as a representative example and is shown in Figure 4 (*, the experiment was performed with 52-week-rotted straw only). However, from both types of amine substituents, i.e., [pip-2,3-^{14}C]ENR (Figure 5) or [pip-2,3-^{14}C]-CIP* and [pyr-7-^{14}C]PRA, and, especially notable, also from the [^{14}COOH]-decylopropyl-1-H congener of ENR*, rates of $^{14}CO_2$ formation were on the order of 25 to 40% per year (data not shown in detail). In general, activities were lower (with higher standard deviations) with straw pre-rotted for only four weeks. Otherwise, no great difference in $^{14}CO_2$ release rates was observed. Material balances were close to 100%.

As the carboxyl group is essential for the antibacterial activity of FQs (*1*, *3*), placing the ^{14}C label at that position had been anticipated to provide an ideal probe for detecting inactivation of ENR. However, the presence of the cyclopropyl substituent in ENR was found to retard $^{14}CO_2$ formation from [^{14}COOH]ENR (note the ranges of activity stated for [^{14}COOH]ENR and its decyclopropyl analog in the previous paragraph). In controls which had received 40 μg of [1-^{14}C]glucose or [U-^{14}C]alanine, 75% of the applied label was released as $^{14}CO_2$ within one week, reaching a total of about 85% after two years (similar to Fig. 6.6 in [*36*]). Obviously, such substrates were unlikely to cause prolonged stimulation of the soil microbial population. $^{14}CO_2$ produced in gamma-sterilized (30 to 50 kGy), 26-week-rotted wheat straw, wetted by adding ENR, may have indicated chemical, possibly irradiation-enhanced decomposition of [4-^{14}C]ENR and [pip-2,3-^{14}C]ENR. However, those rates were in the order of ≤0.1 and 1% per year, respectively.

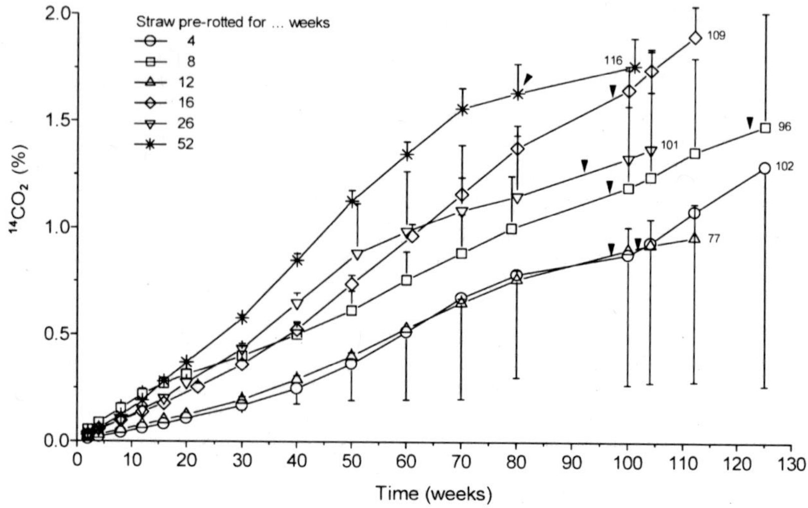

Figure 4. $^{14}CO_2$ formation from [4-^{14}C]enrofloxacin in pre-rotted wheat straw.

Cattle Waste Pile

Diseased cattle are kept straw-bedded inside the barn in most European countries. Therefore, drug residues will be present in waste material, which often will have been stored on piles for prolonged times before being applied to fields. We are not aware of reported FQ concentrations in cattle manure or manure-amended agricultural soils. However; chicken feces may contain up to 15 µg ENR per gram (*38*). During storage, the complex rotting and humification process sets in, often visualized by steaming dung piles. Samples were taken from the surface of a one-week-old waste pile from non-medicated cattle, and aliquots of 2 g were spiked with 3 ml of ^{14}C-labeled FQs, resulting in a final concentration of about 16 µg/g. The rates of $^{14}CO_2$ liberation (Figure 6) were only slightly higher than those obtained for pre-rotted straw or cattle dung pats (see below), amounting to 1.5% from [4-^{14}C]ENR and [4-^{14}C]CIP, 2.6% from [^{14}COOH]-ENR, and 5% from [2-^{14}C]-PRA, but 15 to 30% per year of the ^{14}C applied with the amine labels in ENR, CIP, and PRA (Table 1).

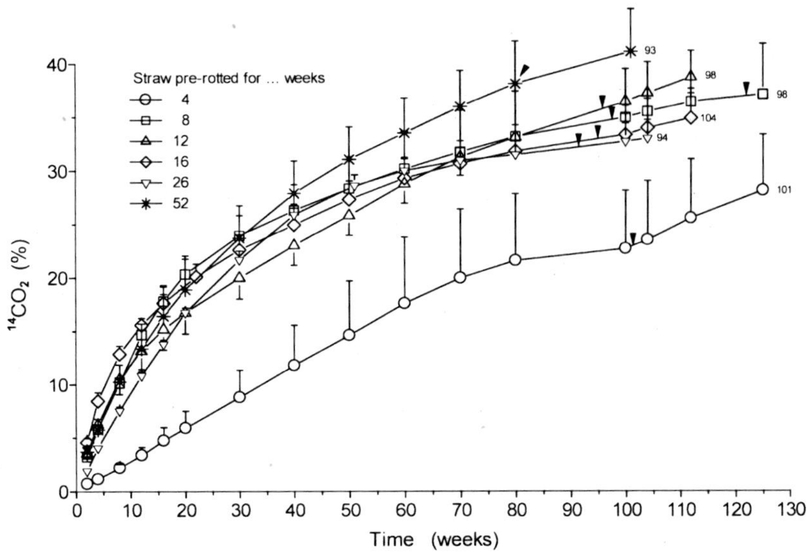

Figure 5. $^{14}CO_2$ *formation from [piperazine-2,3-*^{14}*C]enrofloxacin in pre-rotted wheat straw.*

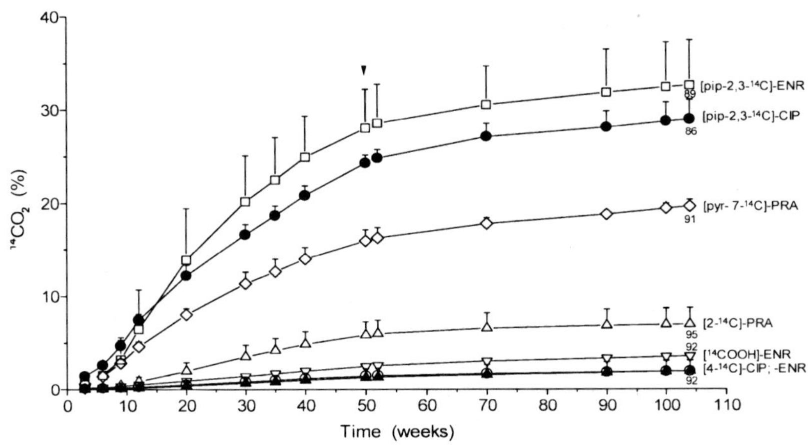

Figure 6. $^{14}CO_2$ *release from various* ^{14}*C label positions in enro-, prado-, cipro-, and moxifloxacin in samples obtained from a cattle waste pile.*

Cattle Dung Pats

Being a premier model ecosystem, cattle dung has been studied extensively (*39*). Here, two groups of Holstein-Friesian heifers with recorded health status, around 15 months of age and not being treated with a FQ, were enrolled. One group was kept inside the barn, feeding on a maize/hay silage. The dung pats were defined as "type A", and $^{14}CO_2$ release kinetics are shown in columns A of

Table 1. Biotransformation of selected fluoroquinolones⁺ in matrices of agricultural relevance

Matrix	$^{14}CO_2$ (%) per year from ^{14}C label at		$T_{1/2}$ of [4-^{14}C]ENR
	Core	Amine substituent	(days)
Wheat straw			
native	40 - 50 *	nt	nt
pre-rotted	0.5 - 1	20 - 30	nt
Cattle dung			
pats	0.5 - 1	10 - 35	84 ± 6
waste pile	2 - 5	20 - 30	nt
Compost	≤0.5	10 - 13	90 - 113
Agricultural soils			
Höfchen I	nt	nt	83
Höfchen II	nt	6.5	≤120
Laacher Hof	nt	7	≤103

⁺ Cipro-, enro-, moxi- and pradofloxacin; * after 8 weeks of incubation; nt, not tested

Figures 7a and 7b. The other group was kept grazing on a pasture near Burscheid, Germany, providing dung of "type B" (see columns B of Figures 7a and 7b). Fresh dung pats from both groups were placed on a pasture at 51°04'33"N, 7°07'48"E and left exposed to the natural decay process (with the cattle fenced off). Upon sampling, equal amounts of material were collected from the top, center and lower part of the dung pats. Already after 8 to 12 weeks, fresh dung pats had undergone transformation into a loose litter layer covering the soil surface. Aliquots of 1 g of cattle dung were supplemented with 3 ml of either [4-^{14}C]ENR, [2-^{14}C]PRA, [pip-2,3-^{14}C]ENR, or [pyr-7-^{14}C]PRA. At selected sampling times, additional labels were included such as [pip-2,3-^{14}C]-CIP, [^{14}COOH]ENR, the [^{14}COOH]-decyclopropyl-1-H congener of ENR, and [3-^{14}C]MFX. The nominal concentration of FQs was 33 µg/g.

Very similar kinetics were obtained with samples drawn after 4, 8, 12 and 20 weeks. Rates of $^{14}CO_2$ production from ^{14}C at core positions fell in the order of 0.5 to 2% per year (Figure 7a), in contrast to 5 to 30% of $^{14}CO_2$ liberated from the amine labels in [pip-2,3-^{14}C]ENR and [pyr-7-^{14}C]PRA (Figure 7b). Rates from [2-^{14}C]PRA were higher than from [4-^{14}C]ENR, throughout. One-year-old samples represented the litter layer of soil rather than cattle dung. For those, the rates of $^{14}CO_2$ formation from [pyr-7-^{14}C]PRA were consistently higher than from [pip-2,3-^{14}C]ENR (Figure 7b). Similar experiments had been performed in three consecutive years, and all findings were in agreement with those depicted in Figure 7, representing the third and most comprehensive data set. The diet had little effect on the degradation rates of FQs, although dung from grazing animals (type B) tended to produce slightly higher amounts of $^{14}CO_2$. Again, rates of $^{14}CO_2$ formation from [^{14}COOH]ENR were much lower than from its decyclopropyl analog, amounting to 0.5 versus 7 to 14% (compare Figures 7a and 7b at 52 weeks). Thus, an unsubstituted N-1, permitting the formation of an enol-tautomer (Figure 1), strongly enhanced decarboxylation.

In 3-hour-old cattle dung of type B, rates of $^{14}CO_2$ liberation from core label positions reached the order of 0.5 to 2% of the applied ^{14}C after the first year (Figure 8); the rates obtained for dung of type A (not depicted) were similar to the four-week-sample shown in Figure 7a. After 2 years of incubation, the amount of $^{14}CO_2$ liberated from [2-^{14}C]PRA was almost five times higher when compared to [4-^{14}C]ENR or [4-^{14}C]CIP, while [3-^{14}C]MFX took an intermediate position. This suggested position C-2 being the primary site of pyridone ring cleavage and explained the different rates of $^{14}CO_2$ formation initially mentioned for [2-^{14}C]sarafloxacin and [4-^{14}C]ENR (≤0.6 vs 0.09% per 8 weeks). Again, $^{14}CO_2$ formation from the carboxyl label was low unless the cyclopropyl substituent of ENR was absent, as discussed above. Rates of $^{14}CO_2$ formation from the amines were in the order of 5 to 25% per year (similar to Figure 7b, t_0).

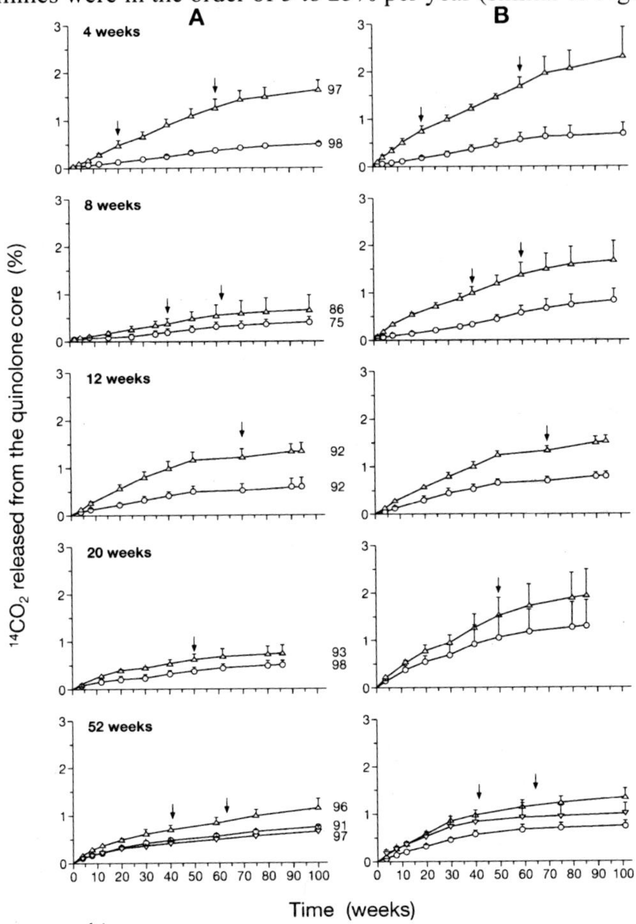

Figure 7a. $^{14}CO_2$ formation from core label positions in enro- and pradofloxacin added to aging cattle dung. [4-^{14}C]enrofloxacin (○),[2-^{14}C]pradofloxacin (Δ) and ^{14}COOH-enrofloxacin (v) were applied at33 μg/g. Dung pats had been obtained from heifers feeding either on a maize/hay silage (type A, column A) or grazing on a pasture (type B, column B) and had been left decaying on the pasture until the indicated sampling time.

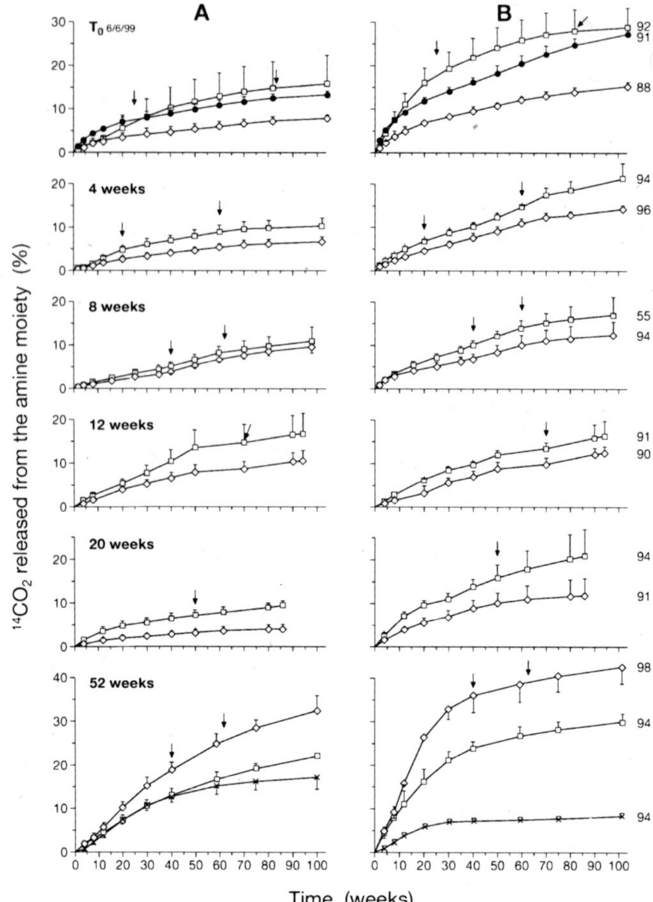

Figure 7b. $^{14}CO_2$ formation from amine label positions in enro- (□), prado- (◊), and ciprofloxacin (●) as well as from the [^{14}COOH]-decyclopropyl-1-H congener of enrofloxacin (x), present in aging cattle dung at 33 µg/g

Compost

To include a matrix supposedly providing a microbial population with maximum humus formation activity (*40*), freshly sieved plant-derived compost at rotting stage V (6 to 7 months of age; water content, 49%; organic C, 18.6% dry weight; total N, 1.2% dry weight; pH 7.6) was collected at the local municipal composting facility at Burscheid-Heiligeneiche near Leverkusen. On the same day, aliquots of 2 g of compost were supplemented with 0.5 ml of either [4-^{14}C]ENR, [^{14}COOH]ENR, [pip-2,3-^{14}C]ENR, [2-^{14}C]PRA, or [pyr-7-^{14}C]-PRA. FQs were applied at a final concentration of 33 µg/g (dry weight). The kinetics of $^{14}CO_2$ formation indicated that 0.3 to 1.4% of $^{14}CO_2$ was produced per year from ^{14}C labels located at core positions, while up to 12% of $^{14}CO_2$ was formed from the amine labels (Figure 9, Table 1). Notably, already after 10 weeks of incubation, biotransformation activity was on the decline, just as in soils (see Figures 10 and 11 below). Only after the addition of 100 mg of ground maize stems to each microcosm at week 32, another cycle of $^{14}CO_2$ production could be elicited from all label positions (the water content had been carefully controlled). Even in this almost ideal organic model matrix, the microbial population, apparently, had responded to a declining nutrient, most likely a specific carbon source, to be re-activated upon the addition of a substrate supporting the entire humification process

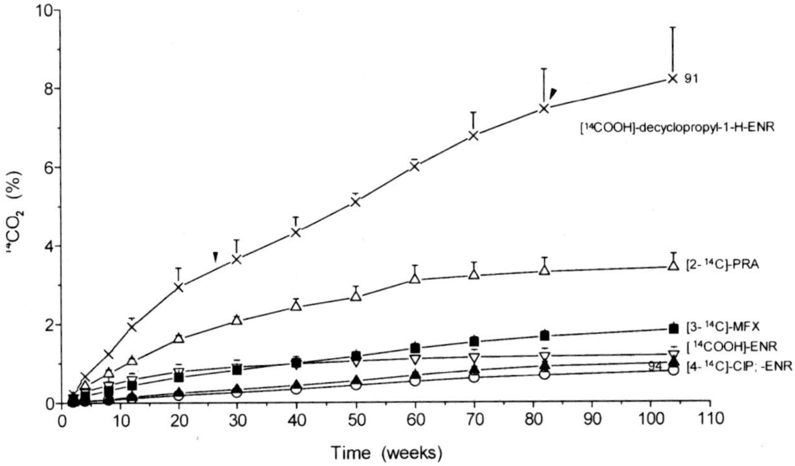

Figure 8. Comparative rates of $^{14}CO_2$ release in fresh cattle dung from core labels of enro-, cipro-, prado-, and moxifloxacin applied at 33 µg/g

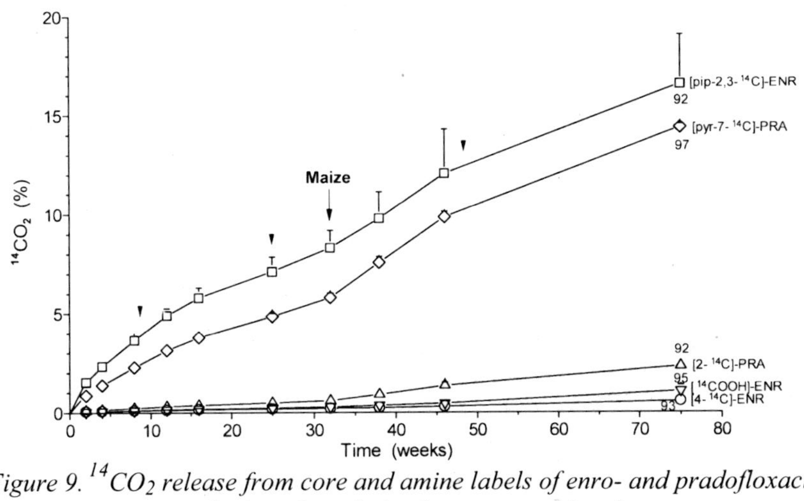

Figure 9. $^{14}CO_2$ release from core and amine labels of enro- and pradofloxacin applied to plant-derived compost at 33 µg/g.

Interim Summary on FQ Degradation in Organic Matrices

The microbial populations encountered in pre-rotted wheat straw, cattle waste, cattle dung pats, and compost catalyzed $^{14}CO_2$ formation at rates on the order of 0.5 to 2% of the initially applied ^{14}C per year, if the label was located in the FQ core. However, up to 20-fold higher rates could be determined in parallel assays, if the label was located in the amine substituent (summarized in Table 1). This indicated (i) the presence of considerable biotransformation activity for FQs in all organic matrices tested, and (ii) a moderate effect of the mineral fraction in limiting biotransformation, due to the similarity of the rates of $^{14}CO_2$ formation in the organic matrices and soils evaluated. Overall, the kinetics of $^{14}CO_2$ formation obtained were unsatisfactory for determining half-lives for the biotransformation of a FQ.

Assessing Biotransformation by Recording the Fate of Parent Drug

In order to demonstrate quantitative biotransformation of a FQ, [4-^{14}C]ENR was selected as model compound. PRA would have been unsuitable because of an inherent lability to hydrolysis of its amine substituent under alkaline conditions, e.g., required for the extraction of drug residues from the organic matrices (see below). Now, quantitative recovery of the ^{14}C label as well as identification of residual ENR among the metabolites by HPLC analysis were prerequisite. As there had been almost no $^{14}CO_2$ liberation from [4-^{14}C]ENR, $^{14}CO_2$ formation was considered negligible. Consequently, an even simpler test system (type III) could be employed, consisting of a sealed sterile 10-ml glass tube holding <1 g of matrix as specified below; soil moisture was adjusted by the ENR solution added to provide the ^{14}C-label at ≥3.7 kBq per microcosm. The gas phase of such vials was exchanged twice a month via syringes, and the moisture content was checked by weighing and supplemented, if required.

A larger number of identical microcosms (16 to 30) were prepared, and at each sampling point, two cultures were analyzed in total to facilitate quantitative recovery of the ^{14}C label. Mass balances comprised ^{14}C label extracted from the matrix, representing ENR and its metabolites; $^{14}CO_2$ recovered from (non-extractable) solid residues upon combustion in a Harvey Oxidizer 500; and small amounts of ^{14}C label obtained by rinsing of the culture vessels. For each extraction, 2 ml of acetonitrile/KOH 0.2 M (3:1, vol/vol) was added to the matrix, and samples were sonicated for 15 min in a water bath kept at room temperature. Depending on the age of the sample, the number of extraction steps increased from 7 (cattle dung) to about 11 (soils) or even 18 (compost). The recovery of ^{14}C label was monitored for each extraction in order to decide on an additional step. Solids were separated by centrifugation, and the combined supernatants were concentrated by evaporation of the acetonitrile (to avoid disturbance of the HPLC gradient). Finally, the volume was adjusted to 10 ml by adding water, and aliquots of 0.5 ml were passed through a membrane with a nominal molecular weight limit of 10,000 (Microcon YM-10, Millipore Corp., Bedford MA). Before HPLC analysis, samples of 0.2 ml were acidified with formic acid to give a final pH of ≤2.5. HPLC analysis of humus and cattle-dung extracts was performed as described before (*8*, *21*, *45*). For soil extracts, an eluent system consisting of 25 mM *ortho*-phosphoric acid pH 2.4 and acetonitrile (*7*) was confirmed to be most appropriate.

Agricultural Soils

A soil sample was drawn in May 2000 from the upper 15 cm of a fallow field at "Höfchen am Hohenseh" (51°04'01"N, 7°06'19"E), which had not been treated with agrochemicals for several years. The soil was slightly dried overnight, then sieved to 2 mm, and stored in plastic bags at 4°C to be used within a week. In preparing 30 microcosms in parallel, each vial (type III) received 0.75 g of "Höfchen" soil I (a silt loam soil, according to the USDA soil classification system [*35*]: 9% sand, 79% silt, 12% clay; 2.4% organic C, 0.25% total N, pH 6.7) and 250 µl of an aqueous solution of [4-^{14}C]ENR, thereby adjusting soil moisture to 90% water-holding capacity. The final concentration of ENR was 44 µg/g (dry weight). Two samples extracted at day zero served as a 100% reference point of the ^{14}C label applied. Upon sampling at selected times between days 27 and 839, two independent microcosms were analyzed for the distribution and chemical identity of the ^{14}C label.

All of the 28 independent microcosms assayed demonstrated biotransformation of ENR (Figure 10). Material balances were ≥92%, throughout. The biotransformation kinetics for ENR revealed: (i) an initial phase (labeled A) with an apparent half-life of 205 days, most likely due to reconstitution of the soil ecosystem disturbed upon sampling; (ii) a second phase between days 82 and 144 (labeled B), during which 30% of the ENR was transformed at a half-life of 83 days; and (iii) ceasing biotransformation activity (labeled C) after about 130 days, as had to be expected (*32*, *33*). The apparent half-life of ENR then would have been calculated to be of the order of 1200 days. However, this phase is interpreted to indicate adaptation of the microbial population for long-term survival in the stationary phase; because (iv) following the addition of 50 mg of ground maize stems to six vials at day 349, after a short phase of reactivation, an additional 15% of ENR underwent biotransformation (labeled D). At the same time, non-extractable residues increased in proportion. Hence, bioavailability of ENR was not severely restricting its transformation, and microbial activity could be stimulated by adding a substrate sustaining the humification process. Again, ENR had been added as an aqueous solution, and some of the drug may have reached soil compartments not readily accessible for microbes.

Notably, the stimulatory effect exerted by maize could be reproduced in the following year with two independent soils (sampled in March), now run at 60% water-holding capacity: the first, "Höfchen" soil II, was similar to the one mentioned above (data not depicted), while the second, collected at "Laacher Hof" (51°05'03"N, 6°55'10"E), represented a loamy sand soil (75% sand, 17% silt, 8% clay, 0.8% organic C, 0.07% total N, pH 5.9). Due to a smaller number of microcosms available, the transformation kinetics obtained (Figure 11) could not reach the degree of resolution attained before. However, at day 388 - and before the addition of 5% (wt/wt) of ground maize roots to two microcosms, the extent of ENR transformation had reached the order of 40 and 70%, respectively; half-lives were estimated at 100 to 120 days (Table 1). Such half-lives resemble decomposition rates of ^{14}C-labeled wheat straw, grass, or maize in agricultural soils, and 30 to 50% of the material will have remained behind after the first growing season (*34*, *35*).

Biotransformation kinetics comparable to those shown in Figures 10 and 11 have been presented before by Hektoen et al. for sarafloxacin and two first-generation quinolones, flumequine and oxolinic acid, when placed in the aerobic part of a sea sediment. The extent of transformation of sarafloxacin had reached about 40 and 70% after 100 and 180 days, respectively (*41*). In another report, half-lives of 87 to 143 days have been discussed for [2-^{14}C]danofloxacin (*26*), though no data were presented. No evidence indicating FQ degradation under anaerobic conditions has yet been reported, either under *in vitro* conditions (*42*), in sludge digesters (*6*, *7*) or in sea sediments (*41*). The susceptibility to FQs of methanogenic bacteria, which are representative of anaerobic ecosystems, has been found to vary greatly: growth of *Methanobrevibacter ruminantium* and *Methanococcus vannielii* was unaffected at ≤44 µg ENR/ml, while *Methano-spirillum hungatei* (at ≤10 µg/ml) and *Methanosarcina barkeri* (at 1 µg/ml) were strongly inhibited, including methanogenesis (*43*). These findings may indicate diversity among the molecular targets of FQs in these genera.

Plant-Derived Compost

Aliquots of 0.75 g of compost (see above) were filled into 18 vessels (test system type III), each receiving 0.5 ml of [4-^{14}C]ENR to give a final drug concentration of 77 µg/g (dry weight). All microcosms indicated biotrans-formation of ENR. The kinetics (Figure 12) revealed an initial phase with half-lives of between 90 to 100 days, and a second phase with a half-life of 113 days (Table 1). Activity ceased when the (remarkably high) concentration level of non-extractable drug residues (20%) was reached. Material balances were >92%, throughout. The increasing number of extraction steps, required towards the end of the experiment, is thought responsible for the slight decline in the recoveries. Nevertheless, quantitative biotransformation of [4-^{14}C]ENR could be proven.

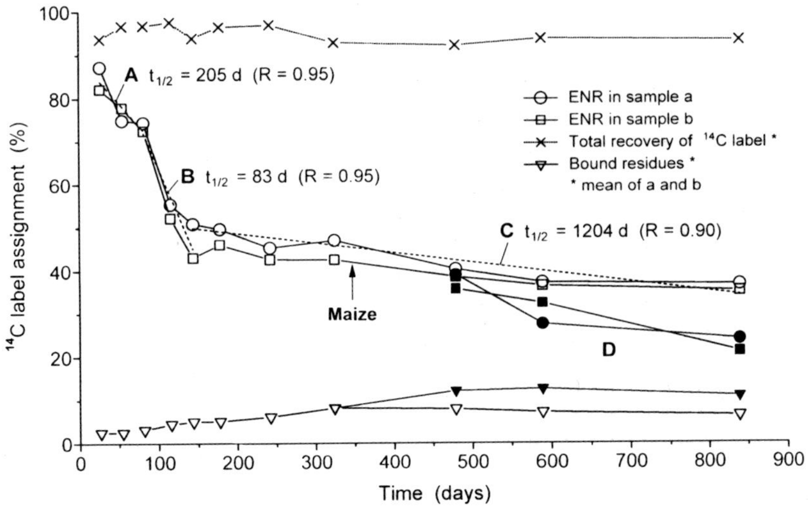

Figure 10. Biotransformation of [4-^{14}C]enrofloxacin in "Höfchen" soil I. After an initial phase of drug metabolism, activity ceased - but could be stimulated in all of six cultures (closed symbols) by the addition of 50 mg of ground maize stems. Enrofloxacin was present at 44 μg/g (dry weight).

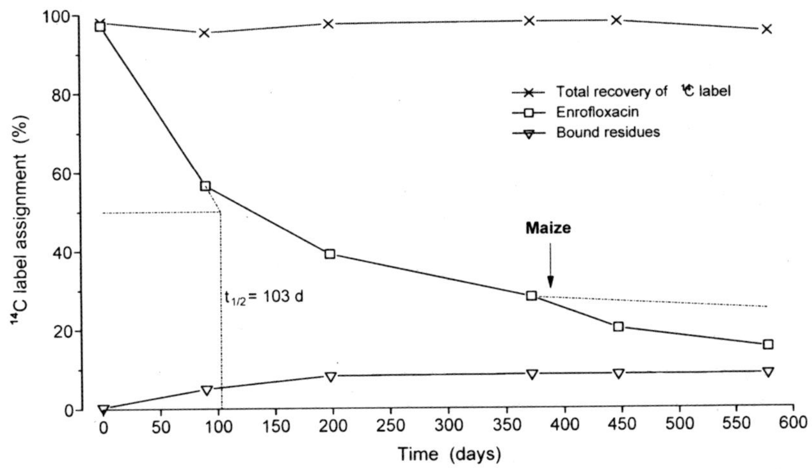

Figure 11. Biotransformation of [4-^{14}C]enrofloxacin in "Laacher Hof" soil. Enrofloxacin was added at 27 μg/g (dry weight).

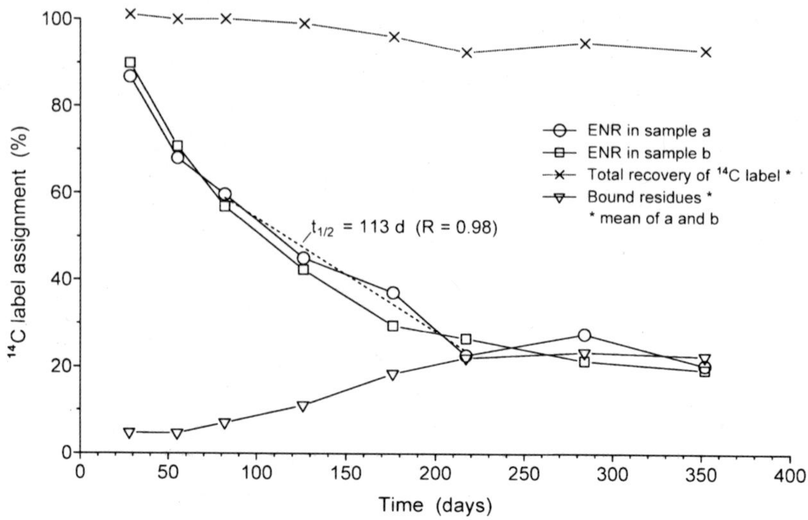

Figure 12. Biotransformation of [4-^{14}C]enrofloxacin in a plant-derived compost. Enrofloxacin was present at 77 µg/g (dry weight). Activity ceased when the concentration level of bound residues was reached, which was higher than in soils or cattle dung.

Cattle Dung

Sixteen vessels (type III) had received 0.5 g of 12-week-old cattle dung of type A (aged on the pasture) and 0.5 ml of [4-^{14}C]ENR (54 kBq). The nominal ENR concentration was 146 µg/g (dry weight). Again, ENR was transformed in all microcosms. The resulting kinetics (Figure 13) indicated that, after an initial activation phase of about 15 days, the concentration of ENR declined at a constant rate with a half-life of 84 ± 6 days (Table 1); apparently, the residual drug concentration did not impact the rate (*44*). Bound residues reached a maximum of just 2.2% at around day 84 and then declined to 1% at day 210.

To verify that ENR had been transformed into metabolites, the radioactivity present in a 0.2 ml sample was compared with the amount of ^{14}C label recovered in fifty 1-min-fractions collected during an HPLC run and analyzed at high precision by LSC (Figure 14). Recoveries of 95 to 101% of the applied ^{14}C label proved that the entire sample had been evaluated and - indicated by the spread of the ^{14}C label throughout the gradient fractions - a high number of metabolites of very different polarity had been produced. Especially notable are the very polar metabolites that eluted before 12 min; these could not be characterized at the time (*45*). Similar results were obtained with "Höfchen" soil I and compost, although the fraction of very polar metabolites was most prominent in cattle dung.

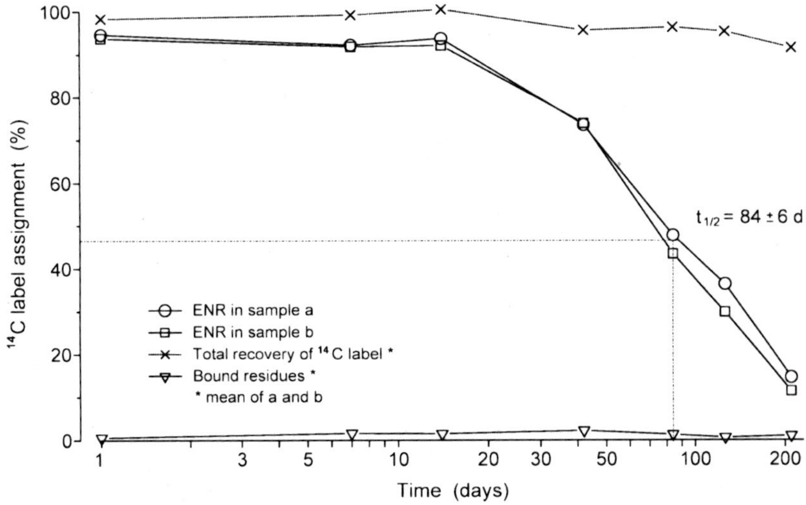

Figure 13. Biotransformation of [4-^{14}C]enrofloxacin in 12-week-old cattle dung. The rate appeared to be independent of the drug concentration. Note the low level of non-extractable drug residues.

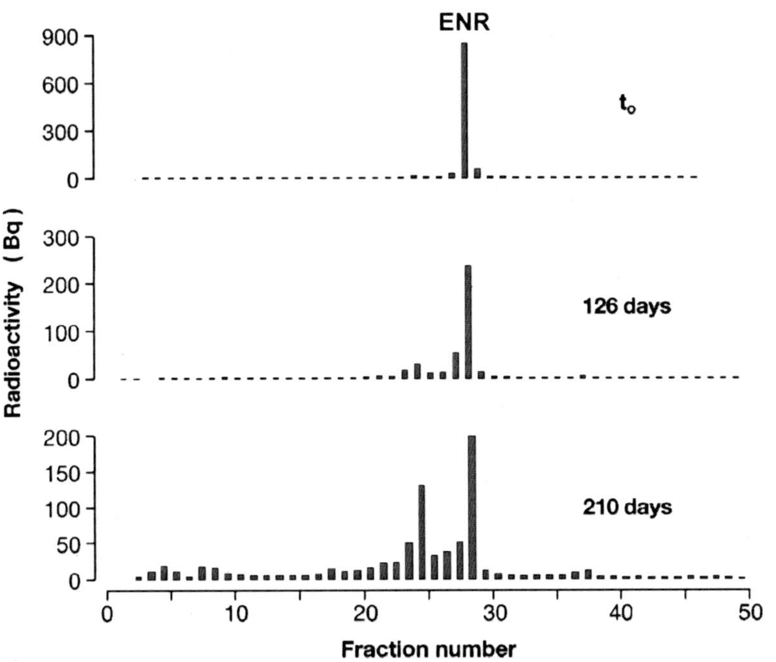

Figure 14. LSC analysis of 1-min-fractions of extracts prepared from cattle dung spiked with [4-^{14}C]enrofloxacin and being collected during HPLC analysis. Initially, radioactivity was detected in about 3 fractions. After 210 days, it was spread throughout the gradient thus indicating the formation of metabolites of very different polarity

Identified Metabolites

In extracts derived from compost, "Höfchen" soil I, and cattle dung pats of type A (analyzed after 126, 182, and 210 days of incubation, respectively) four metabolites could be identified by HPLC/high-resolution mass spectrometry (*45*). In addition to CIP, an active primary congener of ENR also in cattle metabolism (*4*), these metabolites were identified as the *N*-4'-oxide of ENR (F-14), which has also been detected in mammals (*46*), deethylene-ENR (F-4), and the 7-amino congener (F-9), the latter also being described for danofloxacin (*26*); F indicates fungal metabolite. All metabolites are components of the basic degradation schemes for ENR (Figure 15) and CIP in *G. striatum* (*8*, *21*) and have essentially no antibacterial activity, as indicated by minimum inhibitory concentrations determined for *Escherichia coli*, *Staphylococcus aureus* and *Pseudomonas aeruginosa* (Table 2): residual activities were in the order of <1 to 3% that of ENR; the method used has been described elsewhere (*19*, *47*).

Recently, a more detailed degradation scheme for ENR, including the metabolites present at trace concentration level, has been proposed for several basidiomycetous fungi, most species being indigenous to agricultural sites. The chemical structures of 18 metabolites could be defined (14 of which are included in Figure 15), while the structures of 117 metabolites were postulated on the basis of their chemical formulae (*45*, *48*). One of the most significant degradation sequences may be characterized as follows: hydroxylation transforms the piperazine moiety into two types of hemi-aminals, the one involving position C-2' being of primary interest. In its ring-open form, it represents a secondary amine. The aromatic core of ENR will be hydroxylated not only at positions C-5 and C-8: fluorine and the piperazine moiety may also be replaced by a hydroxyl group, so that either catechols or *ortho*-aminophenols are generated. Those are likely to be oxidized first to *ortho*-quinone- and quinone imine-type intermediates and, secondly, to *cis*,*cis*-muconic acid-type metabolites (see Figure 5 in [*45*]). After the elimination of fluorine, which appeared to occur readily (*8*, *21*, *45*, *48*), all metabolites are natural products.

At least 20 of the metabolites identified (*45*, *48*) resembled monomeric building blocks of humus, which is formed by condensation of *ortho*-quinones and amines (*34-36*), i.e., in reverse of the degradation sequence outlined above. Thus, incorporation of quinoid-type congeners of FQs into the humus matrix may be correlated with the availability of amines in the environment. In addition, ^{14}C atoms from core positions appear to be much more likely to be integrated into humus (C-4 > C-3 > C-2) rather than being released as $^{14}CO_2$, as opposed to ^{14}C labels located within saturated amine moieties: congeners of the latter, accordingly, facilitated much higher rates of $^{14}CO_2$ formation.

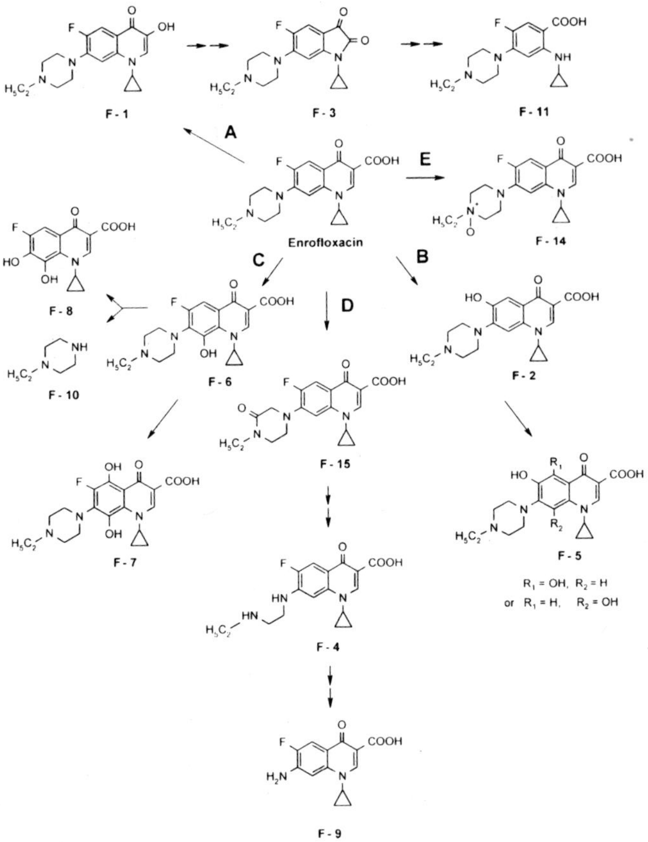

Figure 15. Basic metabolic scheme for enrofloxacin in G. striatum. The major routes of biotransformation comprised: (A) oxidative decarboxylation, which was followed by the elimination of C-2 and C-3, thus resulting in isatin- and anthranilic acid-type metabolites; (B) defluorination and hydroxylation of C-6; (C) hydroxylation of C-8 and C-7, the latter resulting in liberation of the intact piperazine moiety and the formation of a catechol; (D) degradation of the piperazine substituent; and (E) formation of the N-oxide of enrofloxacin, causing immediate inactivation of the drug. (Adapted with permission from reference 21. Copyright 1997 ASM Press.)

Table 2. Minimum inhibitory concentration (MIC) and residual antibacterial activity (RAA) of enrofloxacin and metabolites

Compound	*E. coli* ATCC 8739		*S. aureus* ATCC 6538		*P. aeruginosa* ATCC 13882	
	MIC (µg/ml)	RAA (%)	MIC (µg/ml)	RAA (%)	MIC (µg/ml)	RAA (%)
Enrofloxacin	0.015	100	0.06	100	0.5	100
F-1	>128	<0.01	128	0.05	>128	<0.4
F-2	0.25	6	8	0.8	4	13
F-4	2	0.8	32	0.2	16	3
F-6	0.5	3	2	3	8	6
F-9	0.5	3	8	0.8	64	0.8
F-14	16	0.1	32	0.2	>128	<0.4
F-15	0.25	6	0.25	24	4	13

Conclusions

In several agricultural matrices analyzed, rates of $^{14}CO_2$ formation from ^{14}C labels located at core positions in four FQs were mostly on the order of 0.5 to 2% of the applied ^{14}C per year, which is similar to rates of humus degradation (*34*, *36*, *49*). Thus, recording of $^{14}CO_2$ formation from such labels had little diagnostic potential for indicating FQ biotransformation. Due to the chemical environment in soils, aging cattle dung, and pre-rotted wheat straw highly reactive, quinoid intermediates may be trapped in humus metabolism. In fungal cultures on native wheat straw (*29*), this should be very different because of a low prevalence of amines. In contrast, rates of 10 to 40% of $^{14}CO_2$ per year, formed from ^{14}C labels located in amine substituents, already indicated a rapid biotransformation of FQs with concomitant inactivation of their antibacterial potential, a process being only moderately affected by drug bioavailability.

Regarding *in vitro* biodegradation assays for FQs and other drugs tightly bound by animal manure or soils, in order to simulate *in vivo* conditions, straw or ground maize stems may be used as carriers (low in N) to introduce the test compound, as drug residues will regularly be entering soils bound to waste material. However, extensive supplementation of soil samples, e.g., with liquid manure, may cause divergence of the soil metabolism from the regular "humus turnover" to a heterotrophic, N-sufficient mode for prolonged times, possibly resulting in delayed biotransformation of drug residues as observed in (*7*) - until such time as the eutrophic ecosystem returned to its initial state. Test systems II and III described herein may be readily applicable alternatives to the standard apparatus (*24*), at least for screening purposes.

In order to demonstrate quantitative biotransformation of a FQ, only monitoring the fate of core-labeled parent drug was found to be appropriate. Transformation kinetics obtained with compost, cattle dung, and two types of agricultural soils implied complete turnover of ENR with half-lives of 83 to 113 days. Apparently, this contradicts the widely differing contents of organic C in those matrices. However, as organic C even in soils will be present in large excess over FQ drug residues, the rate of FQ biotransformation may be independent of their concentration, at least in the range of 10 to 144 µg ENR/g dry weight of matrix tested. In contrast, theoretical maximum expected environmental concentrations (*50*) of <0.1µg FQ/g of soil have been calculated for agricultural soils amended with manure from cattle, pigs, and poultry (*7*, *10*). Furthermore, the antibacterial activity (selective pressure) of intact FQs will be attenuated by tight binding to feces and soils. Three key metabolites of ENR, sufficiently stable to be recovered from cattle dung, compost, and a soil were shown to have no significant residual antibacterial activity. Finally, potential for FQ degradation may be present in diverse soils and fungal taxa. Hence, FQ accumulation and long-term persistence in agricultural soils appear to be unlikely, and no relevant selective pressure for bacterial FQ resistance should be exerted.

It has also been reported that the ecological effects of ENR may be more tolerable than those of ivermectin, spiramycin, levamisole, α-cypermethrin, and fenbendazole: ENR residues had the least effect on the rate of decomposition of cattle dung organic matter (*51*). Only recently, a rapid decline of ENR residues in swine sewage and concentrations of 15 mg/kg (*52*) and <1mg/kg (53) were described. So, the rate of $^{14}CO_2$ production from [2-^{14}C]sarafloxacin in soils initially mentioned, <0.6% per 80 days (*23*), indeed had indicated biodegradation - and probably was close to the maximum rate to be expected. However, due to recording of $^{14}CO_2$ formation from a single ^{14}C label position as well as unjustified generalization, these findings have contributed to a perception of "non-biodegradability" of FQs. Hitherto, none of the attributed risks posed to the soil ecosystem by veterinary FQ application has been substantiated experimentally.

Acknowledgements

We would like to acknowledge the exquisite technical assistance of S. Ochtrop, A. Gerhard and A. Seitz; thank our colleagues at various chemistry departments at Bayer AG, namely W. Hallenbach, R. Körmeling, U. Petersen, H. Rast, D. Seidel, and R. Thomas; as well as J. P. E. Anderson, H.-W. Füsers, A. Lagojda, and A. Richter of Bayer CropScience AG.

References

1. *Quinolone Antimicrobial Agents,* 3rd ed.; Hooper, D. C.; Rubinstein, E., Eds.; American Society for Microbiology Press: Washington, DC, 2003.
2. Van Diest, J.; De Jong, A. In *Report and Proceedings of a WHO Meeting: Use of Quinolones in Food Animals and Potential Impact on Human Health;* World Health Organization: Geneva, Switzerland, 1998; pp 97-102.
3. Drlica, K.; Malik, M. *Curr. Topics Med. Chem.* **2003,** *3,* 249-282.
4. Greene, C. E.; Budsberg, S. C. In *Quinolone Antimicrobial Agents*, 2nd ed.; Hooper, D. C.; Wolfson, J. S., Eds.; American Society for Microbiology Press: Washington, DC, 1993; pp 473-488.
5. Boxall, A. B. A.; Kolpin, D. W.; Halling-Sørensen, B.; Tolls, J. *Environ. Sci. Technol.* **2003,** *37,* 286A-294A.
6. Golet, E. M.; Alder, A. C.; Hartmann, A.; Ternes, T. A.; Giger, W. *Anal. Chem.* **2001,** *73,* 3632-3638.

7. Golet, E. M.; Xifra, I.; Siegrist, H.; Alder, A. C.; Giger, W. *Environ. Sci. Technol.* **2003,** *37,* 3243-3249.
8. Wetzstein, H.-G.; Stadler, M.; Tichy, H.-V.; Dalhoff, A.; Karl. W. *Appl. Environ. Microbiol.* **1999,** *65,* 1556-1563.
9. Nowara, A.; Burhenne, J.; Spiteller, M. *J. Agric. Food Chem.* **1997,** *45,* 1459-1463.
10. Thiele-Bruhn, S. *J. Plant. Nutr. Soil Sci.* **2003,** *166,* 145-167.
11. Midtvedt, T. *Scand. J. Infect. Dis.* **1990,** *68*(Suppl.), S7-S13.
12. Levy, S. B. *The Antibiotic Paradox: How Miracle Drugs are Destroying the Miracle;* Plenum Press: New York, 1992; pp 137-181.
13. Ferber, D. *Science* **2000,** *288,* 792-794.
14. Barber, D. A. *JAVMA* **2001,** *218,* 1559-1561.
15. Kümmerer, K. *J. Antimicrob. Chemother.* **2003,** *52,* 5-7.
16. Kümmerer, K. *J. Antimicrob. Chemother.* **2004,** *54,* 311-320.
17. Phillips, I.; Casewell, M.; Cox, T.; De Groot, B.; Friis, C.; Jones, R.; Nightingale, C.; Preston, R.; Waddell, J. *J. Antimicrob. Chemother.* **2004,** *53,* 28-52.
18. Murphy, C. D. *J. Appl. Microbiol.* **2003,** *94,* 539-548.
19. Wetzstein, H.-G. *Antimicrob. Agents Chemother.* **2005,** *49,* 4166-4173.
20. Petersen, U. In *Analogue-based Drug Discovery;* Fischer, J.; Ganellin, C. R., Eds.; Wiley-VCH: Weinheim, Germany, 2006; pp 315-370.
21. Wetzstein, H.-G.; Schmeer, N.; Karl, W. *Appl. Environ. Microbiol.* **1997,** *63,* 4272-4281.
22. Velagaleti, R. R.; Davis, M. L.; O'Brien, G. K.; Stamm, J. M. In *Abstracts of the 205th National Meeting of the American Chemical Society,* Division of Agrochemicals; American Chemical Society: Washington, DC, 1993; abstr. 92.
23. Marengo, J. R.; Kok, R. A.; O`Brien, K.; Velagaleti, R. R.; Stamm, J. M. *Envir. Toxicol. Chem.* **1997,** *16,* 462-471.
24. U. S. Food and Drug Administration. *Environmental Assessment Technical Assistance Handbook;* Document 3.12.; Department of Commerce, National Technical Information Service: Springfield, VA, 1987.
25. Wetzstein, H.-G. *Pharmazie in unserer Zeit* **2001,** *30,* 450-457.
26. Chen, Y.; Rosazza, J. P. N.; Reese, C. P.; Chang, H.-Y.; Nowakowski, M. A.; Kiplinger, J. P. *J. Ind. Microbiol. Biotechnol.* **1997,** *19,* 378-384.
27. Parshikov, I. A.; Freeman, J. P.; Lay, J. O. Jr.; Beger, R. D.; Williams, A. J.; Sutherland, J. B. *Appl. Environ. Microbiol.* **2000,** *66,* 2664-2667.
28. Parshikov, I. A.; Heinze, T. M.; Moody, J. D.; Freeman, J. P.; Williams, A. J.; Sutherland, J. B. *Appl. Microbiol. Biotechnol.* **2001,** *56,* 474-477.
29. Martens, R.; Wetzstein, H.-G.; Zadrazil, F.; Capelari, M.; Hoffmann, P.; Schmeer, N. *Appl. Environ. Microbiol.* **1996,** *62,* 4206-4209.
30. Kerem, Z.; Jensen, K. A.; Hammel, K. E. *FEBS Lett.* **1999,** *446,* 49-54.
31. Jensen, K. A. Jr., Houtman, C. J.; Ryan, Z. C.; Hammel, K. E. *Appl. Environ. Microbiol.* **2001,** *67,* 2705-2711.
32. Anderson, J. P. E. *Soil Biol. Biochem.* **1984,** *16,* 483-489.

33. Anderson, J. P. E. In *Pesticide Effects on Soil Microflora;* Somerville, L.; Greaves, M. P., Eds.; Taylor & Francis: London, 1987; pp 45-60.
34. Wetzstein, H.-G.; Anderson, J. P. E. In *Pesticides, Soil Microbiology and Sustainable Agriculture; Scientific Program and Abstracts of the 3rd International Symposium on Environmental Aspects of Pesticide Microbiology;* Anderson, J. P. E.; Arthur, M. F.; Leake, C. R.; et al., Eds.; Theissen: Monheim am Rhein, Germany, 2000; pp 74-75.
35. Stevenson, F. J. *Humus Chemistry. Genesis, Composition, Reactions;* Wiley: New York, 1994; pp 1-23, 59-95, 188-211.
36. Tate, R. L. *Soil Microbiology;* Wiley: New York, 1995; pp 1-29, 228-253.
37. Paul, E. A.; Clark, F. E. *Soil Microbiology and Biochemistry,* 2nd ed.; Academic Press: San Diego, CA, 1996; pp 129-156.
38. Scheer, M.; De Jong, A.; Froyman, R.; Heinen, E. *J. vet. Pharmacol. Therap.* **1997,** 20(Suppl. 1), S201-S202.
39. Wicklow, D. T. In *The Fungal Community. Its Organization and Role in the Ecosystem,* 2nd ed.; Carroll, G. C.; Wicklow, D. T., Eds.; Marcel Dekker: New York, 1992; pp 715-728.
40. Barker, A.V. In *Agricultural Uses of By-products and Wastes;* Rechcigl, J. E.; MacKinnon, H. C., Eds.; ACS Symposium Series 668. American Chemical Society: Washington, DC, 1997; pp 140-162.
41. Hektoen, H.; Berge, J. A.; Hormazabal, V.; Yndestad, M. *Aquaculture* **1995,** *133,* 175-184.
42. Al-Ahmad, A.; Daschner, F. D.; Kümmerer, K. *Arch. Environ. Contam. Toxicol.* **1999,** *37,* 158-163.
43. Hippe, H.; Wetzstein, H.-G. In *Abstracts of the 96th General Meeting of the American Society for Microbiology;* American Society for Microbiology: Washington, DC, 1996; p 305, abstr. I-014.
44. Wetzstein, H.-G.; Schneider, J.; Karl, W. In *Abstracts of the 102nd General Meeting of the American Society for Microbiology;* American Society for Microbiology: Washington, DC, 2002; p 390, abstr. Q-70.
45. Karl, W.; Schneider, J.; Wetzstein, H.-G. *Appl. Microbiol. Biotechnol.* **2006,** *71,* 101-113.
46. Dalhoff, A.; Bergan, T. In *Quinolone Antibacterials;* Kuhlmann, J.; Dalhoff, A.; Zeiler, H.-J., Eds.; Springer: Berlin, 1998; pp 179-206.
47. Wetzstein, H.-G.; Karl, W.; Hallenbach, W.; Himmler, T.; Petersen U. In *Abstracts of the 100th General Meeting of the American Society for Microbiology;* American Society for Microbiology: Washington, DC, 2000; p 2, abstr. A-8.
48. Wetzstein, H.-G.; Schneider, J.; Karl, W. *Appl. Microbiol. Biotechnol.* **2006,** *71,* 90-100.
49. Martin, J. P.; Haider, K. *Appl. Envir. Microbiol.* **1979,** *38,* 283-289.
50. Spaepen, K. R. I.; Van Leemput, L. J. J.; Wislocki, P. G.; Verschueren, C. Environ. Toxicol. Chem. **1997,** 16, 1977-1982.
51. Sommer, C.; Bibby, B. M. *Eur. J. Soil Biol.* **2002,** *38,* 155-159.
52. Pierini, E.; Famiglini, G.; Mangani, F.; Cappiello, A. *J. Agric. Food Chem.* **2004,** *52,* 3473-3477.
53. Martínez-Carballo, E.; Gonzáles-Barreiro, C.; Scharf, S.; Gans O. *Environ. Pollut.* **2007,** 148, 570-579.

Chapter 7

Environmental fate and chemistry of a veterinary antibiotic—tylosin

Dingfei Hu[1], Keri L.D. Henderson[1], and Joel R. Coats[1]

[1]Department of Entomology, Iowa State University, Ames, IA, USA

Aerobic degradation, photolysis, and mobility of tylosin were investigated in the laboratory. Tylosin A is degraded with a half-life of 200 d in water, while it is stable in the dark. Tylosin C and D are relatively stable except in ultrapure water in the light. Slight increases of tylosin B and formation of two photoreaction products, isotylosin A alcohol (E,Z) and isotylosin A aldol (E,Z), were observed under exposure to light. In soil tylosin A and D has a dissipation half-life of about 1 wk. Sorption and abiotic degradation are the major factors influencing the loss of tylosin in the environment. No biotic degradation was observed at the test concentration of 50 μg/ml or μg/g either in pond water or in an agronomic soil, as determined by comparing dissipation profiles in sterilized and unsterilized conditions. At 7.5 ng/ml, biotransformation may play an important role in degradation of tylosin in water. Tylosin has strong sorption to various soils, and leachbility is dependent on soil properties and manure amendment. Adsorbed tylosin in surface soil might run off to water bodies through soil erosion. In the end, pathways were proposed for tylosin degradation in the environment.

Introduction

Over the last 30 years, intensive animal production has increased globally. As part of these intensive agricultural production practices, veterinary pharmaceuticals, primarily antibiotics, are widely being used as growth promoters or as therapeutic drugs. Many veterinary drugs are poorly adsorbed in the gut of the animal, resulting in as much as 30-90% of the parent compound being excreted (*1*). Manure produced is usually stored in lagoons, and applied to agricultural fields as a fertilizer, which provides a pathway for veterinary pharmaceuticals and their metabolites to enter the environment. In addition, their metabolites can also be bioactive and can be transformed back to the parent compound after excretion (*2*). As a result, there is concern mounting over the over the fate and effects of low levels of veterinary antibiotics in the environment (*3-7*). The utilization of growth-promoting antibiotics in agriculture is one of the primary factors driving increased antibiotic resistance in humans (*8*).

Tylosin is a macrolide antibiotic used for the treatment of disease and growth promotion in cattle, swine, and poultry. Tylosin is also used to control American foulbrood, a wide-spread and devastating disease affecting honey bees (*2*). Tylosin consists of a mixture of tylosin A (Figure 1), tylosin B, tylosin C, and tylosin D. Approximately 80-90% of tylosin is composed of tylosin A (*9*). It has a broad anti-bacterial spectrum against gram positive bacteria and some gram negative bacteria, including vibrio, spirochete, and coccidia. It is also one of the first-choice drugs against infections caused by mycoplasma. Due to the wide-spread in agriculture, tylosin has been detected in 14% of 104 streams tested throughout the United States (*10*). There is concern that the presence of low levels of tylosin in the environment might promote bacterial resistance, which could decrease the effectiveness of erythromycin and other closely related antibiotics used to treat human diseases (*11, 12*).

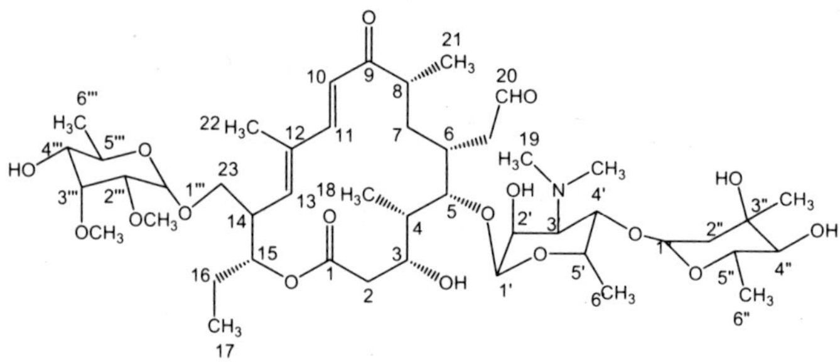

Figure 1. Molecular structure of tylosin A.

Understanding the environmental fate and chemistry of a chemical is a critical component of ecological and human risk assessments. The goal of this study was to characterize the environmental chemistry and fate of the veterinary antibiotic tylosin. Specifically we characterized the dissipation, movement and transformation of tylosin in water and soil. The studies here address: (1) dissipation in water, including the effects of light v. dark, ultrapure water v. pond water v. sterilized pond water, manure v. no manure, and submerged aquatic plant v. no plant; (2) dissipation in sterilized v. non-sterilized soil (one soil type); and (3) leaching through packed soil columns (three soil types).

Materials and Methods

Chemicals

Tylosin tartrate (95.0%, CAS NO. 74610-55-2) was purchased from Sigma Chemical Co. (St. Louis, MO, USA). Acetonitrile and ammonium acetate were purchased from Fisher Scientific (Fair Lawn, NJ, USA). All reagents used during extraction and analysis were analytical reagent grade or better.

Water dissipation study

The methods for the water dissipation study have been previously reported (*13*). Briefly, pond water was collected from the Iowa State University Horticulture Farm Pond (Ames, IA, USA) and was stored in 250-ml French square bottles with perforated lids to maintain aerobic conditions. The pH of the pond water used (in the light and dark studies) was 8.1, the alkalinity is 103 mg/ml, and the total hardness is 190 mg/ml. Tylosin (50 µg/ml) was added to unsterilized pond water, sterilized pond water and ultrapure water, and the water microcosms were incubated. Water microcosms were sampled on days 0, 1, 3, 5, 7, 10, 14, 21, and 30, 60, 180 post-treatment. Tylosin samples were exposed to natural light at Ames, Iowa (42.01° N, 93.97° W) between April 15 and October 15 of 2005. All samples were maintained at a constant temperature of 25 ± 2 °C with 12:12 natural light: dark cycle for 180 days. For the dark experiments, samples were kept in the same conditions but light was excluded by covering the bottles with aluminum foil.

A second water dissipation study (*14*) was conducted at the tylosin concentration of 7.5 ng/ml, which was selected to simulate the detected tylosin level in rivers (*10*). *Ceratophyllum demersum* (coontail) and pond water were collected from the ISU Horticulture Farm, and the manure used in the study was collected from swine on antibiotic-free diets at the ISU Swine Nutrition Farm. There were four microcosm treatments: no manure and no vegetation (pond water only), 0.1% manure (no vegetation), 2.3 g wet wt. coontail (no manure), 0.1% manure and 2.3 g wet wt. coontail. The mixture of Bristol's solution (nutrient-containing water for algal growth), pond water, and ultrapure water (at a ratio of 1:1:4) was placed in French square bottles for the test. The control was 33% Bristol's algal media. After a 3-day acclimation period, solutions were spiked to 7.5 ng/ml tylosin, to simulate tylosin concentrations in rivers. There were 5 replicates per treatment. The containers were maintained at constant temperature of 25 ± 2 °C with 16:8 light (UV specific for plant growth): dark.

Soil dissipation study (*13*)

Soil was collected from an agricultural field at the Iowa State University Agronomy and Agricultural Engineering Research Farm, near Ames, Iowa (from Field 55). The soil was classified as a sandy loam, a Nicollet-Webster complex with 1.6% organic matter, 60% sand, 22% silt, 18% clay and pH 7.0. Soil samples (50 g), after passing through a No. 10 stainless steel sieve, were transferred to 250-ml French square glass bottles. Soil was sterilized by autoclaving at 121 °C for 30 min on three consecutive days. The soil moisture was adjusted to the field capacity water content (1/3 bar) using ultrapure water and was maintained throughout the study. The soil was spiked with 50 μg/g tylosin and incubated at 25°C in light/dark for 30 days. Samples were taken at days 0, 0.25, 0.5, 1, 3, 5, 7, 10, 14 and 21 and 30 post-treatment.

Soil column study

Packed soil columns were used to evaluate the mobility of tylosin in soil. Soil columns consisted of PVC pipe (10-cm diameter and 30-cm length) fitted with a nylon mesh on the bottom. Columns were packed to a bulk density of 1.25 g/cm^3. Columns were housed in upright position at 25°C for a one week prior to treatment and then for a 4-week testing period in an environmental chamber. Prior to treatment soil columns were saturated with 0.01M $CaCl_2$ solution and were allowed to drain overnight. After the treatment was applied, 3-cm rain events were simulated once a week for four weeks. Leachate was collected following each simulated rain event in 500 ml foil-covered glass bottles. At the end of the testing, columns were equally divided into three sections (called top, middle and bottom). Soil from each section was homogenized, and a 60-g subsample of soil was extracted. Three soil types were utilized: Canisteo clay loam, Nicollet-Webster sandy loam, and Hanlon sand; the properties of each soil is presented in Table II. There were three treatments for each soil type, including tylosin without manure, tylosin with manure, and only manure as control. The latter two treatment contained 30 g manure which was spiked with tylosin at 50 μg/g, and the manure was incorporated into about 2-cm soil surface. Four replications were applied for each treatment. Manure was attained from a swine farm feeding antibiotic-free diets, and was also confirmed to contain no tylosin using HPLC method.

Analytical methods

Quantitative analysis of tylosin and its metabolites was achieved by reverse-phase HPLC (*13*). Analysis of the samples was performed using a Hewlett-Packard (Palo Alto, CA, USA) series 1100 HPLC system with a quaternary pump, an autosampler, a thermostatted column compartment, and a Spectroflow 757 absorbance detector (ABI Analytical, Kratos Division, Ramsey, NJ, USA). Data were collected and analyzed using HP Chemstation system software (REV. A.04.01). A Waters Atlantis™ (Milford, MA, USA) dC18 column (4.6×250 mm, 5-μm particle size) was used. Detection was conducted at 290nm, with a flow rate of 1.0 ml/min at room temperature. The mobile phases consisted of acetonitrile and 20 mM ammonium acetate (35:65, v/v, pH=6.0). The same ultraviolet-response capability of all tylosin-related compounds is assumed in this study because of lack of standards except for tylosin A. Soil samples (20g) were extracted, using the mobile phase as the solvent, and the extracts were cleaned up with Oasis hydrophilic-lipophilic balance (HLB) cartridges (6 cc) (Waters, Milford, MA, USA) before HPLC analysis. Filtered water samples were directly injected (*13*). In order to prevent photodegradation, all extracts were stored in the dark and analyzed under minimum lighting.

Results and Discussion

Dissipation profile in water and soil

Tylosin-related compounds enter agricultural fields through application of tylosin-containing manure (*15*). Water dissipation study and soil dissipation study provided insights on persistence of tylosin A and its related compounds, which helps understand the risk posed by those tylosin residues in the environment. As a result, the dissipation profiles in water and soil are listed in Table I. Due to lack of analytical standards except for tylosin A, all tylosin-related compounds were identified using HPLC/MS (*13*).

Tylosin A had a dissipation time (DT_{50}) of 200 d calculated from pooled data of three different treatments in the light, and tylosin A was stable, with less than 6% loss of initial spiked amount in the 6-month study. Tylosin C and D are relatively stable except in ultrapure water in the light. Slight increases of tylosin B after two months were observed in pond water, which is probably due to ionic strength or light exposure. Two photoreaction products were detected although the experimental conditions filtered out a majority of the UVA and UVB wavelengths which tylosin absorbs, and they were proved to be isotylosin alcohol and isotylosin aldol with application of HPLC-ESI-MS and ^{1}H and ^{13}C NMR. The structural elucidation of these two photoreaction products were conducted in our lab (*16*), and the structures of isotylosin alcohol and isotylosin aldol can be seen in Figure 5. Two photo-reaction products both have a (E, Z) configuration at the double-bond conjugated site, and one is isotylosin A alcohol (E, Z) , and the other is isotylosin A aldol (E,Z) with two epimers.

Tylosin A and tylosin D had a DT_{50} of about 1 week in sterilized and unsterilized soil. The concentration of tylosin C increased slightly, and the mechanism of tylosin C formation is unknown, but it is probably facilitated by clay particles. The short DT_{50} of tylosin A in soil does not necessarily mean a rapid degradation. Tylosin might strongly adsorb to soil so that the recovery from the soil using the current extraction method is low. Although 80% of tylosin were able to be recovered right after the spike, the aging of tylosin in soil might remarkable reduce the final recovery from soil. It is evidenced that there are some bound tylosin residues in soils after solvent extraction by demonstrating biological activities of extracted soils (*17*).

Table I. Dissipation profile of tylosin in water and soil in the laboratory

Matrices and Conditions		Dissipation Profile[a] Tylosin A	Tylosin B	Tylosin C	Tylosin D
Ultrapure water	Light	227 d	Increase	200 d	280 d
	Dark	Stable	Stable	Stable	Stable
Pond water	Light	184 d	Stable	Stable	Stable
	Dark	Stable	Stable	Stable	Stable
	Sterilized + light	192 d	Increase	Stable	Stable
	Sterilized + dark	Stable	Increase	Stable	Stable
Soil	Sterilized	7 d	ND	Increase	8 d
	Unsterilized	7 d	ND	Increase	8 d

[a] Dissipation profile is described using DT_{50} (Time to 50% dissipation, listed in days (d) where applicable); Stable: ≤6% loss; Increase: increase in concentration compared to study initiation; ND = not detected.

Although biodegradation of tylosin at concentration of 1 to 100 µg/L was found (*18*), no biodegradation was observed within the experimental periods by comparing the results from sterilized and unsterilized experimental conditions. The difference might be caused by the higher tylosin testing concentration and the lower microbial activity in our test system. The concentration of 30 µg/ml showed a complete inhibition of biodegradation (*18*), and in our studies the worst-case level of tylosin, 50 µg/ml in water or 50 µg/g in soil, was adopted for easy analysis of tylosin and its related compounds.

When the test concentration of tylosin was 7.5 ng/ml, the water dissipation study showed that there were significant differences after day 4 between treatments with manure and those without manure (Figure 2). At initial time point, tylosin concentrations between different treatments were not statistically different, and at later time points microbes from manure might play a role in dissipation or degradation of tylosin at this low level of tylosin. Coontail did not appear to have an effect on tylosin dissipation, indicating bioavailability of tylosin for coontail was low. In this separate water dissipation study, tylosin concentration was detected using a competitive direct enzyme-linked immunosorbent assay (ELISA) (*19*). The ELISA method was not totally specific for tylosin A, but showed cross-reactivity for other tylosin-related compounds. However, the ELISA detection results are in agreement with results from HPLC analysis within a 4-week experimental period, which is shown in another study done in our lab (*16,19*).

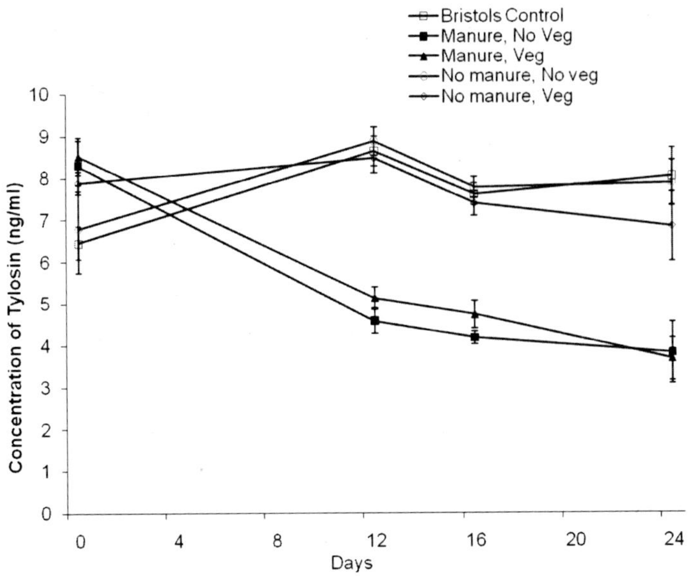

Figure 2. Dissipation of tylosin in surface water at 7.5 ng/ml.

Mobility of tylosin in soil columns

Among three types of soil (Hanlon, Nicollet-Webster, and Canisteo) and two treatments (tylosin with or without manure) for each soil, tylosin A was detected at a range from 0 to 0.27 ng/ml in leachate from packed soil columns. An intact soil column study (*20*)) showed that tylosin concentration was from 0.1 to 2.8 ng/ml in leachate. The average of 0.8ng/ml from intact soil columns is at least 3 times higher than tylosin levels in leachate from packed soil columns, although the amount of spiked tylosin of 60 μg (5 μg/g in manure) on intact soil columns is 25 times lower than 1,500 μg (50 μg/g in manure) packed soil columns. Various factors might contribute to the difference such as soil properties, microbial bioactivity, etc. One important factor is the pore size of soils. All those intact soil columns were selected with a pre-leaching test in order to determine drainage, and columns with less than 24-48 h drainage were discarded. Packed soil columns would not have macropores inside which could drain water with chemicals quickly through soil columns with a low chance of adsorption to soil.

The distribution of recovered tylosin in packed soil columns is illustrated in Figure 3. Total recoveries of tylosin were below 4% of the initial spiked amount. Another study reported remarkably different recoveries of 61%-81% when soil column sections were extracted immediately after one rainfall event (*21*).

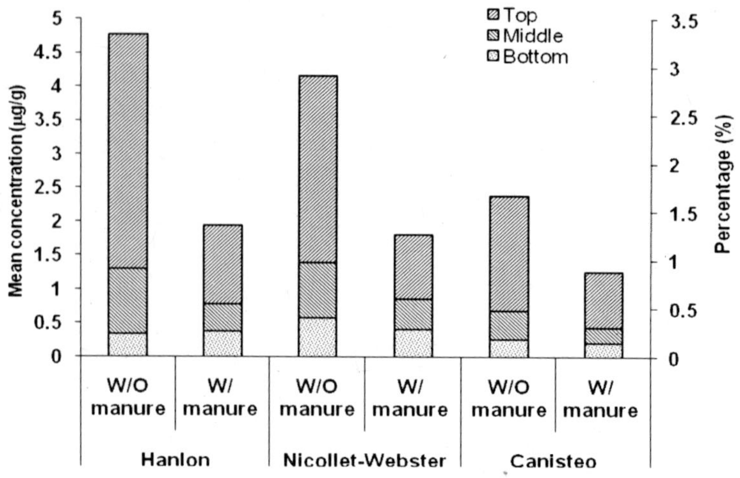

Figure 3. Distribution of recovered tylosin in packed soil columns

Table II. Soil properties and sorption coefficient (Kd) to tylosin

Soil types	texture	Sand%	Silt%	Clay%	%OM[a]	pH	CEC[b]	Kd (ml/g)
Canisteo	Clay loam	36	34	30	4.0	6.0	20.1	65
Nicollet-Webster	Sandy loam	47	36	17	2.6	6.8	14.9	42
Hanlon	Sand	90	6	4	0.5	8.1	6.9	24

[a] Organic matter content of soil

[b] Cation exchange capacity (Meq/100g)

In the one-month period of our soil column study, the majority of the tylosin might have dissipated before the soil extraction, since the DT_{50} in the soil was about 7 d. Decreased recovery from aged soils could be another factor, although the recovery of tylosin is more than 80% at the initial time point. It seems that there are more tylosin residues in the top section of soil columns than the middle and bottom sections. When tylosin is tightly bound to soil particles in highly adsorptive surface soil, soil adsorbed tylosin can be lost by erosion in runoff.

Soil properties such as organic matter, clay, C.E.C., etc. have been shown to influence the environmental behavior of a chemical (*22*). In this study, the three soils were selected to contain a wide range in composition to represent various soils in reality, and soil properties for each soil are listed in Table II. In Figure 3, the greatest leachability was observed in the Hanlon soil with no manure amendment. That soil contained highest sand content (90%), the lowest organic matter content (0.5%), and lowest C.E.C. (6.9 Meq/100 g) among the three soil types. Less or no tylosin was detected in the leachate from the other soil columns. Correlation analysis (Figure 4) of soil properties with leachability indicates that soils high in organic matter, clay or C.E.C are the most adsorptive of tylosin, and manure amendment also decreased the leaching of tylosin through soil columns. In Hanlon and Nicollet-Webster soils amended with manure, tylosin concentrations in leachate were lower than in those without manure. This difference might be explained by higher sorption to manure which is evidenced by higher Kd value (285 ml/g) in manure compared to soils. Another possibility is that the swine manure contributed additional biodegraders to the test system.

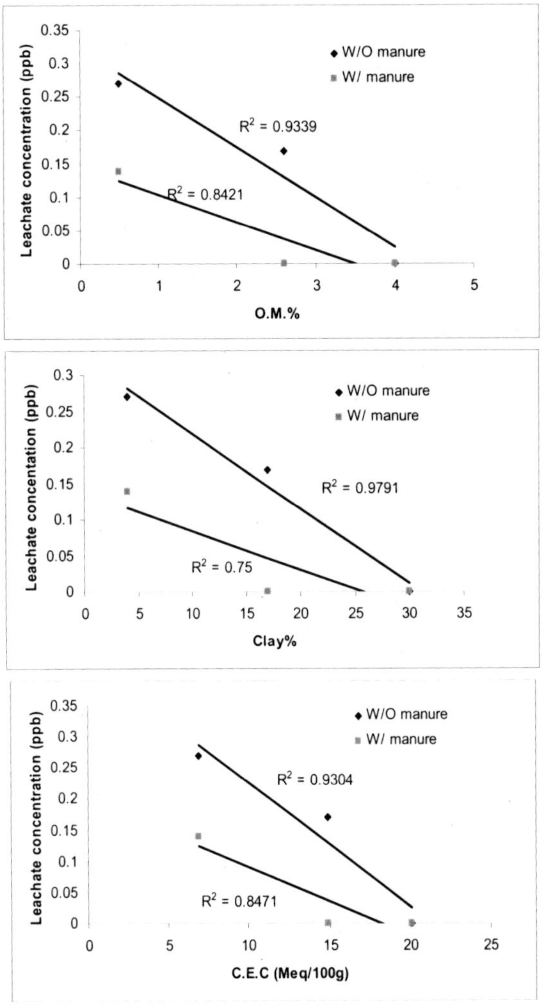

Figure 4. Correlation of leachability with OM%, clay%, and cation exchange capacity

Proposed degradation pathways

Based on literature reports and our experimental findings, degradation pathways of tylosin in the environment are proposed and illustrated in Figure 5. Through cleavage of the mycarose sugar, tylosin B is formed from tylosin A in an acidic environment and light exposure (*1,13,23*). In neutral and alkaline conditions, tylosin aldol and some polar degradation products were detected (*23*). Tylosin D was found the major metabolite in swine and rat (technical sheet from Lilly Research Laboratories, Greenfield, IN, USA). Tylosin D was also detected in manure-containing test systems (*1,15*), which might be formed through microbial biotransformation, although it has yet to be determined. The soil dissipation study shows that tylosin C might form through abiotic transformation with facilitation by clay particles. Tylosin A associated with NH_4^+ was found in our testing system, however, it is unknown if the formation appeared in the environmental matrices or in the HPLC mobile phase, which contained ammonium acetate. Under light exposure, isotylosin A aldol (E, Z) can be transformed from tylosin A or from the intermediate, tylosin A aldol. Isotylosin A alcohol (E, Z) can be formed from tylosin A or from tylosin D which is usually present with tylosin A, and it may possibly be biologically formed from tylosin A. Further degradation and mineralization of tylosin was observed (*15,24*).

Tylosin in the environment dissipates rapidly to very low concentrations which might not pose acute toxicity to living organisms in the ecosystem. However, persistence of low concentrations and their potential impact on the microbial community is not yet fully investigated and understood (*14*). For example, tylosin-resistant bacteria were isolated from cornfields (*12*). However, whether these tylosin-resistant bacteria were produced in the environment or brought from applied manure is still not clear. The impacts might include disturbance of soil respiration (*25*), soil enzyme activity (*26*), soil microbial communities (*27,28*), and soil nitrogen cycling (*29*), and analysis of effects on microbial community structure has been investigated using molecular genetic techniques (*30-32*). Recently, it has been shown that unextractable (bound) antibiotic residues, including bound tylosin residues, can still execute biological activity through unclear mechanisms (*17*), which may put selective pressure on bacteria in the terrestrial environment. Impact of low-level or bound residues of chemicals on the ecosystem, especially on the microbial community, should be a critical part of a complete risk assessment to ensure their safety to the environment and humans.

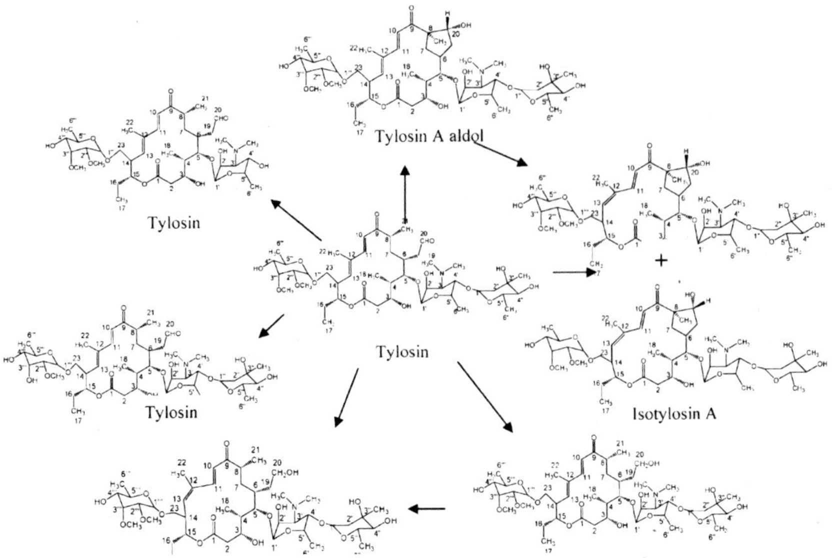

Figure 5. Possible transformation pathway of tylosin in the environment

Acknowledgments

This project was funded by the Center for Health Effects of Environmental Contamination, Iowa City, Iowa, and by the USDA National Research Initiative grant 2006-35102. It is a publication of the Iowa Agriculture and Home Economics Experiment Station Project No. 5091.

References

1. Loke, M. L.; Ingerslev, F.; Halling-Sorensen, B.; Tjornelund, J. *Chemosphere* **2000,** 40, 759-765.
2. Elzen, P. J.; Westervelt, D.; Causey, D.; Ellis, J.; Hepburn, H. R.; Neumann, P. *J. Econ. Entomol.* **2002,** 95, 1119-1122.
3. Halling-Sorensen, B.; Nielsen, S. N.; Lanzky, P. F.; Ingerslev, F.; Lutzhoft, H. C. H.; Jorgensen, S. E. *Chemosphere* **1998,** 36, 357-394.
4. Daughton, C. G.; Ternes, T. A. *Environ.l Health Perspect.* **1999,** 107, 907-938.
5. Boxall, A. B. A.; Kolpin, D. W.; Halling-Sorensen, B.; Tolls, J. *Environ.l Sci. Technol.* **2003,** 37, 286a-294a.
6. Boxall, A. B. A.; Fogg, L. A.; Blackwell, P. A.; Kay, P.; Pemberton, E. J.; Croxford, A. Veterinary medicines in the environment. *Reviews of Environmental Contamination and Toxicology, Vol 180* **2004,** 180, 1-91.
7. Boxall, A.; Long, C. *Environ. Toxicol. Chem.* **2005,** 24, 759-760.
8. Chopra, I.; Roberts, M. *Microbiol. Mol. Biol. Rev.* **2001,** 65, 232-60 ; second page, table of contents.
9. Sarmah, A. K.; Meyer, M. T.; Boxall, A. B. A. *Chemosphere* **2006,** 65, 725-759.

10. Kolpin, D. W.; Furlong, E. T.; Meyer, M. T.; Thurman, E. M.; Zaugg, S. D.; Barber, L. B.; Buxton, H. T. *Environ. Sci. Technol.* **2002,** 36, 1202-1211.
11. Aarestrup, F. M.; Carstensen, B. *Microb. Drug Resist.* **1998,** 4, 307-12.
12. Onan, L. J.; LaPara, T. M. *FEMS Microbiol. Lett.* **2003,** 220, 15-20.
13. Hu, D.; Coats, J. R. *Environ. Toxicol. Chem.* **2007,** 26, 884-9.
14. Henderson, K. L. Impact of veterinary antibiotics in the environment. Iowa State University, Ames, 2008.
15. Kolz, A. C.; Moorman, T. B.; Ong, S. K.; Scoggin, K. D.; Douglass, E. A. *Water Environ. Res.* **2005,** 77, 49-56.
16. Hu, D. Environmental fate and chemistry of a veterinary antibiotic-tylosin and monoterpenoid pesticides-thymol and phenethyl propionate. Iowa State University, Ames, 2007.
17. Chander, Y.; Kumar, K.; Goyal, S. M.; Gupta, S. C. *J. Environ. Qual.* **2005,** 34, 1952-1957.
18. Ingerslev, F.; Torang, L.; Loke, M. L.; Halling-Sorensen, B.; Nyholm, N. *Chemosphere* **2001,** 44, 865-872.
19. Hu, D.; Fulton, B.; Henderson, K. L.; Coats, J. R. *Environ. Sci. Technol.* **2008,** 42, 2982-2987.
20. Henderson, K. L., T.B. Moorman, and J.R. Coats. , Mobility of tylosin and enteric bacteria in soil columns. In *Fate of pharmaceuticals in the environment and in water treatment systems*, Aga, D., Ed. Boca Raton, FL, 2008.
21. Rabolle, M.; Spliid, N. H. *Chemosphere* **2000,** 40, 715-722.
22. Somasundaram, L.; Coats, J. R.; Racke, K. D.; Shanbhag, V. M. *Environ. Toxicol. Chem.* **1991,** 10, 185-194.
23. Paesen, J.; Cypers, W.; Pauwels, K.; Roets, E.; Hoogmartens, J. *J. Pharmaceut. Biomed. Anal.* **1995,** 13, 1153-1159.
24. Ingerslev, F.; Halling-Sorensen, B. *Ecotoxicol. Environ. Safety* **2001,** 48, 311-320.
25. Gigliotti, C.; Allievi, L.; Salardi, C.; Ferrari, F.; Farini, A. *J. Environ. Sci. Health B* **1998,** 33, 381-398.
26. Engelen, B.; Meinken, K.; von Wintzintgerode, F.; Heuer, H.; Malkomes, H. P.; Backhaus, H. *Appl. Environ. Microbiol.* **1998,** 64, 2814-2821.
27. Hu, S.; vanBruggen, A. H. C.; Wakeman, R. J.; Grunwald, N. J. *Plant Soil* **1997,** 195, 43-52.
28. Wardle, D. A.; Parkinson, D. *Plant Soil* **1990,** 122, 21-28.
29. Goring, C. A. I.; Laskowski, D. A., *The effects of pesticides on nitrogen transformations in soils*. American Society of Agronomy, Inc., Soil Society of America, Inc., and Crop Science Society of America, Inc.: Madison, WI, 1982; p 689-720.
30. Bej, A. K.; Perlin, M.; Atlas, R. M. *Fems Microbiol. Ecol.* **1991,** 86, 169-175.
31. Eichner, C. A.; Erb, R. W.; Timmis, K. N.; Wagner-Dobler, I. *Appl. Environ. Microbiol.* **1999,** 65, 102-109.
32. Griffiths, B. S.; Ritz, K.; Ebblewhite, N.; Dobson, G. *Soil Biol. Biochem.* **1999,** 31, 145-153.

Chapter 8

An overview of the fate and effects of antimicrobials used in aquaculture

Yong-Hak Kim[1] and Carl E. Cerniglia[2]

[1]Functional Proteomics Center, Korea Institute of Science and Technology, Seongbuk, Seoul 136-791, Republic of Korea
[2]National Center for Toxicological Research, US Food and Drug Administration, 3900 NCTR Road, Jefferson, AR 72079, USA

Introduction

The use of antimicrobial drugs in food animal production, including the aquaculture industry, is of global concern due to potential impact of residues on the environment and to the spread of antimicrobial resistance microorganisms. Scientific evaluation of a drug's effectiveness in the target animals and of human food and environmental safety is required before drug approval and product marketing.

The use of antimicrobial agents in aquaculture may result in an increase in the prevalence of resistant bacteria that can be transmitted to, and cause infections in humans. Such direct spread of resistance from aquatic environments to humans may occur owing to (*i*) consumption of aquaculture derived food products or drinking water, and (*ii*) direct contact with process effluents, excreted wastes or aquatic organisms.

The United States Food and Drug Administration (US FDA) works with various government agencies and aquaculture associations to increase the number of safe and effective drugs that can be used by the aquaculture industry. Currently, four antibiotics are approved by the FDA Center for Veterinary Medicine (CVM) for use in US aquaculture: sulfamerazine, oxytetracycline, florfenicol, and a potentiated sulfonamide (Romet-30®, sulfadimethoxine: ormetoprim). They are used to treat particular diseases in some aquatic animals, such as channel catfish, trout and other salmonids, other fin fish, and lobster, but cannot be used as animal growth promoters or prophylactic drugs. About 25,000 to 34,000 kg of active ingredient were sold to the US aquaculture industry in the late 1990s, which accounted for approximately 0.3-0.4 % of all the antibiotics used in US agriculture (*9, 45*).

Table I. A preliminary list of classes of antimicrobial agents used in world aquaculture

Classes	Antimicrobial agents	Target Organisms	Aquacultural use in countries*
Tetracyclines	chlortetracycline doxycycline oxytetracycline tetracycline	*Aeromonas hydrophila* *Chlamydia* spp. *Rickettsia* spp.	Canada, Chile, EU (12), Norway, India, Philippines, USA
Sulfonamides and combinations	sulfamerazine sulfadimethoxine-ormetoprim trimethoprim-sulfadiazine	*Enterobacteriaceae*	Canada, EU (7), Norway, Philippines, USA
Phenicols	chloramphenicol florfenicol		Chile, EU (5), Norway, Philippines, USA
Quinolones	flumequine nalidixic acid oxolinic acid sarafloxacin	*Campylobacter* spp. *Enterobacteriaceae* *Salmonella* spp. *Shigella* spp.	Canada, Chile, EU (12), India, Norway, Philippines
Penicillins, Aminopenicillins	amoxicillin ampicillin penicillin	*Listeria* spp. *Enterococcus* spp.	EU (3)
Macrolides	erythromycin oleandomycin rifampicin tylosin	*Chlamydia pneumoniae* *Mycobacterium* spp. *Streptococcus pneumoniae*	Philippines
Aminoglycosides	gentamicin streptomycin		Philippines
Nitrofurans	nitrofurazolidone nitrofurantoin		Norway, Philippines

*Adapted from Arthur et al. (*4*); Grave et al. (*26*); Lunestad & Grave (*41*); Markestad & Grave (*42*); WHO (*80*)

Twenty European countries have approved eight antibiotics for use in aquaculture (*80*): oxytetracycline (12 countries), trimethoprim-sulfadiazine (7), oxolinic acid (6), flumequine (5), florfenicol (5), amoxicillin (3), chlortetracycline (1), and sarafloxacin (1). The numbers registered for each country vary from 1 to 5 antibiotics because of different drug approval and regulatory requirements. In Japan, 29 individual or combination antibiotics have been approved for use in aquatic animals (*49*). Although it is difficult to estimate a total used in world aquaculture, a preliminary list shows the use of not only veterinary-use but also human-use antimicrobials, such as amoxicillin, ampicillin, chloramphenicol, erythromycin, and others that may be used illegitimately (Table I).

Regulatory safety guidances are being written to harmonize technical requirements for drug approval processes and ensure that the drugs used in all animal agriculture throughout the world are evaluated under similar guidelines. The World Health Organization (WHO) has classified 'critically important human antimicrobials' to set up risk management strategies for non-human use (*79*). The WHO, the Food and Agricultural Organization (FAO), and the World Organization for Animal Health (OIE) are working to develop a common approach dealing with the risk of antimicrobial use in agriculture and aquaculture (*71*, *73–80*).

Unapproved antimicrobial drugs used in aquaculture will likely cause adverse impacts on the environment and human health. The use of broadly effective, clinically important drugs such as tetracyclines and erythromycins is of particular concern because the antimicrobial residues in fish tissues and in the environment can facilitate development and transfer of antibiotic resistance to human pathogens. To control potential hazards of antimicrobial drugs used in aquaculture, the use of new animal drugs must be conducted in compliance with the various risk assessment approaches developed by national and international regulatory authorities.

Bacterial pathogens in aquatic environments

Bacteria are pathogenic if they are associated with diseases in aquatic animals, terrestrial animals or humans. For diseases of aquatic animals, there is international cooperation on the protocols of the Clinical and Laboratory Standards Institute (http://www.clsi.org/) to make details of test conditions or relevant control data available for a number of important and more fastidious organisms. Bacterial diseases that adversely affect both aquatic and terrestrial animals are much less known than those that affect both terrestrial animals and humans, e.g., *Salmonella* spp. Zoonotic pathogens transmitted by aquaculture practices have recently been brought into public consideration by a joint FAO/NACA/WHO study group (Table II; reference *73*).

Table II. Pathogenic bacteria transmitted by aquaculture that are associated with human diseases

Pathogenic bacteria	Diseases and Descriptions	Clinical Therapeutics
Indigenous		
Vibrio parahaemolyticus	gastroenteritis, associated with consumption of raw marine crustaceans and fish	tetracyclines or quinolones in cases of extended diarrhea
Vibrio vulnificus	septicemia and gastrointestinal disease, associated with ingestion of raw bivalves and crabs	tetracyclines (1st) ceftazidime or ciprofloxacin (2nd)
Vibrio cholerae	acute diarrhea, low risk with eating fish, normally non-O1/O139 serotypes in aquatic environments	
Aeromonas hydrophila	septicemia, low risk in farmed fish	
Plesiomonas shigelloides	diarrhea, low risk in farmed fish	
Listeria monocytogenes	listeriosis, potential risk with consumption of raw fish produced in temperate inland aquaculture	ampicillin (1st) potentiated sulfonamide (2nd)
Edwardsiella tarda	inflammatory bowel disease	ampicillin
Postharvest contamination		
Salmonella spp.	salmonellosis, low risk with aquaculture products	ampicillin, amoxicillin, potentiated sulfonamides
Other *Enterobacteriaceae*	low risk of infection associated with consumption of fish product	
Campylobacter spp.	campylobacteriosis, low risk of infection associated with consumption of fish products.	erythromycin
Clostridium botulinum	botulism, an intoxication associated with fish and fish products contaminated by *Clostridium botulinum*	
Contact/wound infection		
Erysipelothrix rhusiopathiae	erysipeloid, frequently associated with fish	penicillin
Leptospira interrogans	leptospirosis, caused by exposure to water contaminated with infected animal urine	doxycycline or penicillin
Mycobacterium marinum	significant morbidity in immunocompetent patients, infected by exposure to freshwater and saltwater, or working with tropical fish tanks	rifampicin and ethambutol
Photobacterium damsela	pseudotuberculosis or fish pasteurellosis, associated with disease in fish, wound infection in tropical environments	ceftazidime (or cefotaxime) and fluoroquinolones or tetracyclines
Streptococcus iniae	associated with serious infections of farmed fish, low risk to healthy humans	erythromycin
Vibrio vulnificus	septicemia, skin infection when open wounds are exposed to warm seawater	doxycycline and cephalosporins

Bacterial pathogens in aquaculture are either indigenous in aquatic environments or introduced by human activities. Indigenous bacteria may be divided again into aquatic animal pathogens and other natural inhabitants. It is thought that few fish pathogens in temperate climates are infectious to humans, but that the risks are greater in warmer climates, where *Aeromonas hydrophila* and *Edwardsiella* spp. are important fish pathogens (*73*). Human pathogens are occasionally introduced through discharges of feces or post-harvest processing of aquaculture products. Human infection can occur not only via aquaculture products but also from contaminated drinking water, by skin contact, or by wound infection by pathogenic bacteria, such as *Erysipelothrix rhusiopathiae* (*22*), *Leptospira interrogans* (*6*), *Mycobacterium marinum* (*38*), *Photobacterium (Vibrio) damsela* (*81*), *Streptococcus iniae* (*60*), and *Vibrio vulnificus* (*40*).

Antibiotic resistance

A major impact of antimicrobial use in aquaculture is an increase in the frequency of drug resistance in the microfloras of aquatic animals and the environment. Generally, two methods are used for determination of drug resistance frequency in the microflora. One is to isolate bacteria from samples, using non-selective media at first, and then to do antibiotic susceptibility tests according to standard laboratory protocols using selective media. The other method is to enumerate resistant populations by differential media or replica-plating techniques without isolating individual bacteria. Every test requires appropriate controls for comparison of the results, especially those from environmental samples. Significantly higher levels of antibiotic resistance can result from sampling antibiotic-untreated sites without knowing the background levels (*29*, *44*). Even well-designed experiments using appropriate controls could fail to interpret the impact of antimicrobial use on the resistance frequency. They are usually subject to the selection and cultivation of antibiotic resistant microorganisms under laboratory conditions, but there may be a lot of unknown factors that influence the development and propagation of antibiotic resistance in the environment. For example, oxolinic acid used in fish farms has produced conflicting results about the impact on resistance frequency in the pond outflows (*24*, *27*). The impact of antimicrobial treatments used near or in fish farming on antimicrobial susceptibility among bacteria has been clearly shown in species- or genus-specific approaches in diversified habitats with more than one bacterial group (*1*, *25*, *50*, *59*).

Impacts of antimicrobial use in aquaculture are not readily determined because bacterial populations are constrained by many resistance determinants and biogeochemical factors influencing the abundance and distribution of resistance genes. Most bacterial species in nature have not yet been cultured in defined media. In addition, antibiotic resistance of intestinal microflora can vary with therapeutic regimes and species. Oxytetracycline administered to catfish and goldfish results in significant changes in the resistance frequencies of intestinal microflora (*18*), but intestinal bacteria of salmon fed with oxytetracycline-medicated feeds do not show any significant change in the frequency of resistance (*30*). Oxolinic acid also results in different frequencies of resistance after administration to goldfish (*66*) or sea bass (*24*).

One of the major questions of antimicrobial use in aquaculture is how to safeguard humans and the environment from enrichment of resistance genes, because the genes, not the bacteria, can be transferred to human pathogenic bacteria. Advances in molecular biology and bioinformatics allow easy access to the total DNA encoding antibiotic resistance factors of intestinal or environmental microflora. The molecular genetic evidence show that antibiotic resistance genes are widely spread in the vicinity of aquaculture (*17*, *19*, *56-58*, *65*) and that some mobile genes coding for resistance are similar to those of clinical isolates from humans (*23*, *56*). DePaola et al. (*17*) reported that oxytetracycline- and tetracycline-resistant *Aeromonas hydrophila* strains (Octr Tet^r) became prevalent in water and sediment as well as in intestinal contents of channel catfish (*Ictalurus punctatus*) by the use of oxytetracycline in catfish culture ponds, and that some of their Tet^r determinants were able to transfer to *Escherichia coli* without the help of any plasmid. In addition, Sandaa et al. (*58*) reported that 7 strains of 34 multiple drug resistant isolates obtained from two oxytetracycline-treated aquaculture sediments were able to transfer resistance to *E. coli* HB101 by means of transferable plasmids. Spangaard et al. (*65*) showed no significant difference in the resistance profiles against oxytetracycline and oxolinic acid between Gram-negative isolates among three different trout farms exposed to different environmental stresses. This means, horizontal gene transfers of those resistances likely occur independent of species or groups among Gram-negative strains at the vicinity of antibiotic-treated sites. Some Tet^r determinants, such as TetA, TetE and TetH, are able to transfer to the recipient by using a self-mobile element like transposons Tn*1721* and Tn*5706* (*17*, *47*, *56*), whereas some multiple drug resistance are transferable at a high frequency by using a promiscuous plasmid like pRAS1 of fish-pathogenic *Aeromonas salmonicida* (*57*). Recently, quantitative (real-time) polymerase chain reactions (PCR) was used to determine the concentrations of antibiotic resistance genes (normalized to bacterial 16S ribosomal RNA genes) at various routes of the spread to humans (*54*).

The presence of zoonotic pathogens in the environmental habitats is significant, although it is presently regarded as a minor cause of human diseases (*39*). Some bacteria are frequently resistant to multiple antimicrobials and may act as reservoirs of genetic materials for more virulent bacteria, if associated with broad host ranges and certain degrees of population dynamics and dispersal of the hosts. The ecological aspects of the gene pool provide us with an insight on gene flow from aquatic organisms to terrestrial animals and human-associated bacteria, and *vice versa*. The direction of gene flow has been argued over a decade for phenicol resistance encoded by the *floR* genes of *Pasteurella piscicida* (*31*), *Klebsiella pneumoniae* (*10*), and *Salmonella* spp. (*7*, *8*, *46*). It is as yet uncertain whether the gene flow is related to antimicrobial use in aquaculture. Even if genetic probing and sequence analysis provide simple techniques to obtain genetic information about antibiotic resistance, ecological factors affecting population dynamics and dispersal in relation to human activities and infection remain to be studied.

Mechanisms of antibiotic resistance and lateral gene transfer

Mechanisms of antibiotic resistance have been intensively studied and identified in three main categories. One of them is a DNA mutation leading to structural change of a target biomolecule at the antibiotic-binding site. Secondly, active transport of antibiotic molecules to the outside of the cell is associated with non-specific efflux pumps. Thirdly, structural modifications (or degradation) of antibiotic molecules by catabolic enzymes inhibit binding to the target biomolecules. It is worth noting that some bacterial variants have phenotypic persistence mechanisms involved in antibiotic resistance without having any resistance genes or chromosomal mutations (*5*).

For the purpose of microbiological risk assessment of antimicrobial agents in aquaculture, a hazard is characterized as biological, chemical or physical agent in, or condition of food associated with an adverse health effect (*11*). The hazard may be quantitatively and qualitatively related to antimicrobial residues and antibiotic resistant microorganisms (*2*). Antimicrobials will exert their effects in fish or in fish farms via contamination of antimicrobial residues (hazard 1). Microorganisms, which are resistant to those substances, are of most significance if they have potential to associate with human, animal or fish diseases. This hazard can involve resistance mechanisms in vertical transmission during the duplication of chromosomal genes of resistant bacteria (hazard 2) or in lateral transmission via horizontal transfer of resistance genes (hazard 3) located on plasmids or associated with transposons or integrons. Because the horizontal gene transfer results in a wide spread of resistance factors in the microbial community, the ecological risk of hazard 3 is more significant than that of hazard 2. Not all resistance genes are mobile in the bacterial community. Chromosomally located β-lactamases in mesophilic *Aeromonas* spp. are thought to be non-transferable, intrinsic resistances. So, they are listed under hazard 2 for resistant bacteria, but not under hazard 3 for resistance genes. In contrast, *Aeromonas* spp., which harbor *tetA* genes associated with IncU plasmids capable of transfer to a wide variety of Gram-negative bacteria, are classified to hazard 3 (*56*, *64*).

Plasmids likely mediate lateral gene transfer of antibiotic resistance. Although they are not the only mechanism for DNA exchange between bacteria, they play an eminent role in lateral gene transfer of resistance in the environment. Plasmids are capable of transferring resistance genes either by passive mobilization, with the help of other genetic elements (*63*), or by self-mobilization by harboring transposons or integrons for insertion and excision of plasmid or chromosomal DNA. Self-mobilization not only increases the efficiency of lateral gene transfer between bacteria but also facilitates development and transfer of new gene combinations in bacterial populations. Likely, plasmids not only confer resistance but also influence intrinsic rates of growth or mortality of bacterial populations. Transferred or inherited DNA is not always stable in all bacteria because the incompatibility influences plasmid stability in the host. A prolonged use of antimicrobials in aquaculture may have an effect on stable inheritance of resistance, as did erythromycin esterases of *Pseudomonas* spp. in an erythromycin-treated trout hatchery site (*33*). The *ereA* gene sequences of *Pseudomonas* spp. are nearly identical to those located in integron cassettes of human and aquatic *Enterobacteriaceae* and *Vibrionaceae* (*32*).

Otherwise, the *qnr*-type mechanisms in aquatic *Enterobacteriaceae* and *Vibrionaceae* are responsible for lateral gene transfer and widespread distribution of low-level resistance to fluoroquinolones (*48*, *52*). The Qnr protein would protect DNA gyrase and topoisomerase IV from inhibition by fluoroquinolones (*21*). A transferable plasmid pMG252 containing a *gnr* gene was first found in a strain of *Klebsiella pneumoniae* (*43*, *69*). The plasmid-mediated *qnr*-like gene transfer may confer the fluoroquinolone resistance to other Gram-positive and Gram-negative strains (*3*).

Fate of antimicrobials used in aquaculture

The fate of antimicrobials used in aquaculture has been studied since the late 1980s. Most antimicrobial components disperse along an entire water column and flow throughout the water, so the residual components are normally low and rarely reach the minimum inhibitory concentrations necessary for selection of resistant bacteria (*61*, *62*). A relatively high concentration may occur in the under-cage sediments in shallow waters without bioturbation or in deep water sediments under anoxic conditions (*13*, *14*, *28*). The half-lives of residual antimicrobials in sediments are determined by a variety of biological and environmental factors affecting biodegradation, chemical decomposition, and sorption processes (*32*). In aquatic animals, Reimschuessel et al. (*55*) compiled a searchable database for pharmacokinetics of antimicrobials, but there are considerable differences in the half-lives of antimicrobial residues, even for the same agents in the same animals. So it is difficult to estimate the elimination times of antimicrobial residues in fish tissues. Moreover, the limited data or lack of data makes it difficult to evaluate the environmental fate of antimicrobials used in aquaculture or to do the risk analysis for the safety of humans and the environment.

To conduct risk analysis of antimicrobial use in aquaculture, not only the impacts on aquatic and terrestrial animals and humans but also the reduction, via inactivation or elimination, of biological activity by biological and chemical processes in the environment should be taken into account. The ecological risk of antimicrobials used in aquaculture can be evaluated in the process of risk assessment consisting of hazard identification, hazard characterization, exposure assessment, and risk characterization. This process provides numerical and/or descriptive estimates of risk and indication of an adverse effect on the environment and human health, and will definitely support the risk communication for interactive exchange of information and opinions among regulators and other stakeholders. The risk management, if necessary, is to make a decision or a policy for selecting and implementing proper control options and regulatory guidelines. The risk analysis is the most general term of risk-based disciplines that comprise the four pillars of risk analysis (*15*).

- Comparative risk analysis is to compare two or more types of risk, principally a tool for policy makers to decide on resource allocation.
- Risk assessment is a set of analytical techniques for estimating the frequency of undesired events and their consequences (damage or injury), and is properly accompanied by a description of uncertainty in the assessment process.
- Risk management, in contrast to risk assessment, considers social, economic and political factors to determine the acceptability of damage and what action can be taken to mitigate it.
- Risk communication is about conveying information about risk.

In addition, a new Australian Commonwealth Scientific and Industrial Research Organisation (CSIRO) research program adds

- Monitoring, to detect the impact of hazards at an early stage (although purists might include such monitoring under the heading of risk management) or to provide data to refine future risk assessments.

However, not every environmental issue can be addressed, and priorities need to be set by governmental agencies and risk managers.

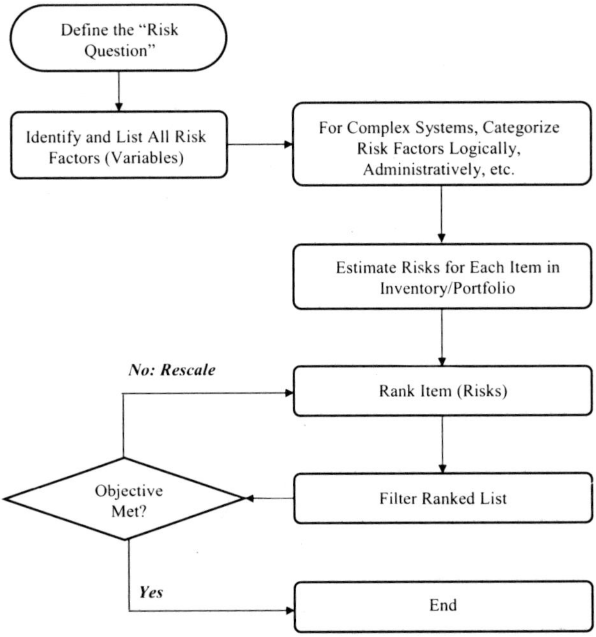

Figure 1. General process for risk ranking. A high-level schematic is shown: the process includes numerous intermediate administrative activities, such as organizing teams of experts and determining data needs under each factor (reproduced from reference 70).

Risk ranking is a risk management tool for comparing and prioritizing risks from a long track record among governmental agencies that prioritize work on portfolios of hazards that fall within their regulatory purview and that is often driven by a disparity between obligations to manage, mitigate, or reduce an array of risks, or many sources of a given type of risk, and available resources (*70*). Formal risk ranking is based on well-defined analytical processes and enhances the quality, transparency, and, potentially, the performance of risk management programs, particularly in situations in which the portfolio of risks and the underlying consequences to be managed are diverse and difficult to compare using a single tool. As shown in *Figure 1*, this process is comprised with three typical elements: hazard identification, risk estimation, and risk filtering that are to identify potential hazards and risks; characterize factors that can be used as variables for quantifying risk; and mathematically combine the variables to yield an overall risk score for risk ranking, respectively.

Environmental fate of erythromycin A

Erythromycin A is widely used for the treatment and prevention of infectious bacterial diseases in humans, livestock, and poultry (*53*), and used in food fish under an INAD program of the US FDA/CVM. Kim and his colleagues (*32*, *34-36*) have extensively studied the fate of erythromycin A in aquaculture sediment microcosms, since it is a clinically important drug for the treatment of respiratory tract infections or as an alternative to penicillin in patients who are allergic to that antibiotic. In veterinary medicine, it is also used for the treatment of mastitis in cows, chronic respiratory diseases due to mycoplasmas in poultry and other infectious diseases due to erythromycin-sensitive bacteria. Owing to the broad efficacy and applications, it has been widely contaminated in soils, water and sediments through discharges of drugs, manure and other wastes, and most frequently detected in U.S. streams (*37*, *68*). Erythromycin is not yet registered for its use in aquaculture in most European countries. The Council Directive (EEC) no. 82/2001 established a general withdrawal time of 500 °C-day (water temperature × days) only for the off-label drug use in fishes (*12*). Esposito et al. (*20*) reported that erythromycin residues administered for 21 days to rainbow trout (*Oncorhynchus mykiss*) with a therapeutic dose of 100 mg per kg-body weight per day showed a withdrawal time of 255 °C-day that is less than the general value. Since antimicrobial residues in water and sediments can serve as potential reservoirs of antibiotic-resistant microorganisms, it is important to evaluate their fate in water and sediments to access the risks of resistance developments in zoonotic pathogens or resistance transfer to human pathogens (*16*, *51*, *53*, *67*, *72*).

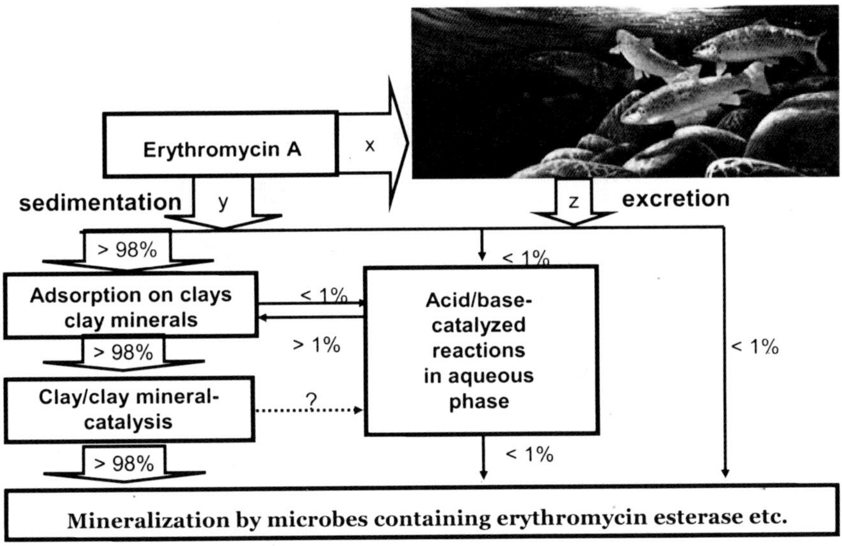

Figure 2. A scheme of the fate of erythromycin A used in aquaculture. Erythomycin (x) enters the water system from sedimentation of uningested feed (y) and through discharge of fish excretes (z). The bioavailability of erythromycin in medicated feed will vary with many factors yet to be determined. Erythromycin in the aqueous phase and sediments is degraded mainly by clay/clay mineral-catalyzed reactions, acid/base-catalyzed reactions, and microbial degradation and mineralization. The percent values of those ways contributing to degradation or inactivation of erythromycin A at neutral pH conditions are estimated from the observed data in references 34-36.

Erythromycin A is likely degraded by general acid/base-catalyzed reactions, leading to the formation of erythromycin A enol ether, anhydroerythromycin A and pseudoerythromycin A enol ether that mostly lose the original antimicrobial activity (*34*). The sorption process in sediments would enhance the degradation rate of erythromycin A by two to four orders of magnitude due to the likely occurrence of clay/clay mineral-catalyzed reactions (*35*). In addition to chemical and geochemical processes, the degradation of erythromycin A is facilitated by such a constitutive erythromycin-resistance gene product, erythromycin esterase, which is able to hydrolyze the lactone rings of some 14-membered macrolides, including oleandomycin, erythromycin A, and erythromycin A enol ether (*33*). The enzymatic turnover rates (k_{cat}) of erythromycin A (1.36×10^4 s^{-1}) and its derivative erythromycin A enol ether (4.61×10^3 s^{-1}) appear high compared to the acid/base-catalyzed reactions of erythromycin A ($k_{obs} = 3 \times 10^{-7}$ s^{-1}) in neutral conditions. But, the microbial degradation is very limited by the presence and density of metabolizing populations and their cell volumes (*32, 36*). As illustrated in *Figure 2*, erythromycin A and its derivatives are partially or completely mineralized by sediment microorganisms, albeit slowly due to the low initial densities of metabolizing populations. Even though the microbial degradation is slower than the whole geochemical degradation in sediments, it is the only feasible way to mineralize the applied antimicrobial drug to carbon dioxide.

Summary

National and international authorities, in conjunction with animal and public health officials, have been proactive to promote the prudent use of veterinary antimicrobial drugs to minimize development of resistant bacteria and transfer of resistance genes. To allow food animal producers to use beneficial antimicrobials in the treatment and/or prevention of animal diseases or in the improved production of animals and fish, the safety and effectiveness of drugs should be ensured with certainty that products will be safe, wholesome and free of drug residues when the products reach the consumer. The legitimate use of antimicrobials in aquaculture necessitates careful use of drugs, not only for therapeutic purposes but also, for the safety of humans and the environment, to minimize the development and spread of antimicrobial resistance from exposure to low concentrations.

The primary goal of antimicrobial use is effective therapy for pathogen infection of aquatic animals with as narrow a range of drug spectrum as possible. The use of clinically important broad-spectrum antimicrobials like erythromycin A is considered to have potential for development of resistant bacteria and transfer of resistance genes to non-resistant bacteria, especially human pathogens. The adverse impacts result not only from the increase in the frequency of resistance in aquatic animals and environments, but also from acceleration of gene flow between aquatic and terrestrial animals and humans.

Antibiotic-exposed environments may serve as reservoirs of antibiotic resistance if an undigested or poorly absorbed portion of medicated feed exists in water and sediments for a long period or for intervals of time during repeated use. Waste treatments, such as sedimentation or filtration through soil, will help reduce concentrations of antimicrobial residues at the outflows. In the case of erythromycin A in sediments, the biological activity is almost completely lost within a few days by soil sorption and clay/clay mineral-catalyzed degradation processes.

Disclaimer

The use of trade names is for identification purposes only, and does not imply endorsement by the U.S. Food and Drug Administration or the U.S. Department of Health and Human Services. Views presented in this manuscript do not necessarily reflect those of the FDA.

References

1. Agersø, Y.; Peterson, A. *J. Antimicrob. Chemother.* **2006**, 59, 23-27.
2. Alday, V.; Guichard, B.; Smith, P.; Uhland, C. *Towards a risk analysis of antimicrobial use in aquaculture. Joint FAO/OIE/WHO expert consultation on antimicrobial use in aquaculture and antimicrobial resistance*, Seoul, Republic of Korea, 2006.
3. Arsene, S.; Leclercq, R. *Antimicrob. Agents Chemother.* **2007**, 51, 3254-3258.
4. Arthur, J. R.; Lavilla-Pitogo, C. R.; Subasinghe, R. P. *Use of chemicals in aquaculture in Asia.* Proceedings of the Meeting on the Use of Chemicals in Aquaculture in Asia, 20-22 May 1996, 2000, Tigbauan, Iloilo, Philippines.
5. Balaban, N. Q.; Merrin, J.; Chait, R.; Kowalik, L.; Leibler, S. *Science* **2004**, 305, 1622-1625.
6. Baranton, G.; Postic, D. *Int. J. Infect. Dis.*, **2006**, 10, 162-170.
7. Bolton, L. F.; Kelley, L. C.; Lee, M. D.; Fedorka-Cray, P. I.; Maurer, H. *J. Clin. Microbiol.* **1999**, 37, 1348-1351.
8. Briggs, C. E.; Fratamico, P. M. *Antimicrob. Agents Chemother.* **1999**, 43, 846-849.
9. Carnevale, R. *Practical considerations in the collection of antibiotic use data in animals. A US perspective.* Animal Health Institute, Washington, D.C., 2001.
10. Cloeckaert, A.; Baucheron, S.; Chaslus-Dancla, E. *Antimicrob. Agents Chemother.* **2001**, 45, 2381-2382.
11. Codex Alimentarius Commission. *Principles and Guidelines for the conduct of microbiological risk assessment, CAC/GL-30*, 1999.
12. Council Directive (EEC) no. 82/2001. *Official Journal of the European Communities L Ser.* **2001**, 311, 0001-0066.
13. Coyne, R.; Hiney, M.; O'Connor, B.; Kerry, J.; Cazabon, D.; Smith, P. *Aquaculture* **1994**, 123, 31-42.
14. Coyne, R.; Smith, P.; Moriarty, C. *Marine Environ. Health* **2001**, Series 3, 1-24.
15. Davies, J. C., Ed. *Comparing environmental risks. Tools for setting government priorities.* Washington, DC: Resources for the Future, 1996.
16. Davies, J. E. In *Antibiotic resistance: origins, evolution, selection and spread*; S. Levy et al., Ed.; Ciba Foundation Symp. 207. John Wiley & Sons, NY, 1997, pp 15-27.
17. DePaola, A.; Flynn, P. A.; McPhearson, R. M.; Levy, S. B. *Appl. Environ. Microbiol.* **1988**, 54, 1861-1863.
18. DePaola, A.; Peeler, J. T.; Rodrick, G. E. *Appl. Environ. Microbiol.* **1995**, 61, 2335-2340.
19. DePaola, A.; Roberts, M. C. *Mol. Cell. Probes* **1995**, 9, 311-313.
20. Esposito, A.; Fabrizi, L.; Lucchetti, D.; Marvasi, L.; Coni, E.; Guandalini, E. *Antimicrob. Agents Chemother.* **2007**, 51, 1043-1047.
21. Ferrero, L.; Cameron, B.; Manse, B.; Lagneaux, D.; Crouzet, J.; Famechon, A.; Blanche, F. *Mol. Microbiol.* **1994**, 13, 641-653.
22. Fidalgo, S. G.; Wang, Q.; Riley, T. V. *Appl. Environ. Microbiol.* **2000**, 66,

2066-2070.

23. Furushita, M.; Shiba, T.; Maeda, T.; Yahata, M.; Kaneoka, A.; Takahashi, Y.; Torii, K.; Hasegawa, H.; Ohta, M. *Appl. Environ. Microbiol.* **2003**, 69, 5336-5342.
24. Giraud, E.; Douet, D.-G.; Le Bris, H.; Bouju-Albert, A.; Donnay-Mreno, C.; Thorin, C.; Pouliquen, H. *FEMS Microbiol. Ecol.* **2006**, 55, 439-448.
25. Gordon, L.; Giraud, E.; Ganière, J. P.; Armand, F.; Bouju-Albert, A.; de la Cotte, N.; Mangion, C.; Le Bris, H. *J. Appl. Microbiol.* **2007**, 102, 1167-1176.
26. Grave, K.; Engelstad, M.; Søli, N. E.; Håstein, T. *Aquaculture* **1990**, 86, 347-358.
27. Guardabassi, L.; Dalsgaard, A.; Raffatellu, M.; Olsen, J. E. Aquaculture **2000**, 188, 205-218.
28. Hektoen, H.; Berge, J. A.; Hormazabal, V.; Yndestad, M. *Aquaculture* **1995**, 133, 175-184.
29. Kelch, W. J.; Lee, J. S. *Appl. Environ. Microbiol.*, **1978**, 36, 450-456.
30. Kerry, J.; NicGabhainn, S.; Smith, P. *Aquaculture* **1997**, 157, 187-195.
31. Kim, E. H.; Aoki, T. *Microbiol. Immunol.* **1993**, 37, 103-109.
32. Kim, Y.-H.; Cerniglia, C. E. *Aquat. Toxicol.* **2005**, 73, 230-241.
33. Kim, Y.-H.; Cha, C. J.; Cerniglia, C. E. *FEMS Microbiol. Lett.* **2002**, 210, 239-244.
34. Kim, Y.-H.; Heinze, T. M.; Beger, R.; Pothuluri, J. V.; Cerniglia, C. E. *Int. J. Pharmaceut.* **2004a**, 271, 63-76.
35. Kim, Y.-H.; Heinze, T. M.; Kim, S. J.; Cerniglia, C. E. *J. Environ. Qual.* **2004b**, 33, 257-264.
36. Kim, Y.-H.; Pak, K.; Pothuluri, J. V.; Cerniglia, C. E. *FEMS Microbiol. Lett.* **2004c**, 234, 169-175.
37. Kolpin, D. W.; Furlong, E. T.; Meyer, M. T.; Thurman, E. M.; Zaugg, S. D.; Barber, L. B.; Buxton, H. T. *Environ. Sci. Technol.* **2002**, 36, 1202-1211.
38. Lahey, T. *Emerg. Infect. Dis.*, **2003**, 9, 1496-1498.
39. Leatherbarrow, A. J. H.; Griffiths, R.; Hart, C. A.; Kemp, R.; Williams, N. J.; Diggle, P. J.; Wright, E. J.; Sutherst, J.; Houghton, P.; French, N. P. *Environ. Microbiol.* **2007**, 9, 1772-1779.
40. Lehane, L.; Rawlin, G. T. *Med. J. Austral.*, **2001**, 174, 480-481.
41. Lunestad, B. T.; Grave, K. *Bull. EAFP* 2005, 25, 284-290.
42. Markestad, A., and Grave, K. In *Fish vaccinology: development in biological standardization*; Guddings, R.; Lillehaug, A.; Midlyng, A. J.; Brown, F., Eds.; Basel, Karger, 1997, Vol. 90, pp 365-369.
43. Martinez-Martinez, L.; Pascual, A.; Jacoby, G. A. *Lancet* **1998**, 351, 797-799.
44. McKeon, D. M.; Calabrese, J. P.; Bissonnettes, G. K. *Water Res.* **1995**, 29, 1902-1908.
45. Mellon, M.; Benbrook, C.; Benbrook, K. L. *Hogging it: estimates of antimicrobial abuse in livestock.* Union of Concerned Scientists, Cambridge, MA, 2001.
46. Meunier, D.; Baucheron, S.; Chaslus-Dancla, E.; Martel, J.-L.; Cloeckaert, A. *J. Antimicrob. Chemother.* **2003**, 51, 1007-1009.

47. Miranda, C. D.; Kehrenberg, C.; Ulep, C.; Schwarz, S.; Roberts, M. C. *Antimicrob. Agents Chemother.* **2003**, 47, 883-888.
48. Nordmann, P.; Poirel, L. *J. Antimicrob. Chemother.* **2005**, 56, 463-469.
49. Okamoto, A. In *Chemotherapy in aquaculture: from theory to reality*; Michel, C.; Alderman D. J., Eds.; Office International de Epizooties, Paris, 1992, pp 109-114.
50. Petersen, A.; Andersen, J. S.; Kaewmak, T.; Somsiri, T.; Dalsgaard, A. *Appl. Environ. Microbiol.* **2002**, 68, 6036-6042.
51. Piddock, L. J. V. *J. Antimicrob. Chemother.* **1996**, 38, 1-3.
52. Poirel, L.; Liard, A.; Rodriguez-Martinez, J.-M.; Nordmann, P. *J. Antimicrob. Chemother.* **2005**, 56, 1118-1121.
53. Pothuluri, J. V.; Nawaz, M.; Cerniglia, C. E. In *Bioremediation: principles and practice. Biodegradation technology developments*; Sikdar, S. K.; Irvine, R. L., Eds.; Technomic Publ., Lancaster, PA, 1998, Vol. II, pp 221-248.
54. Pruden, A.; Pei, R.; Storteboom, H.; Carlson, K. H. *Environ. Sci. Technol.* **2006**, 40, 7445-7450.
55. Reimschuessel, R.; Stewart, L.; Squibb, E.; Hirokawa, K.; Brady, T.; Brooks, D.; Shaikh, B.; Hodsdon, C. *AAPS J.* **2005**, 7, E288-E327.
56. Rhodes, G.; Huys, G.; Swings, J.; McGann, P.; Hiney, M.; Smith, P.; Pickup, R. W. *Appl. Environ. Microbiol.* **2000**, 66, 3883-3890.
57. Sandaa, R. A.; Enger, Ø. *Appl. Environ. Microbiol.* **1994**, 60, 4234-4238.
58. Sandaa, R. A.; Torsvik, V.; Goksøyr, J. *Can. J. Microbiol.* **1992**, 38, 1061-1065.
59. Schmidt, A. S.; Bruun, M. S.; Dalsgaard, I.; Pedersen, K.; Larsen, J. L. *Appl. Environ. Microbiol.* **2000**, 66, 4908-4915.
60. Shoemaker, C.; Klesius, P. In *Tilapia aquaculture*; Fitzsimmons, K., Ed.; NREAES 106, Ithaca, 1997, Vol. 2, pp 671-680.
61. Smith, P.; Samuelsen, O. B. *Aquaculture* **1996**, 144, 17-26.
62. Smith, P.; Donlon, J.; Coyne, R.; Cazabon, D. J. *Aquaculture* **1994**, 120, 319-325.
63. Sørum, H. In *Antimicrobial resistance in bacteria of animal origin*; Aarestrup, F. M., Ed.; ASM Press, Washington, D.C., 2006, pp 213-238.
64. Sørum, H., L'Abée-Lund, T.M., Solberg, A., and Wold, A. *Antimicrob. Agents Chemother.* **2003**, 47, 1285-1290.
65. Spangaard, B., Jørgensen, F., Gram, L., and Huss, H. H. *Aquaculture* **1993**, 115, 195-207.
66. Sugita, H.; Miyajima, C.; Fukumoto, M.; Koyama, H.; Deguchi, Y. *Aquaculture* **1989**, 80, 163-174.
67. Tollefson, L.; Fedorka-Cray, P.; Marano, N.; Angulo, F. U. S. *National antimicrobial resistance monitoring system for enteric bacteria. Informal information meeting on antimicrobial resistance surveillance in foodborne pathogens*. World Health Organization, Geneva, Switzerland, 1999.
68. Tolls, J. *Environ. Sci. Technol.* **2001**, 35, 3397-3406.
69. Tran, J. H.; Jacoby, G. A. *Proc Natl Acad Sci USA* **2002**, 99, 5638-5642.
70. US FDA. *Risk-based method for prioritizing cGMP inspections of pharmaceutical manufacturing sites - a pilot risk ranking model.* Department of Health and Human Services, US Food and Drug

Administration, 2004.

71. WHO. *The medical impact of the use of antimicrobials in food animals. Report of a WHO meeting*, Berlin, Germany, 1997.
72. WHO. *WHO's Fact Sheet no. 194*. World Health Organization, Geneva, Switzerland, 1998.
73. WHO. *Food safety issues associated with products from aquaculture*. World Health Organization, Geneva, Switzerland, 1999.
74. WHO. *WHO global principles for the containment of antimicrobial resistance in animals intended for food. Report of a WHO consultation with the participation of the Food and Agriculture Organization of the United Nations and the Office International des Epizooties*, World Health Organization, Geneva, Switzerland, 2000.
75. WHO. *Monitoring antimicrobial usage in food animals for the protection of human health. Report of a WHO consultation*, Oslo, Norway, 2001.
76. WHO. *Beyond antimicrobial growth promoters in food animal production. Working papers for the WHO international review panels evaluation*, World Health Organization, Foulum, Denmark, 2002.
77. WHO. *Background document for the joint WHO/FAO/OIE expert workshop on non-human antimicrobial usage and antimicrobial resistance: scientific assessment*. World Health Organization, Geneva, Switzerland, 2003.
78. WHO. *Second joint WHO/FAO/OIE expert workshop on non-human antimicrobial usage and antimicrobial resistance: management options*. Oslo, Norway, 2004.
79. WHO. *Critically important antibacterial agents for human medicine for risk management strategies of non human use. Report of a WHO working group consultation*, Canberra, Australia, 2005.
80. WHO. *Antimicrobial use in aquaculture and antimicrobial resistance. Report of a joint FAO/OIE/WHO expert consultation on antimicrobial use in aquaculture and antimicrobial resistance*, Seoul, Republic of Korea, 2006.
81. Yamane, K.; Asato, J.; Kawade, N.; Takahashi, H.; Kimura, B.; Arakawa, Y. *J. Clin. Microbiol.*, **2004**, 42, 1370-1372.

Chapter 9

Fate and bioavailability of sulfamethazine in freshwater ecosystems

Keri L.D. Henderson[1], Thomas B. Moorman[2], and Joel R. Coats[1]

[1]Department of Entomology, Iowa State University, Ames, IA, USA
[2]U.S. Department of Agriculture, Agricultural Research Service, National Soil Tilth Laboratory, Ames, IA, USA

The antibiotic sulfamethazine can be transported from manured fields to surface water bodies. We investigated the degradation, fate, and bioavailability of sulfamethazine in surface water using ^{14}C-U-phenyl-sulfamethazine in small pond water microcosms. Sulfamethazine dissipated exponentially from the water column, with the majority of loss occurring via movement into the sediment phase. Manure input significantly increased sorption and binding of sulfamethazine residues to the sediment. These results indicate sediment is a potential sink for sulfamethazine and sulfamethazine-related residues, which could have important implications for benthic organisms. Understanding the bioavailability of pharmaceuticals in environmental matrices is particularly important considering they are often in a bioactive form. The bioavailability of sulfamethazine in surface water microcosms was evaluated using *Lumbriculus variegatus* in a bioassay. Bioconcentration factors (BCFs) were calculated, and a log BCF >2 was observed during aquatic exposure (0.05 mg/l). Interestingly, a significant inverse relationship between exposure concentration and BCF was also noted. Our results indicate the need for further assessment of the bioaccumulation potential of SMZ residues as a result of sediment exposure of benthic invertebrates.

Introduction

Recent monitoring studies have confirmed the presence of veterinary antibiotics, including sulfamethazine (SMZ), in surface water and sediment throughout the U.S. and abroad (*1-4*). Antibiotics could potentially alter bacterial populations in sediment and water, thus affecting nutrient cycling, water quality, and food webs via direct or indirect effects on non-target organisms. In particular, small ponds and wetlands that support invertebrate communities can receive significant amounts of contaminated agricultural runoff, which could contain antibiotic residues (*5,6*). It is believed that these drugs may have an affinity for clay particles in soil or sediment, which could affect their availability for degradation; this is also likely important for bioavailability and uptake for higher organisms (*7*). An understanding of the fate and bioavailability of antibiotics in ponds is important because of widespread use of the compounds in livestock production in the U.S., and concurrent application of manure to agricultural lands dotted with small ponds and wetlands.

The sulfonamides are a class of antibiotics frequently detected in the aforementioned monitoring studies. These drugs have been used extensively in human and veterinary medicine applications for decades. The sulfonamide mode of action is prevention of folic acid synthesis in bacteria by acting as a structural analog of p-aminobenzoic acid, and they are mostly active against gram-positive, but also some gram-negative bacteria; in the present study, sulfamethazine was chosen as a representative sulfonamide because of its frequent detection in aquatic systems (*1-4,8*). Sulfamethazine (Figure 1) is used in cattle and swine production as a therapeutic and as a growth promoter. Concentrations can exceed 5 mg/kg in swine manure (*9*).

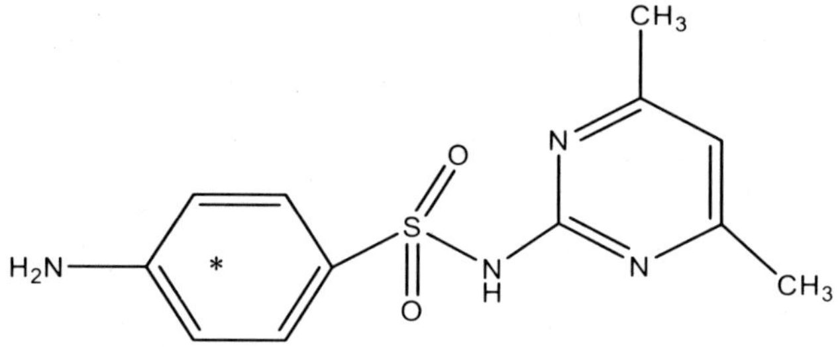

Figure 1. Chemical structure of sulfamethazine, with 14C-uniformly labeled in the phenyl ring (*).

Previous soil fate studies have indicated that SMZ is not strongly sorbed to soils and is likely to be highly mobile in the aqueous portion of runoff, thus being likely to reach surface water bodies, e.g. small ponds (*5,10-12*). Though its fate has been extensively examined in soil, SMZ fate in surface water, and sediment in particular, has not been extensively studied.

Published reports have confirmed that very low levels (e.g. <5 μg/l) of antibiotic residues are detected in water and sediment monitoring studies (*1,3,4*) but the significance of those residues will depend, to a great extent, on their biological availability. Although acute effects in traditional test organisms are unlikely at these concentrations (*3,4,13,14*), the potential biological significance of antibiotics lies in the possibility that they could induce resistance in bacterial populations, maintain resistant populations, allow for easier transfer of resistance, or alter beneficial bacterial populations, such as gut symbionts or those bacteria responsible for nutrient cycles or biodegradation of contaminants.

Bioavailability is defined as the fraction of a compound that is immediately available to an organism, as indicated by the compound's tendency to be transferred from the habitat (e.g., sediment or water) and subsequently cross the organism's cellular membrane. Bioavailability has traditionally been studied using bioassays by examining the amount of a chemical taken up into an organism's tissues following exposure to a known concentration. Simplistically, a bioconcentration factor (BCF) or bioaccumulation factor (BAF) is calculated by determining the ratio between the concentration of a compound detected in an organism and the concentration in the matrix. Bioavailability studies serve as a crucial link to understanding the likelihood of environmental residues impacting a species in a negative way.

The objective of the studies presented here was to assess the environmental fate of sulfamethazine in aquatic microcosms, and subsequent uptake and bioaccumulation of SMZ residues by *L. variegatus*, a common freshwater sediment-dwelling oligochaete worm. The methods and results from those studies on the fate and bioavailability of SMZ are presented in this chapter.

Materials and Methods

Environmental fate study

Pond water (surface 20 cm) and sediment were collected from the Iowa State University Horticulture Research Station pond (Ames, IA). Prior to use, sediment was thoroughly mixed, and moisture content was determined to be 46.2%. The sediment was a sandy loam (60% sand, 28% silt, 12% clay) with 2.0% organic matter, and a pH of 8.1. The pH of the pond water was 8.1, the total hardness was 150 mg/ml, and the alkalinity was 103 mg/ml.

Fifty grams (dry wt) of sediment was measured into wide-mouth pint jars, and topped with 177 ml pond water, to equal 200 ml water per replicate. There were four replicates at each timepoint per treatment group. Treatment groups were fresh pond water with sediment (PWS), fresh pond water with sediment plus manure (PWS+M), autoclaved pond water with sediment (Auto PWS), and pond water only (PW).

Stock solutions of labeled and non-labeled sulfamethazine were prepared to make a final treatment solution to be added to each microcosm. The final concentration of SMZ in pond water was 5 mg/l, and 0.2 μCi ^{14}C-U-labeled phenyl ring-SMZ/jar.

For the manure treatment (PWS+M), a manure slurry was added to obtain 0.1% manure in pond water, which was enough manure to make the pond water slightly murky in appearance. The slurry consisted of a 33% w/v solution of fresh manure from adult hogs on an antibiotic-free diet (Iowa State University Swine Nutrition Facility; 33 g wet wt (22% dry mass) in 100 ml distilled water). Microcosms were maintained in environmental chambers at 22°C in a 12:12 photoperiod for the 63-day period; pH was monitored weekly in the surface water of the microcosms, and did not significantly change during the test period.

To track mineralization of ^{14}C-SMZ added to the microcosms, NaOH traps for CO_2 were included in the systems. Traps consisted of 10 ml 0.5M NaOH, and were collected at days 3, 7, 14, 21, 28, 35, 42, 49, and 56 and counted for radioactivity using a Packard Tri-Carb 1900 (PerkinElmer, Waltham, MA).

At each timepoint (7, 14, 28, or 63 d), the appropriate replicates were sacrificed. Pond water was siphoned from the top of the sediment, and a 2-ml aliquot was mixed with 12 ml Ultima Gold™ XR cocktail (PerkinElmer, Waltham, MA) and counted for radioactivity. Water was filtered through glass fiber filters and extracted using HLB solid phase extraction cartridges (6 cc, Oasis HLB®, Waters Corporation, Milford, MA); cartridges were conditioned with methanol, 0.5N HCl, and ultrapure water prior to sample loading. After water samples were passed through, cartridges were washed with ultrapure water, and finally eluted with methanol. Eluates were brought to a final volume of 5.0 ml using ultrapure water, and stored in a freezer until analysis.

For sediment extraction, sediment was transferred to Teflon centrifuge bottles using a 70% methanol solution. The bottles were then shaken 80 min and allowed to settle overnight at room temperature. Bottles were centrifuged and decanted, and a second extraction was performed. Extracts were then concentrated by evaporation under nitrogen flow (15 psi, 40°C), and enriched using the SPE method previously described.

Extracted sediments were dried in a fume hood overnight, were processed to remove rocks and debris, and ground using mortar and pestle to ensure thorough mixing. To determine the amount or radioactivity remaining bound in the sediment following extraction, aliquots were combusted using an OX-600 biological oxidizer (RJ Harvey Instrument Co., Hillsdale, NJ), and radioactivity was determined by LSC as previously described.

A reverse-phase high-performance liquid chromatography (HPLC) method was used to quantify parent sulfamethazine and to identify related metabolites. Analysis of the samples was performed using a Hewlett-Packard (Palo Alto, CA, USA) series 1100 HPLC system with a quaternary pump, an autosampler, a thermostatted column compartment, and UV and β-RAM detectors. A Waters Atlantis™ (Milford, MA, USA) dC18 column (4.6 × 250 mm, 5-μm particle size) was used with a flow rate of 1.0 ml/min at 30° C, and 200μl injection volume. The mobile phase consisted of 30% methanol. Radiolabeled and non-labeled standards were used to quantify parent sulfamethazine in samples. A Beta-RAM radiodetector (IN/US Systems, Tampa, FL) was also used to detect radioactivity in the samples.

Bioavailability study

To assess bioavailability of SMZ in aquatic ecosystems, the aquatic oligochaete *Lumbriculus variegatus* was chosen; these worms are commonly used in sediment bioaccumulation assays (*15*). Worms were exposed to a range of SMZ concentrations either in water only, or in sediment-containing microcosms.

For the aquatic exposure assay, test chambers consisted of 200-ml glass jars filled with 50 ml ultrapure water spiked with a mixture of ^{14}C- and non-labeled sulfamethazine to achieve the following concentrations: 0.05, 0.5, and 5 mg/l in test water. Specific activity of the labeled sulfamethazine within test chambers was 0.72 μCi/mg for the 0.05 mg/l treatment, and 0.52 μCi/mg for the 0.5 and 5 mg/l treatments. At day 0, five adult *Lumbriculus variegatus*, weighing approximately 8 mg each, were placed in test chambers. Worms were sacrificed at days 1, 3, 7, and 14, and worm tissue from a single replicate was pooled for analysis, with 4 replicates per treatment and timepoint. Worms were removed from treatment water and placed in 5 ml fresh ultrapure water for 6 hours to allow for gut clearance. Clearance water was removed for quantification of ^{14}C residues using liquid scintillation counting (LSC; Beckman Coulter 6500, Fullerton, CA). Next, worms were homogenized in 5 ml methanol for 10 min; extracts including homogenized tissue were shaken at 300 rpm for 15 min on an orbital shaker and then centrifuged at 2000 rpm for 5 min.

Treatment water was extracted using the previously described solid-phase extraction technique. Aliquots of worm extract supernatant or water eluate were immediately counted for radioactivity using LSC, and remaining extract was stored at -20° C until analysis using high pressure liquid chromatography with radiodetection, as previously described.

The sediment exposure studies were run in parallel with the previously described environmental fate experiment. Pond water with sediment (PWS) microcosms were constructed as previously described; sediment was allowed to settle for 1 h, and 10 adult *L. variegatus* (approximately 8 mg each) were added to each replicate test chamber. Worms were allowed to acclimate for 1 h prior to addition of the ^{14}C-SMZ spiking solution. The final spiking solution was 0.425 mg/ml non-labeled SMZ and 0.085 μCi/ml ^{14}C-SMZ in 10% methanol, as in the fate study. At the end of the 7-day period, pond water was removed from the top of the sediment and the remaining water and sediment were carefully sieved through a 2-mm mesh to remove worms. Worms were then rinsed with ultrapure water, and allowed the 6-h clearance period described for the aquatic exposure study. Worms were ground and extracted, and extracts analyzed as previously described.

Bioconcentration factors were calculated based on quantification of total radioactivity (SMZ equivalents) in worm extracts, and parent sulfamethazine in water (C_w) for the aquatic exposure study, and in pond water and sediment for the sediment exposure (C_{sedOC}). Concentrations in worm tissue (C_{lipid}) were calculated as ng SMZ equivalents/g lipid. Sediment concentrations were also normalized by 2% organic carbon (mg SMZ/kg OC). Bioconcentration factors were calculated using the normalized data.

Results and Discussion

Environmental fate of sulfamethazine

Mean ^{14}C-mass balances for pond water (PW) systems were >90% for each timepoint. All sediment-containing systems (PWS, PWS+M, and Autoclaved PWS) exceeded 95%, with the exception of the PWS+M treatment at day 63, with a mean ^{14}C mass balance of 84±2%. Figure 2 displays the distribution of ^{14}C in pond water and sediment at day 7 and day 63. Clear differences exist between day 7 and day 63 for sediment binding and amount of radiolabel remaining in water ($p<0.001$ for all treatments).

Sulfamethazine dissipated rapidly from surface water in all treatments, with DT_{50}s (time to 50% dissipation) ranging from 2.7 to 17.8 days. The most rapid loss from the water column occurred in the non-autoclaved sediment-containing treatments. The slowest dissipation rate occurred in the autoclaved treatment, indicating the importance of biodegradation of sulfamethazine residues in freshwater ecosystems. Sulfamethazine moved from the water column into the sediment within the first 14 days of the study. Parent SMZ detected in the sediment peaked at 7-14 days, and then showed a slow decline, which corresponded with an increase in bound residue detected in all sediment-containing treatments. Figure 3 simultaneously displays dissipation of parent SMZ from the water column, and its movement into and subsequent dissipation in sediment.

^{14}C-bound (unextractable) residue concentrations increased exponentially over time, and appeared to plateau toward the end of the 63-day study. In the manure-containing treatment, 62% of applied radioactivity was bound to sediment at the end of the study, while the other two treatments had approximately 40% of applied radioactivity as bound residue. The manure-containing treatment (PWS+M) had significantly more bound residues than PWS or the autoclaved treatment throughout the study ($p<0.04$ and <0.01; Fig. 2). It is possible that SMZ sorbed to manure in the pond water phase in our study, and then settled onto the sediment. In a study of the effect of hog manure slurry on sulfonamide sorption in soil, Thiele-Bruhn and Aust (*16*) found an increase in the nondesorbable (bound) fraction for SMZ in systems with hog manure input.

The amount of sorption observed in this study is relatively consistent with previously reported sorption values (*5,10,11,17,18*). Our results showing increased binding in the treatment with highest organic carbon content (PWS+M) are in agreement with Bialk et al. (*19*) and Thiel-Bruhn et al. (*11*); however, at pH 8.1 in pond water and sediment, a significant portion of SMZ is likely to be ionized, therefore some alternative sorption mechanisms may come in to play. These mechanisms could include cation exchange or bridging, surface complexes, or hydrogen bonding (*10,11,19,20*)

In a sediment monitoring study, Kim and Carlson (*3*) detected SMZ in 25% of river sediment samples, with mean of 4.7 μg/kg and maximum concentration detected of 13.7 μg/kg; overlying water samples from that study were typically <0.1 μg/l, pointing to the significance of sediment in the fate of SMZ in aquatic ecosystems. Our results indicate that SMZ or metabolite residues may be transported from the aqueous phase into the sediment via diffusion, residues become adsorbed, and a fraction of those residues become bound over time; these data further support Kim and Carlson's work identifying sediment as a potential sink for SMZ residues. The potential implications of sediment-associated residues for benthic-dwelling organisms need to be further explored.

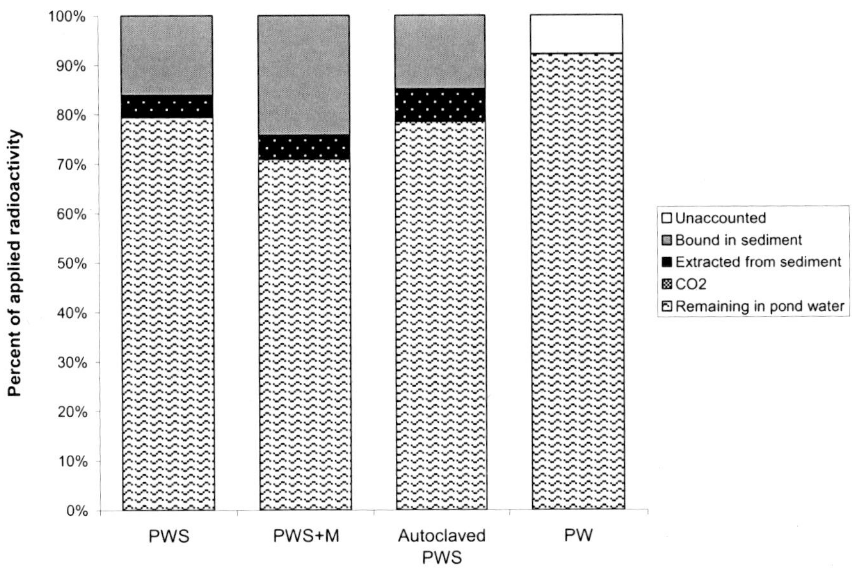

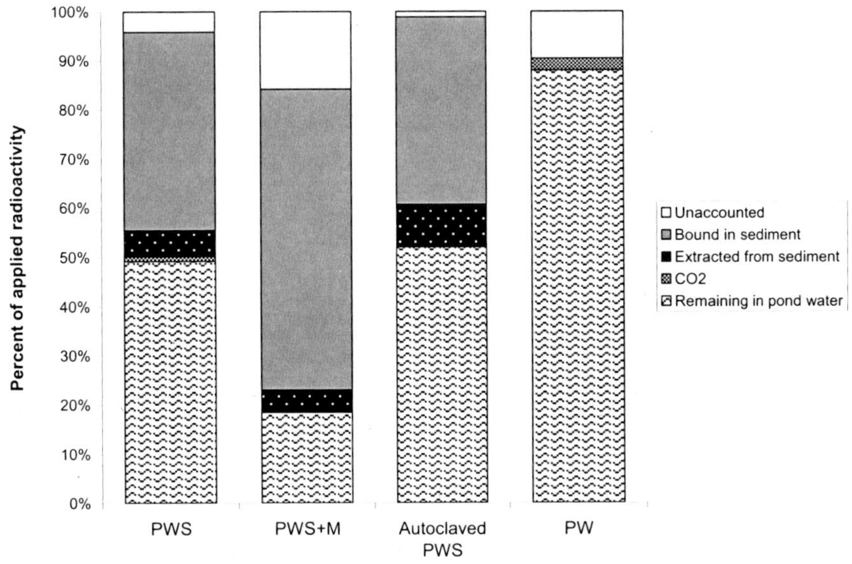

Figure 2. Mass balance of SMZ-residues in each treatment at day 7 (top) and day 63 (bottom).

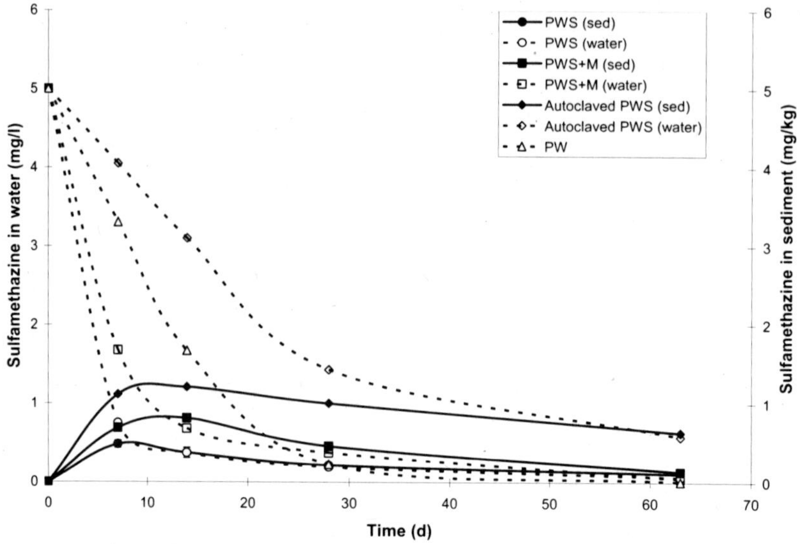

Figure 3. Dissipation of parent sulfamethazine from pond water and movement into sediment.

Bioavailability of sulfamethazine

During the aquatic exposure bioassay, no toxicity of sulfamethazine to *L. variegatus* was noted during the study. The lipid content of adult *L. variegatus* was 12.7% as determined by acid hydrolysis (AOAC Method 4.5.02 16th ed; Midwest Laboratories, Inc., Omaha, NE). Mean total uptake of SMZ-equivalents was 0.05, 0.02, and 0.10 micrograms for the 0.05, 0.5, and 5 mg/l exposure concentrations, respectively. This corresponds to ≤2% uptake of the total available ^{14}C-residue. Log BCFs were calculated for each treatment concentration at equilibrium (~ 7 days), and significant treatment effects were noted ($p<0.001$), indicating an inverse relationship between amount of uptake and exposure concentration in the water (Table I). The mean log BCF for the 0.05 mg/l exposure concentration was 2.17, which is of the same magnitude as BCFs expected for persistent hydrophobic compounds such as PCBs (e.g. Aroclors 1242 and 1254) or PAHs (e.g. phenanthrene and fluoranthene), which was not expected given the relatively low K_{ow} for SMZ (*21-23*). The inverse concentration-dependent accumulation appears to be unique for an organic compound; a brief review of the literature found similar relationships in metal bioaccumulation studies (*24,25*). Reasons for an inverse relationship in metals are proposed to be related to saturated uptake conditions at higher exposure concentrations (*26*) and active regulation of metal uptake. Since SMZ may be ionized at environmental pHs, as in the present study, it is possible that it was similarly regulated. The inverse relationship points to a need for further study of bioavailability of sublethal levels of pharmaceuticals in the environment.

At the end of the 7-day sediment exposure study, 68% of worms were recovered from the four replicates, and all calculations were based off of recovered tissue mass. Less than 0.5% of total ^{14}C was taken up by the worms; however the mean log BCF (or BSAF) was calculated to be 1.89 (Table 1). Significantly more uptake (in µg) was observed in the sediment assay compared to all of the aquatic exposure concentrations, with a mean difference of approximately 1.5 µg ($p<0.0001$). Interestingly, there was no significant difference between log BCFs for the sediment exposure and the lowest aquatic exposure concentration (0.05 mg/l; $p=0.8145$), though the exposure concentrations differed by nearly two orders of magnitude. This observation may indicate the significance of feeding behavior as a route of uptake resulting from sediment exposure for *L. variegatus*. The results presented herein indicate that sulfamethazine and SMZ-related residues bioaccumulate in *L. variegatus* tissues.

Table I. Bioaccumulation of sulfamethazine in the aquatic worm L. variegatus.

Treatment exposure concentration (mg/l)	Mean uptake[1] (µg) (sd)	Log BCF[2] (g OC/g lipid) (sd)
Aquatic exposure		
0.05	0.052 (0.050)	2.17 (0.45)
0.5	0.016 (0.016)	0.55 (0.64)
5.0	0.101 (0.045)	0.56 (0.21)
Sediment exposure[3]		
3.03	1.614 (0.489)	1.89 (0.12)

[1] Uptake was calculated as SMZ equivalents based on total radioactivity in worms.
[2] Log bioconcentration factors normalized by worm lipid content (12.7% wet wt.) and sediment organic carbon content (2.0%). BCFs for sediment exposure are the same as BSAFs.
[3] Weighted sediment exposure concentration = 0.9(mean SMZ concentration in overlying water) + 0.1(mean SMZ concentration in sediment organic carbon). Based on 90% of ^{14}C residues in overlying water during days 0-7, and on observed worm behavior.

Conclusions

Our results indicate that antibiotics should not be studied alone; manure inputs had a significant effect on fate in soil and sediment, and sediment may serve as a sink for antibiotic residues in surface water ecosystems. Our results also indicate that the low concentrations of antibiotics detected in the environment could have biological significance, based on the bioaccumulation seen in organisms exposed to sulfamethazine. Future studies should examine potential effects on benthic invertebrates and their gut symbionts following exposure in chronic studies, rather than the acute (e.g. 48-hr) toxicity studies that are traditionally performed for ecological risk assessments. Risk assessments of antibiotics may require new methodologies, including extended study lengths based on possibility for chronic exposures that result from continual inputs into water bodies, and new modes of action, compared to the risk assessment paradigm established for pesticides.

Acknowledgements

The authors would like to thank Beth Douglass, Amy Morrow, and Ashley Jessick for their technical assistance, and Dr. David Smith (USDA-ARS, Fargo, ND) for donation of metabolite standards. Funding for this study was provided by a U.S. EPA STAR Fellowship and USDA-CSREES-NRI Grant #IOW05091. This chapter is a contribution from the Iowa Agricultural Experiment Station, Project No. 5091.

References

1. Kolpin, D.W.; Furlong, E.T.; Meyer, M.T.; Thurman, E.M.; Zaugg, S.D.; Barber, L.B.; Buxton, H.T. *Environ. Sci. Technol.* **2002,** *36*, 1202-11.
2. Managaki, S.; Murata, A.; Takada, H.; Tuyen, B.C.; Chiem, N.H. *Environ. Sci. Technol.* **2007,** *41*, 8004-10.
3. Kim, S.C.; Carlson, K. *Anal. Bioanal. Chem.* **2007**, *387*, 1301-15.
4. Kim, S.C.; Carlson, K. *Environ. Sci. Technol.* **2007**, *41*, 50-7.
5. Davis, J.G.; Truman, C.C.; Kim, S.C.; Ascough II, J.C.; Carlson, K. *J. Environ. Qual.* **2006**, *35*, 2250-60.
6. Lahr, J. *Arch. Environ. Contam. Toxicol.* **1997,** *32*, 50-57.
7. Rabølle, M.; Spliid, N.H. *Chemosphere* **2000**, *40*, 715-22.
8. Batt, A.L.; Snow, D.D.; Aga, D.S. *Chemosphere* **2006**, *64*, 1963-71.
9. Haller, M.Y.; Müller, S.R.; McArdell, C.S.; Alder, A.C.; Suter, M.J. *J. Chromatogr. A.* **2002**, *952*, 111-20.
10. Boxall, A.B.A.; Blackwell, P.; Cavallo, R.; Kay, P.; Tolls, J. *Toxicol. Lett.* **2002**, *131*, 19-28.
11. Thiele-Bruhn, S.; Seibicke, T.; Schulten, H.R.; Leinweber, P. *J. Environ. Qual.* **2004**, *33*, 1331-42.
12. Coats, J.R.; Metcalf, R.L.; Lu, P.Y.; Brown, D.D.; Williams, J.F.; Hansen, L.G. *Environ. Health Perspect.* **1976**, *18*, 167-79.
13. Baguer, A.J.; Jensen, J.; Krogh, P.H. *Chemosphere* **2000**, *40*, 751-7.
14. Wollenberger, L.; Halling-Sorensen, B.; Kusk, K.O. *Chemosphere* **2000**, *40*, 723-30.
15. U.S. Environmental Protection Agency (USEPA). *Methods for measuring the toxicity and bioaccumulation of sediment-associated contaminants with freshwater invertebrates.* EPA 600/R-99/064, 2000; 2nd ed., pp. 63–71.

16. Thiele-Bruhn, S.; Aust, M.O. *Arch. Environ. Contam. Toxicol.* **2004**, *47*, 31-9.
17. Sukul, P.; Spiteller, M. *Rev. Environ. Contam. Toxicol.* **2006**, *187*, 67-101.
18. Tolls, J. *Environ. Sci. Technol.* **2001**, *35*, 3397-3406.
19. Bialk, H.; Simpson, A.J.; Pedersen, J.A. *Environ. Sci. Technol.* **2005**, *65*, 2352-7.
20. Gao, J.; Pedersen, J.A. *Environ. Sci. Technol.* **2005**, *39*, 9509-16.
21. Trimble, T.A.; You, J.; Lydy, M.J. *Chemosphere* **2008**, *71*, 337-44.
22. Krauss, M.; Wilcke, W. *Environ. Sci. Technol.* **2001**, *35*, 3931-5.
23. Kraaij, R.; Mayer, P.; Busser, F.J.M.; van het Bolscher, M.; Seinen, W.; Belfroid, A.C.; Tolls, J. *Environ. Sci. Technol.* **2003**, *37*, 268-74.
24. McGeer, J.C.; Brix, K.V.; Skeaf, J.M.; DeForest, D.K.; Brigham, S.I.; Adams, W.J.; Green, A. *Environ. Toxicol. Chem.* **2003**, *22*, 1017–37.
25. DeForest, D.K.; Brix, K.V.; Adams, W.J. *Aquat. Toxicol.* **2007**, *84*, 236-46.
26. Simkiss, K.; Taylor, M.G. *Rev. Aquat. Sci.* **1989**, *1*, 173–88.

Assessing Potential Environmental Impacts

Chapter 10

Livestock Hormones in Aquatic Ecosystems

Yue-wern Huang[1], Dev K. Niyogi[1], Paul K. Nam[2], and Janet M. Bandeff[1]

[1] Department of Biological Sciences and Environmental Research Center, Missouri University of Science and Technology, 105 Schrenk Hall, 1870 Miner Circle, Rolla, MO 65409
[2] Department of Chemistry and Center for Environmental Science and Technology, Missouri University of Science and Technology, 142 Schrenk Hall, 1870 Miner Circle, Rolla, MO 65409

We begin this article with a general description of sources of hormones followed by their transport and fate in the environment. We then focus on occurrences and effects of livestock estrogens in aquatic ecosystems. Two case studies from our research are presented in details to contrast levels of estrogens in a concentrated animal farming area in northern Missouri with those in a relatively pristine area in southern Missouri. Estimation of human exposure to estrogenic compounds from various sources would be informative to the general public. Thus, we estimate exposures from environmental estrogens, medicinal estrogens, and dietary phytoestrogens. Estimation of exposure via these three sources was made using 17β-estradiol equivalent concentrations. Finally, we discuss the limited available data on other livestock hormones and then conclude with future research that should be conducted to further understand the impact of livestock hormones on aquatic ecosystems.

Introduction

Humans and wildlife are exposed to numerous chemicals that have the potential to modulate their endocrine systems (*1-6*). Some of the most potent of the identified endocrine modulators are natural and synthetic estrogens. These modulators are especially potent because of their strong binding affinity to estrogen receptors, which leads to biological effects even at very low concentrations (*7*). Additionally, they may exert their estrogenic effects via estrogen receptor-independent pathways, most of which are not well understood. Natural and synthetic hormones may reach the environment through confined animal feedlot effluent, sewage treatment plant effluent, and run-off from land-applied manure. Aquatic wildlife species face the greatest threat of exposure, although terrestrial wildlife may be exposed through soil and humans may be exposed through drinking water contamination and food sources.

One of the best examples of endocrine disruption in wildlife from natural and synthetic hormones is in the United Kingdom, where sewage effluent entering rivers was causing feminization of fish. The chemicals responsible for this endocrine disruption were found to be 17α-ethinyl estradiol (a synthetic estrogen in contraceptives), 17β-estradiol, and estrone (*8-11*). Due to these and other studies, concern has been raised about possible contamination of surface waters by hormones from livestock. The situation with livestock differs from that of human sewage in that while large scale livestock operations are required to treat animal excretion before release into the environment, land applied manure may enter surface waters via run-off. Furthermore, the cumulative effect of animal excretion from traditional family-run livestock farms can not be ignored because they are not required to have an animal waste treatment system. Additionally, livestock manure differs from human sewage because most livestock animals raised in the United States are treated with growth hormones (*12*). Therefore, it is necessary to assess the possibility of contamination and effects on humans and wildlife with these issues in mind.

Hormones in Livestock

The three major types of livestock raised in major operations are cattle, swine, and poultry. These animals all naturally produce the estrogens 17α- and 17β-estradiol, estrone, estriol, and progesterone, which are responsible for the development of female secondary sex characteristics and regulation of the menstrual cycle, pregnancy, and embryogenesis. To date, six hormones are approved for use in the United States as growth hormones in livestock: 17β-estradiol and the synthetic estrogen zeranol, progesterone and the synthetic progesterone melengestrol acetate (used to synchronize or induce estrus), and testosterone and the synthetic androgen trenbolone acetate. These hormones are administered as subcutaneous implants or feed additives. Typical treatments with growth hormones increase growth rates by 5-20% and feeding efficiency by 3-10% (*13*). Hormones are not approved for use in the European Union.

Excretion of natural sex steroid hormones occurs via the feces or urine. Lange et al. (*14*) estimate that yearly excretion of estrogens, progesterones, and androgens by livestock in the U.S. to be 49, 4.4, and 279 tons, respectively. Before or after excretion, hormones may be metabolized. The major metabolite of zeranol is zearalanone and the minor metabolite is taleranol. Progesterone is typically excreted as metabolites including pregnanediones, prenanolones, and prenanediols (*15*). Melengesterol acetate is excreted largely unaltered. Trenbolone acetate is mainly excreted as the metabolites 17α- and 17β-trenbolone. It has been estimated that approximately 8-12% of added growth hormones is excreted (*14*). Little is known about the total amount of growth hormones excreted on a yearly basis compared to natural hormones.

A portion of the cattle, swine, and poultry in the U.S. are raised in concentrated animal feeding operations (CAFOs). These facilities contain large numbers of livestock in small areas. Concerns about such operations include problems with manure storage and land application, *Escherichia coli* outbreaks, groundwater and surface water contamination, excess nutrients, fish kills, air quality, and the use of antibiotics and growth hormones. Such facilities are considered point sources for hormones, nutrients, and other pollutants.

Transport and Fate in the Environment

Natural and synthetic steroid hormones from livestock excretion reach the environment through feedlot effluent and runoff from land applied manure. Factors affecting the concentrations of hormones in the environment include inputs of hormones to the system, sorption to soils and sediments, degradation, and dilution (*14-17*). Inputs are controlled by the number, age, sex, reproductive status, and type of the livestock, as well as the runoff potential of the soil and the amount of precipitation in the region. Sorption is controlled by soil characteristics such as particle size and organic matter content, as well as the solubility of the hormone. Degradation is affected by temperature and light, as well as nutrient availability and biological activity of the soil (*18*). Dilution is affected by precipitation, stream discharge, and other stream characteristics. Based on these factors, hormones may accumulate either in soil or surface water.

Natural estrogens are rapidly degraded under aerobic conditions, often within several hours to several days (*18-22*). Estradiol will be degraded to estrone under aerobic conditions (*23*), but estrone will not be broken down further. Estrogens also sorb strongly to soils and sediments, which may serve as a sink in the environment (*24-28*). Testosterone easily leaches from the soil because it does not sorb strongly compared to estrogens (*12, 14*). Although testosterone is fairly mobile, it degrades even faster than natural estrogens. Metabolites of testosterone include 4-andostene-3,17-dione, 5 α-androstane 3,17-dione, and 1,4-androstadiene-3,17-dione (*12, 18, 20, 21,23*). Generally, progesterone is excreted as a biologically inactive metabolite. Little is known about its fate in the environment (*15*).

Synthetic hormones are considered to be more persistent than natural steroid hormones. For example, soils that received manure application had detectable concentrations of melengestrol acetate for the entire growing season and trenbolone for 8 weeks. The half-life of trenbolone was approximately 260 days (*29*). Both compounds adsorb strongly to organic matter and are unlikely to leach in large quantities (*30*). There is no information regarding the fate of zeranol in the environment.

Based on sorption and degradation potentials, the occurrence of hormones in the environment might be expected to be low. However, the abundance of liquid and solid manure from concentrated animal feedlot operations and the widespread use of growth hormones increase the likelihood of short-term and even chronic contamination of surface waters by hormones. Additionally, hormones are the most potent of endocrine disrupting chemicals, and even low concentrations in the environment might have detrimental effects on humans and wildlife, particularly if contamination occurs during crucial developmental periods affected by such hormones.

Occurrence and Effects of Livestock Estrogens in the Environment

Several studies have measured the concentrations of natural and synthetic estrogens from livestock in the water. The results of these studies are summarized in **Table 1**. Concentrations of 17α- and 17β-estradiol in water ranged from not detected to 2,530 ng/L. Relatively little is known about the occurrence of hormones in drinking water and meat. Concentrations in meat would be expected to vary based on time of treatment with hormones, when or if the subcutaneous hormone implant was removed, and the sex and reproductive status of the animal.

Table 1. Concentrations of natural and synthetic estrogens from livestock. Concentrations in water are in ng/L; nd is not detected.

Chemical	*Source*	*Livestock Type*	*Min*	*Mean/Median*	*Max*	*Citation*
17α-estradiol	Stream	Cattle	nd		0.026	a
17β-estradiol	Run-off	Broiler chickens	50		150	b
17β-estradiol	Pond	Cattle	5		23	a
17β-estradiol	River	Dairy cows	nd		0.6	c
17β-estradiol	Run-off	Poultry			1,280	d
17β-estradiol	Run-off	Broiler chickens	20		2,530	b
17β-estradiol	Springs	unknown	6		66	e
17β-estradiol	Stream	Cattle	nd		nd	a
Estradiol	Streams	Cattle	<0.5		5	f
E2 equivalents	Streams	Dairy cows	0	2	26.5	g
Estrone	River	Dairy cows	nd		0.9	d
Estrone	Stream	Cattle	0.246		7.7	a

[a]Soto et al. 2004 (*33*). [b]Finlay-Moore et al. 2000 (*52*). [c]Kolodziej et al. 2004 (*53*). [d]Nichols et al. 1997 (*59*). [e]Petersen et al. 2000 (*60*). [f]Shore et al. 1995 (*61*). [g]Matthiessen et al. 2006 (*62*).

Effects on Wildlife

Studies on the effects of natural estrogens on wildlife species have found a wide array of responses, including feminization, intersex, decreased sperm count, and vitellogenin induction (*31*). Synthetic hormones are often created to be more potent than natural estrogens, which combined with their increased persistence in the environment, makes them especially of great concern. Most studies performed to assess the effects of concentrated animal feedlot effluent on wildlife species have focused on fish. Orlando et al. (*32*) found that female fathead minnows downstream of a cattle feedlot effluent were defeminized, with decreased estrogen to androgen ratios, while male fathead minnows were demasculinized, with lowered testosterone synthesis and smaller testis. The authors suggest that either there are potent androgenic substances in the FLE [feedlot effluent], and/or there is complex mixture of androgenic and estrogenic substances that alter the hypothalamic-pituitary-gonadal axis, inhibiting the release of gonadotropin-releasing hormone or gonadotropins (*32*). A companion study (*33*) detected estrogenic activity in the surface water using E-SCREEN bioassays. Estrone and 17β-estradiol were all detected at least once, although only estrone concentrations were high enough to account for a substantial component (3-46%) of the measured hormonal activity of the water (*33*).

Effects on Humans

Hormones from livestock may affect humans either through dietary intake of hormones remaining in meat and milk products or through contamination of drinking water. Most of the information regarding the effects of human exposure to livestock hormones has been obtained by investigating the effects on surrogate mammal species. For example, zarenol and its metabolites cause accelerated puberty and vaginal opening, abnormal estrus, and anovulatory ovary in prepubertal mice fed 10 mg/kg of zarenol daily (*34-35*). In utero exposure to zeranol by feeding pregnant mice 10 mg/kg of zeranol had earlier and abnormal testicular differentiation in offspring (*36*). The no observed effect concentration of zeranol in monkey species was determined to be 0.225 mg/kg (*37*). More information is needed on the occurrence and concentrations of zeranol and other hormones in meat to begin to determine the effects of human consumption.

Case Studies of the Occurrence of Estrogenic Compounds in the Environment

Long term studies on temporal and spatial scales of hormone contamination are needed. Thus, we undertook two monitoring studies with the objective of quantifying concentrations of hormones and other organic chemicals in surface waters. The first study was performed in an agricultural area of northern Missouri, USA, and the second study was conducted in a relatively pristine recreational area in southern Missouri.

Occurrence of Organic Chemicals in the Grand River Watershed, Northern Missouri

The Grand River Watershed is located in a glaciated plain in northern Missouri (**Figure 1**). Approximately 60 to 70% of the area is covered by native grasses. The major soil types are GARA-22-E2, Armstrong-24C2, Lamoni-15C2, and Adair-14C2. Constant soil erosion is characteristic of the region, particularly during spring and late autumn. Row crops include soybean and corn. Pastured cattle farms are spread over the region (**Table 2**), and there are approximately one million hogs raised by a corporate hog farm and by individual farmers. Streams within the watershed may be affected by runoff from confined hog operations and subsequent land application of manure, as well as run-off from pastured cattle and corn and soybean fields. This has led to concern over the potential for endocrine disruption of aquatic species via natural or synthetic hormones, as well as industrial chemicals and pesticides. Therefore, a monitoring study was undertaken between June 2003 and September 2005 to determine occurrence and concentrations of estrogens and other organic chemicals that might be estrogenic. Water samples were collected monthly during the growing season from two streams. Two sampling sites were established on each of the streams. Automatic sampling devices (ISCO model 3710, Teledyne ISCO, Lincoln, Nebraska, USA) were used to collect 24-h composite samples in a time-proportional mode. Eight liters of water sampled from each site were collected monthly from June to November 2003, May to November 2004, and June to September 2005. The samples were filtered through 1.0 μm glass fiber filters (Whatman, NJ, USA). Solid phase extraction was then performed using C-18 cartridges. The extract was dried and reacted with Sylon BFT (Supelco, Bellefonte, PA, USA) to produce silyl derivatives. Concentrations of selected organic chemicals were determined by gas chromatography/mass spectrometry (**Figure 2**) (*38*). Mean recoveries of all analytes in the laboratory spike from GC-MS analysis generally exceeded 70% at all spike levels. Percent recoveries of the steroids were between 56% and 79% with recoveries of the phenolic compounds between 70% and 92%. The linearity of calibration curves (R^2) was greater than 0.97, with the exception of simazine at 0.87. Limits of detections (LODs) ranged from 0.1 to 3.4 ng/L. Nutrients and several physicochemical parameters of the water were determined by Hach methods (Hach Company, Loveland, Colorado) that are endorsed by USEPA.

During the three year study period, temperature ranged from 5.5 to 32.7 °C and pH values ranged from 4.5 to 8.9. Median turbidity was 29 NTU. Median total nitrogen and total phosphorus concentrations were 0.78 mg/L N and 0.14 mg/L, respectively, and median total organic carbon was 12.8 mg/L. Thirteen of the nineteen organic chemicals were detected at least once between June 2003 and September 2005 (**Table 3**). 17α-Estradiol was detected three times, with a maximum concentration of 54 ng/L. Estrone was detected twice with a maximum concentration of 29 ng/L. Both cattle and swine farming activities are likely to contribute the source of 17α-estradiol and estrone. Studies have shown that cattle mainly excrete 17α-estradiol instead of 17β-estradiol (*16*). The most commonly detected chemicals were dibutyl phthalate, metolachlor, nonylphenol, bisphenol A, atrazine, and benzyl butyl phthalate. All these chemicals have been shown to be estrogenic to various biological endpoints (*7*) with the exception of metolachlor, which has not been tested. Concentrations of metolachlor and atrazine decreased throughout the growing season, while concentrations of industrial chemicals showed no clear pattern. Currently we are conducting a field enclosure study of amphibian tadpoles to evaluate total estrogenic effects of these compounds.

Table 2. Land-use upstream of the four sampling sites in Little Medicine Creek (LMC) and West Locust Creek (WLC) (data obtained from MoRAP GIS data set). Zero percent of the land-use is classified as urban and less than or equal to 0.1% is classified as water.

Site	*Latitude*	*Longitude*	*Drainage Area (ha)*	*% Row Crop*	*% Forest*	*% Grassland*
LMC1	N 40° 29'	W 93° 24'	4,278	23	9	67
LMC2	N 40° 20'	W 93° 23'	16,023	23	17	60
WLC1	N 40° 29'	W 93° 11'	8,699	17	11	72
WLC2	N 40° 25'	W 93° 10'	12,320	15	12	73

LMC1: north site in LMC; LMC2: south site in LMC; WLC1: north site in WLC; WLC2: south site in WLC.

Table 3. Percent detection, median, and maximum concentrations of 19 selected chemicals found in filtered streamwater samples from Little Medicine Creek and West Locust Creek during 2003-5. Data from both creeks were pooled (n = 65). "nd"= not detected; "-" = not applicable.

Chemical	*Detection Limit (ng/L)*	*3-yr Ave. Detection (%)*	*Median conc. (ng/L)*	*Maximum conc. (ng/L)*
Dibutyl phthalate	0.3	78	27	1,127
Metolachlor	0.4	75	9	472
Nonylphenol	1.7	72	43	725
Bisphenol A	0.1	66	6	1,986
Atrazine	0.9	65	12	536
Benzyl butyl phthalate	0.8	63	5	145
Simazine	2.4	34	<DL	375
p,p'-DDE	0.6	23	<DL	2,247
4-Octylphenol	2.0	22	<DL	812
Tebuthiuron	3.4	15	<DL	373
Bioallethrin	0.8	3	<DL	28
17α-Estradiol	1.3	3	<DL	54
Estrone	0.9	2	<DL	29
Diethylstilbestrol	0.6	nd	-	-
Estriol	0.8	nd	-	-
17α-Ethinyl estradiol	0.8	nd	-	-
17β-Estradiol	1.3	nd	-	-
Permethrin	0.9	nd	-	-
Tamoxifen	1.1	nd	-	-

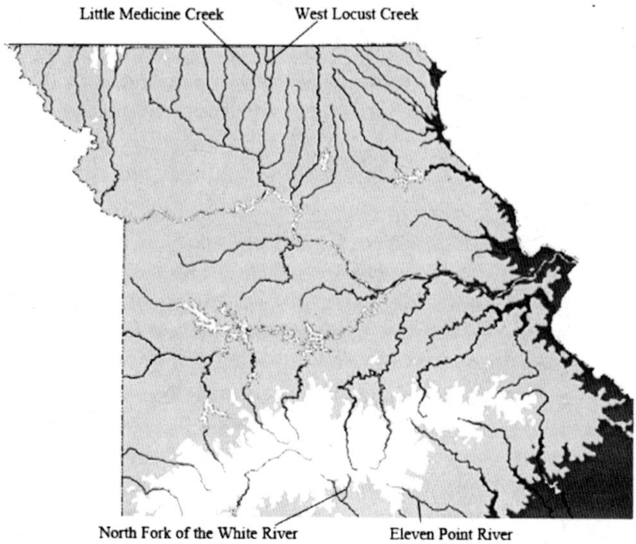

Figure 1. Study sites in northern and southern Missouri. Two study sites are located in Little Medicine Creek and West Locust Creek in northern Missouri. Another two study sites are located in North Fork of the White River and Eleven Point River in southern Missouri

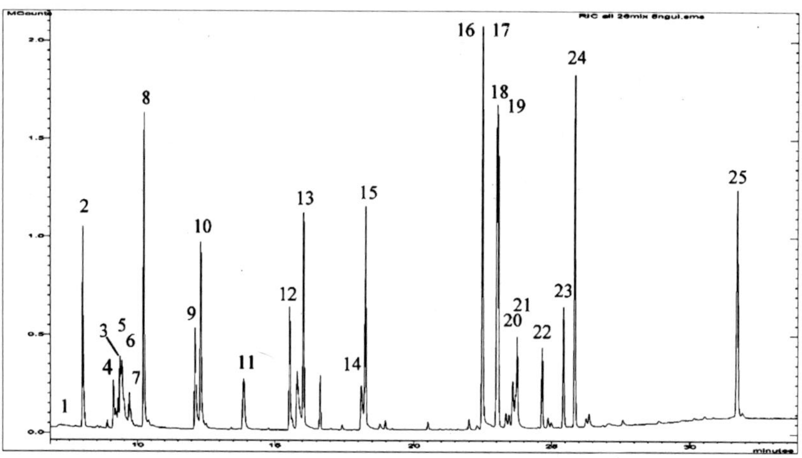

Figure 2. Total ion chromatogram from GC/MS analysis of trimethylsilyl (TMS) derivatives of environmental contaminants monitored in river waters. 1 = Tebuthiuron, 2 = 4-tert-octyl phenol, 3 = simazine, 4 = nonyl phenol, 5 = d_5-atrazine (internal std), 6 = atrazine, 7 = 4-octyl phenol, 8 = d_{10}-anthracene (internal std), 9 = dibutyl phthalate, 10 = metolachlor, 11 = bioallethrin, 12 = p,p'-DDE, 13 = bisphenol A, 14 = benzyl butyl phthalate, 15 = diethylstilbestrol, 16 = estrone, 17 = 17α-estradiol, 18 = d_5-estradiol (internal standard), 19 = 17β-estradiol, 20 = permethrin, 21 = tamoxifen, 22 = 17α-ethinyl estradiol, 23 = cholestane (internal std), 24 = estriol, 25 = β-sitosterol.

Occurrence of Estrogens in Two Rivers in Southern Missouri

A two-year monitoring study was conducted in two relatively pristine rivers in southern Missouri to identify possible estrogenic chemicals in the rivers. Study sites were established in the lower sections of the Eleven Point and the North Fork of the White Rivers (**Figure 1**). The stretches of both rivers are fast moving and spring fed with numerous pools and riffles. The land use adjacent to the rivers is deciduous forest with pasture and grassland in the upland areas. Traditional family-run cattle farms are located throughout the watersheds. Most of the areas adjacent to the riparian zones of the two rivers were intact without much crop cultivation. Canoeing in the summertime drew large crowds, particularly during weekends and holidays. Twenty-four hour composite water samples were collected monthly from August 2003 to November 2004. Water samples were analyzed with the same protocols used in the northern Missouri study.

Data from both rivers were very similar. For regional comparison with northern Missouri, data from Eleven Point River and North Fork of the White River are pooled. Temperature ranged from 7.4 to 23.3°C and pH values ranged from 5.7 to 8.6. Median turbidity was 1.9 NTU. Median total nitrogen and total phosphorus concentrations were 0.68 mg/L N and 0.03 mg/L, respectively. Median total organic carbon concentration was 1.6 mg/L. No natural estrogens from humans or livestock were detected throughout the study period (**Table 4**). However, several estrogenic chemicals were present. For instance, benzyl butyl phthalate, dibutyl phthalate, and bisphenol A were all detected in the majority of samples (87 to 97%) with median concentrations ranging from 18 to 164 ng/L. Occurrences of dibutyl phthalate, benzyl butyl phthalate, and nonylphenol were all highly correlated, which may indicate a common source or process controlling their concentrations in the rivers, but further study would be needed to determine such mechanisms. Additionally, the correlation between nonylphenol and metolachlor is consistent with nonylphenol being used as an inert ingredient in herbicides (*39*). Dibutyl phthalate and bisphenol A have both been shown to be estrogenic in amphibians, causing deformities and impaired spermatogenesis (*40*), and altered sex ratios (*41*), although at concentrations ranging from 100-10,000 μg/L for dibutyl phthalate and 2.3 to 228 μg/L for bisphenol A. Nonylphenol has been shown to alter sex ratios in *Rana pipiens* and *R. sylvatica* tadpoles at concentrations as low as 10 μg/L (*42*). The plant sterol β-sitosterol, a weak binder of estrogen receptors, was detected in the Eleven Point and North Fork of the White River (*38*). Metolachlor, an herbicide generally used in soybean fields and in mixtures with atrazine on corn fields, was the most commonly detected herbicide. Little evidence is available related to its estrogenicity. Tebuthiuron, an herbicide typically applied to roadsides and other non-cropland areas, was also detected. Its estrogenicity is also unknown. Atrazine, an herbicide commonly applied to corn fields, was not detected. Though atrazine is considered estrogenic, its effects and potency on wildlife are still under vigorous debate.

Despite the prevalence of livestock in the Grand River Watershed in northern Missouri, few occurrences of estrogens were found. Our findings of low frequency of presence of estrogens are in agreement with those of other studies in areas downstream of cattle, hog, and poultry operations (Table 1). In our study, the rare detection of natural and synthetic estrogens in may be attributed to several factors such as the efficiency of the waste management system, rapid breakdown, and sorption to soils and sediments. Additionally, it should be noted that bed and suspended sediments in the streams were not measured. Due to the high turbidity, it is also likely that most of estrogenic chemicals are adsorbed in the organic matrix of the water. A special analytical protocol needs to be developed and validated to analyze the water with such high turbidity

In contrast to the Grand River Watershed, the Eleven Point River and the North Fork of the White Rivers in southern Missouri are located in relatively isolated areas with much fewer livestock operations. It is thus not a surprise that estrogens were below detection limits.

Table 4. Detection frequency, median, and maximum concentrations of organic chemicals in Eleven Point and North Fork of the White Rivers during 2003-4. Data were pooled from two rivers (n=31).

Chemical	*DL[a] (ng/L)*	*Detection Frequency (%)*	*Median (ng/L)*	*Maximum (ng/L)*
Benzyl butyl phthalate	0.8	97	56	248
Dibutyl phthalate	0.3	97	154	2,753
β-Sitosterol	1.0	88	190	220
Bisphenol A	0.1	87	21	135
Nonylphenol	1.7	46	<DL	3,186
Metolachlor	0.4	36	<DL	912
Tebuthiuron	3.4	30	<DL	163
4-Octylphenol	0.3	10	-	51
17α-Ethinylestradiol	0.8	nd	-	-
17α-Estradiol	1.3	nd	-	-
17β-Estradiol	1.3	nd	-	-
Estriol	0.8	nd	-	-
Estrone	0.9	nd	-	-
Diethylstilbestrol	0.6	nd	-	-
Tamoxifen	1.1	nd	-	-
Atrazine	0.9	nd	-	-
p,p'-DDE	0.6	nd	-	-
Bioallethrin	0.8	nd	-	-
Permethrin	0.9	nd	-	-
Simazine	2.4	nd	-	-

[a] DL = detection limit

Levels of Human Exposure to Environmental, Dietary, and Medicinal Estrogens

We have discussed environmental concentrations of natural and synthetic estrogens as well as exposure of aquatic organisms to them. Estimation of human exposure to estrogenic compounds would be informative to the general public. In addition to being potentially exposed to estrogenic chemicals in drinking water and contaminants in the diet (categorically "environmental"), humans may be exposed to medicinal estrogens and dietary phytoestrogens. To understand contributions from these three sources, we estimate possible maximum exposure to each source. Exposure levels and potency factors of each source of estrogens were adopted from several studies. Clinical data show that medicinal estrogens in female serum/plasma range between 0.032 and 12.3 ng/mL (**Table 5**) (*43-48*). The Third National Health and Nutrition Examination Survey from 1988-1994 showed that the phytoestrogen levels in human serum/plasma ranged from non-detectable to 373.2 ng/mL (**Table 6**) (*49*). Many studies have investigated serum levels of estrogenic organic chemicals from the environment. However, the data were normalized by fat contents and thus were not comparable here. To our knowledge, only one study estimated serum/plasma levels of DDT and metabolites, aldrin, dieldrin, and *p,p'*-methoxychlor in the control subjects of the Long Island Breast Cancer Study Project. The concentration range was between 2.39 and 6.85 ng/mL (**Table 6**) (*50*). The serum/plasma estrogen levels acquired from consuming the livestock with administered estrogens were unknown and difficult to estimate.

The potency factors for medicinal estrogens and livestock estrogens were assumed to be 1.0 because those estrogens have binding affinities and elicit biological effects similar to 17β-estradiol. According to the available receptor binding affinity and cell proliferation data, the potency factors of phytoestrogens range from 10^{-5} to 10^{-2}, and the highest number was adopted (*7*). The potency factors of environmental estrogens (e.g., DDT and its metabolites, certain PCBs) range between 10^{-6} and 10^{-3} (**Table 7**) (*7*).

Table 7 summarizes human exposure levels to medicinal estrogens, dietary phytoestrogens, environmental estrogens, and estrogens from the release of livestock. 17β-Estradiol equivalent concentrations are the product of estrogens in serum/plasma and their potency factors. Accordingly, the ranking order of 17β-estradiol equivalent concentrations is medicinal estrogens > phytoestrogens >> environmental estrogens. Unfortunately, the 17β-estradiol equivalent concentrations from exposure to the release of livestock estrogens can not be calculated and ranked because of the lack of adequate exposure data. It should be noted that the calculation of 17β-estradiol equivalent concentrations can be much improved when more chemicals are tested for additional critical toxicological endpoints.

Table 5. Steady-state concentrations of medicinal estrogens in plasma or serum.

Brand Name (form)	*Types of estrogens*	*Serum level (ng/ml)*
Premarin® (tablet)	Estrone, equilin, 17α- and 17β-dihydroequilin, 17α- and 17β-estradiol	12.3
Prempro® (tablet)	Conjugated estrogens found in Premarin, medroxyprogesterone	11.7
Cenestin® (tablets)	Conjugated estrogens found in Premarin tablets, 17β-estradiol, equilenin, 17α- and 17β-dihydroequilenin	7.7
Alora® (transdermal patch)	17β-Estradiol	0.092 - 0.144
Climara® (tablet)	17β-Estradiol	0.032 - 0.147
Vivelle® (transdermal patch)	17α –Ethinyl estradiol	0.046 - 0.133

Table 6. Levels (ng/ mL^{-1}) of phytoestrogens and environmental estrogens detectedin human serum or plasma. Ranges are given followed by means in parentheses.n.d. = not detectable.

Chemicals	*Levels*[a]
Daidzein	n.d. – 162 (3.9)
Equol	n.d. – 8.2 (LOD)
Genistein	n.d. – 203 (4.7)
Total isoflavones	n.d. – 373.2 (8.6)
Enterolactone	n.d. – 112 (3.6)
Matairesinol	n.d. – 3.3 (LOD)
Enterodiol	n.d. – 19 (1.8)
Total lignans	n.d. – 134.3 (5.4)
Total environmental estrogenic chemicals	2.39 – 6.85[b]

[a] The Third National Health and Nutrition Examination Survey from 1988-1994 (*49*).
[b] DDT and metabolites, aldrin, dieldrin, and *p,p'*-methoxychlor in the control subjects of the Long Island Breast Cancer Study Project (*50*).

Other Livestock Hormones

In addition to estrogens, the levels and effects of progesterone, 17α-trenbolone, testosterone, and trendione were reported in several studies (**Table 8**). One study found that progesterone was below the detection limit (0.2 ng/mL) in any of the groundwater samples (*51*). Levels of testosterone ranged from not detected to 1,830 ng/L, with the highest concentration in run-off from land-applied broiler chicken manure (*52*). The most commonly measured synthetic hormones were 17α- and 17β-trenbolone, which ranged in concentration from not detected to 120 ng/L. Studies focusing on fish and the effects of trenbolone acetate metabolites have found that concentrations ranging from 9-193 ng/L cause reduced fecundity and plasma vitellogenin and steroid concentrations (*53-56*). In male fish, concentrations of trenbolone acetate metabolites ranging from 50-41,000 ng/L caused decreased 11-ketotestosterone concentrations and increased testicular area and sperm percentage (*53, 55*). Holbech et al. (*57*), in a study using juvenile zebra fish, found that concentrations of trenbolone of 9.7 ng/L led to the formation of all male populations. In a modeling study of population viability of fathead minnows, concentrations of 27 ng/L trenbolone acetate metabolites caused a 50% reduction in population size after two years of exposure. (*58*).

Table 7. Human exposure levels of estrogens (ng/mL) via three different sources. Estrogens from the environment can range from estrogenic chemicals with very low potency (e.g., DDT and its metabolites, certain PCBs) to those with very high potency like livestock estrogens. Data sources were stated in the text.

Category	*Estrogens in serum/plasma*	*Potency factor*	*17β-estradiol equivalent concentration*
Medicinal estrogens (female)	0.032 – 12.3	1.0	0.032 – 12.3
Phytoestrogens	n.d. – 373.2	10^{-2}	n.d. – 3.73
Environmental estrogens	2.39 – 6.85	10^{-3}	0.00239 – 0.00685
Estrogens from livestock areas	N/A	10^{-6} – 1.0	N/A

Table 8. Concentrations of natural and synthetic estrogens in streams from different livestock types (ng/L). nd is not detected. Mean or median was not available.

Chemical	*Source*	*Livestock Type*	*Min*	*Max*	*Citation*
Progesterone	River	Dairy cows	nd	nd	a
17α trenbolone	Discharge	Cattle	nd	120	b
17α trenbolone	Stream	Cattle	0.0016	0.035	c
17α trenbolone	Stream	Cattle	nd	50	b
17β trenbolone	Discharge	Cattle	nd	20	b
17β trenbolone	Stream	Cattle	<0.0004	0.0015	c
17β trenbolone	Stream	Cattle	nd	7	b
Testosterone	Run-off	Broiler chickens	15	125	d
Testosterone	River	Dairy cows	nd	0.6	a
Testosterone	River	Cattle and fish	nd	6	e
Testosterone	Run-off	Broiler chickens	10	1,830	d

[a]Kolodziej et al. 2004 (*51*). [b]Durhan et al. 2006 (*63*). [c]Soto et al. 2004 (*33*). [d]Finlay-Moore et al. 2000 (*52*). [e]Shore et al. 2004 (*64*).

Conclusions

The natural hormones 17α- and 17β-estradiol, estrone, estriol, testosterone, and progesterone, as well as the growth hormones 17β-estradiol, testosterone, progesterone, zeranol, trenbolone acetate, and melengestrol acetate, are all excreted in large quantities by livestock. These hormones have the potential to reach the environment through run-off from confined animal feedlots and land-applied manure. Natural hormones are rapidly degraded under aerobic conditions and sorb strongly to soils and sediments, which may decrease their tendency to accumulate in the environment. Synthetic hormones are more persistent in the environment. Overall, the detection frequencies and levels of estrogens in aquatic ecosystems seem to be low. However, their effects on aquatic organisms can not be overlooked due to their high estrogenic potency.

There are numerous knowledge gaps that need to be filled to make informed decisions about the effects of current livestock production practices on environmental quality. Research needs include, but are not limited to: 1) determination of acute toxicity to aquatic organisms due to overland flow, especially during storm events, as higher concentrations of hormones were observed in many studies, 2) assessment of the effects of hormones on soil organisms, given that both natural and synthetic growth hormones bind strongly to soils, 3) study of sorption to sediments in surface waters as an environmental sink for hormones, of which toxic effects to benthic organisms are unknown, 4) quantification of the amount of livestock hormones in meat to get a better understanding of human exposure levels, and 5) study of effects of testosterone, which is understudied compared to estrogens.

Acknowledgements

The authors thank the Missouri Department of Natural Resources, the Missouri Department of Conservation, the U.S. Department of the Interior Fish and Wildlife Service, and the Saint Louis Zoo for financial support. Our results have not been subjected to agency review and therefore do not necessarily reflect the views of our funding agencies and no official endorsement should be inferred.

References

1. Howell WM, Black DA, Bortone SA. 1980. Abnormal expression of secondary sex characters in a population of mosquitofish, *Gambusia affinis holbrooki*: evidence for environmentally induced masculinization. Copeia 4:676-681.
2. Colborn TF, vom Saal S, Soto AM. 1993. Developmental effects of endocrine-disrupting chemicals in wildlife and humans. Environ Health Persp 101:378-384.

3. Guillette LJ, Crain DA, Rooney AA, Pickford DB. 1995. Organization versus activation: the role of endocrine-disrupting contaminants (EDCs) during embryonic development in wildlife. Environ Health Persp 103:157-164.
4. Tyler CR, Jobling S, Sumpter JP. 1998. Endocrine disruption in wildlife: a critical review of the evidence. Crit Rev Toxicol 28:319-361.
5. Belfroid A, Van Der Horst A, Vethaak AD, Schafer AJ, Rijs GB, Wegener J, Cofino WP. 1999. Analysis and occurrence of estrogenic hormones and their glucuronides in surface water and waste water in The Netherlands. Sci Total Environ 225:101-108.
6. Körner W, Bolz U, Sussmuth W, Hiller G, Schuller W, Hanf V, Hagenmaier H. 2000. Input/output balance of estrogenic active compounds in a major municipal sewage plant in Germany. Chemosphere 40:1131-1142.
7. Huang YW, Hunter LD, Phillips JR. 2007. Human exposure to medicinal, dietary, and environmental estrogens. Toxicol Environ Chem 89:141-160.
8. Purdom CE, Hardimon PA, Bye VJ, Enu NC, Tyler CR, Sumpter JP. 1994. Estrogenic effects of effluents from sewage treatment works. Chem Ecol 8:275-285.
9. Harries JE, Sheahan DA, Jobling S, Mattiessen P, Neall P, Sumpter JP, Tylor T, Zaman N. 1997. Estrogenic activity in five United Kingdom rivers detected by measurement of vitellogenesis in caged male trout. Environ Toxicol Chem 16:534-542.
10. Desbrow C, Routledge EJ, Brighty GC, Sumpter JP, Waldock M. 1998. Identification of estrogenic chemicals in STW effluent, 1. Chemical fractionation and in vitro biological screening. Environ Sci Technol 32:1549-1558.
11. Jobling SMN, Tyler CR, Brighty G, Sumpter JP. 1998. Widespread Sexual Disruption in Wild Fish. Environ Sci Technol 32:2498-2506.
12. Casey FXM, Hakk H, Simunek J, Larsen GL. 2004. Fate and transport of testosterone in agricultural soils. Environ Sci Technol 38:790-798.
13. Birkelo CP. 2003. Pharamaceuticals, direct-fed microbials, and enzymes for enhancing growth and feed efficiency of beef. Veterinary Clinics of North America-Food and Animal Practice 19:599.
14. Lange IG, Daxenberger A, Schiffer B, Witters H, Ibarreta D, Meyer HHD. 2002. Sex hormones originating from different livestock production systems: fate and potential disrupting activity in the environment. Analytica Chimica Acta 473:27-37.
15. Lorenzen A, Hendel JG, Conn KL, Bittman S, Kwabiah AB, Lazarovitz G, Masse D, McAllister TA, Topp E. 2004. Survey of hormone activities in municipal biosolids and animal manures. Environ Toxicol 19:216-225.
16. Hanselman TA, Graetz DA, Wilkie AC. 2003. Manure-borne estrogens as potential environmental contaminants: a review. Environ Sci Technol 37:5471-5478.
17. Lissemore L, Hao CY, Yang P, Sibley PK, Mabury S, Solomon KR. 2006. An exposure assessment for selected pharmaceuticals within a watershed in southern Ontario. Chemosphere 64:717-729.

18. Das BS, Lee LS, Rao PSC, Hultgren RP. 2004. Sorption and degradation of steroid hormones in soils during transport: column studies and model evaluation. Environ Sci Technol 38:1460-1470.
19. Colucci MS, Bork H, Topp E. 2001. Persistence of estrogenic hormones in agricultural soils: I. 17β-estradiol and estrone. J Environ Qual 30:2070-2076.
20. Jacobsen AM, Lorenzen A, Chapman R, Topp E. 2005. Persistence of testosterone and 17β-estradiol in soils receiving swine manure of municipal biosolids. J Environ Qual 34:861-871.
21. Lee LS, Stock TJ, Sarmah AK, Rao PSC. 2003. Sorption and dissipation of testosterone, estrogens, and their primary transportation products in soils and sediment. Environ Sci Technol 37:4098-4105.
22. Lorenzen A, Burnison K, Servos M, Topp E. 2006. Persistence of endocrine disrupting chemicals in agricultural soils. J Environ Eng and Sci 5:211-219.
23. Czaijka CP, Londry KL. 2006. Anaerobic biotransformation of estrogens. Sci Total Environ 367:932-941.
24. Braga O, Smythe GA, Schafer AI, Feitz AJ. 2005. Fate of steroidal estrogens in Australian inland and coastal wastewater treatment plants. Environ Sci Technol 39:3351-3358.
25. Holthaus KIE, Johnson AC, Jurgens MD, Williams RJ. Smith JJL, Carter JE. 2002. The potential for estradiol and ethinylestradiol to sorb to suspended and bed sediments in some English rivers. Environ Toxicol Chem 21:2526-2535.
26. Peck M, Gobson RW, Kortenkamp A, Hill EM. 2004. Sediments are major sinks of steroidal estrogens in two United Kingdom rivers. Environ Toxicol Chem 23:945-952.
27. Yu Z, Xiao B, Huang W, Peng P. 2004. Sorption of steroid estrogens to soils and sediments. Environ Toxicol Chem 23:531-539.
28. Ying GG, Kookana RS. 2005. Sorption and degradation of estrogen-like endocrine disrupting chemicals in soil. Environ Toxicol Chem 24:2640-2645.
29. Schiffer B, Daxenberger A, Meyer K, Meyer HHD. 2001. The fate of trenbolone acetate and melengestrol acetate after application as growth promoters in cattle: environmental studies. Environ Health Persp 109:1145-1151.
30. Schiffer B, Totsche KU, Jann S, Kogel-Knabner I, Meyer K, Meyer HHD. 2004. Mobility of the growth promoters trenbolone and melengestrol acetate in agricultural soils: column studies. Sci Total Environ 326:225-237.
31. Tyler CR, Routledge EJ. 1998. Oestrogenic effects in fish in English rivers with evidence of their causation. Pure Appl Chem 70:1795-1804.
32. Orlando EF, Kolok AS, Binzcik GA, Gates JL, Horton MK, Lambright CS, Gray LE, Soto AM, Guillette LJ. 2004. Endocrine-disrupting effects of cattle feedlot effluent on an aquatic sentinel species, the fathead minnow. Environ Health Persp 112:353-358.
33. Soto AM, Calabro JM, Prechtl NV, Yau AY, Orlando EF, Daxenberger A, Kolok AS, Guillette LJ, le Bizec B, Lange IG, Sonnesnschein C. 2004. Androgenic and estrogenic activity in water bodies receiving cattle feedlot effluent in eastern Nebraska, USA. Environ Health Persp 112:346-352.

34. Nikaido Y, Danbara N, Tsujita-Kyutoku M, Yuri T, Uehara N, Tsubura A. 2005. Effects of prepubertal exposure to xenoestrogen on development of estrogen target organs in female CD-1 mice. In Vivo 19:487-494.
35. Yuri T, Nikaido Y, Shimano N, Uehara N, Shikata N, Tsubura A. 2004. Effects of prepubetal zeranol exposure on estrogen target organs and N-methyl-N-nitrogsourea-induced mammary tumorigenesis in female Sprague-Dawley rats. In Vivo 18:775-761.
36. Perez-Martinez C, Garcia-Ingelsias MJ, Ferreras-Estrada MC, Bravo-Moral AM, Espinosa-Alvarez J, Escudero-Diez A. 1996. Effects of in-utero exposure to zeranol or diethylstilboestrol on morphological development of the fetal testis in mice. J Comp Pathol 114:407-418.
37. Baldwin RS, Williams RD, Terry MK. 1983. Zeranol: a review of the metabolism, toxicology, and analytical methods for detection of tissue residues. Regul Toxicol Pharmacol 3:9-25.
38. Solis M, Liu C, Bandeff J, Nam P, Niyogi D, Huang Y. 2007. Occurrence of organic chemicals in two rivers inhabited by Ozark hellbenders (*Cryptobranchus alleganiensis bishopi*). Arch Environ Contam Toxicol 53:426-434.
39. USEPA 2006. Animal Feeding Operations. Accessed October 19, 2006. Available on-line at http://cfpub.epa.gov/npdes/home.cfm?program_id=7.
40. Lee SK, Veeramachaneni DN. 2005. Subchronic exposure to low concentrations of di-n-butyl phthalate disrupts spermatogenesis in *Xenopus laevis* frogs. Toxicol Sci 84:394-407.
41. Levy G, Lutz I, Krüger A, Kloas W. 2004. Bisphenol A induces feminization in *Xenopus laevis* tadpoles, Environ Res 94:102-111.
42. MacKenzie CA, Berril, M, Metcalfe C, Pauli, BD. 2003. Gonadal differentiation in frogs exposed to estrogenic and antiestrogenic compounds. Environ Toxicol Chem 22:2466-2475.
43. U.S. Food and Drug Administration. 1999. Cenestin label [Online]. Available by http://www.fda.gov/cder/foi/label/1999/20992lbl.pdf.
44. U.S. Food and Drug Administration. 2003. Prempro label [Online]. Available by http://www.fda.gov/cder/foi/label/2004/020527S034035lbl.pdf.
45. U.S. Food and Drug Administration. 2003. Premarin lable [Online]. Available by http://www.fda.gov/cder/foi/label/2003/04782s129lbl.pdf.
46. U.S. Food and Drug Administration. 2004. Vivelle label [Online]. Available by http://www.fda.gov/cder/foi/label/2004/20323slr017,025,030_vivelle_lbl.pdf
47. U.S. Food and Drug Administration. 2005. Climara label [Online]. Available by http://www.fda.gov/cder/foi/label/2005/021258s003lbl.pdf.
48. U.S. Food and Drug Administration. 2005. Alora label [Online]. Available by http://www.fda.gov/cder/foi/label/2005/020655s011lbl.pdf.
49. Valentin-Blasini, L., B.C. Blount, S.P. Caudill, and L.L. Needham. 2003. Urinary and serum concentrations of seven phytoestrogens in a human reference population subset. Journal of Expo Anal Environ Epidemiol 13:276-282.
50. Gammon, M.D., M.S. Wolff, A.I. Neugut, S.M. Eng, S.L. Teitelbaum, et. al. 2002. Environmental toxins and breast cancer on Long Island. II.

Organochlorine compound levels in blood. Cancer Epidemiol, Biomarkers Prev 11:686-697.

51. Kolodziej EP, Harter T, Sedlak DL. 2004. Dairy wastewater, aquaculture, and spawning fish as sources of steroid hormones in the aquatic environment. Environ Sci Technol 38:6377-6384.
52. Finlay-Moore O, Hartel PG, Cabrera ML. 2000. 17β-Estradiol and testosterone in soil and run-off from grasslands amended with broiler litter. J Environ Qual 29:1604-1611.
53. Ankley GT, Jensen KM, Makynen EA, Kalh MD, Korte JJ, Hornung MW, Henry TR, Denny JS, Leino RL, Wilson VS, CArdon MC, Hartig PC, Gray LE. 2003. Effects of the androgenic growth promoter 17-beta-trenbolone on fecundity and reproductive endocrinology of the fathead minnow. Environ Toxicol Chem 22:1350-1360.
54. Jensen KM, Makynen EA, Kahl MC, Ankley GT. 2006. Effects of the feedlot contaminant 17α-trenbolone on reproductive endocrinology of the fathead minnow. Environ Sci Technol 40:3112-3117.
55. Orn S, Yamini S, Norrgren L. 2006. Comparison of vitellogenin induction, sex ratio, and gonad morphology between zebrafish and Japanese medaka after exposure to 17 alpha-ethinyl estradiol and 17 beta-trenbolone. Archives of Environmental Contamination and Toxicology 51:237-243.
56. Seki M, Fuijishima S, Nozaka T, Maeda M, Kobayashi K. 2006. Comparison of response to 17 beta-estradiol and 17 beta-trenbolone among three small fish species. Environ Toxicol Chem 25:2742-2752.
57. Holbech H, Kinnberg K, Petersen GI, Jackson P, Hylland K, Norrgren L, Bjerregaard P. 2006. Detection of endocrine disrupters: evaluation of a fish sexual development test (FSDT). Comp Biochem Physiol C-Toxicol Pharmacol 144:57-66.
58. Miller DH, Ankley GT. 2004. Modeling impacts on populations: fathead minnow (*Pimephales promelas)* exposure to the endocrine disruptor 17 beta-trenbolone as a case study. Ecotoxicol Environ Saf 59:1-9.
59. Nichols DJ, Daniel TC, Moore PA, Edwards DR, Pote DH. 1997. Run-off of estrogen hormones 17β-estradiol from poultry litter applied to pasture. J Environ Qual 26:1002-1006.
60. Peterson EW, Davis RK, Orndorff HA. 2000. 17β-estradiol as an indicator of animal waste contamination in mantled karst aquifers. J Environ Qual 29:826-834.
61. Shore LS, Correel D, Chakroborty PK. 1995. In Animal Waste and the Land-Water Interface, K Steele (Ed.) pp 49-56. Lewis Publishers, Boca Raton, FL, USA.
62. Matthiessen P, Arnold D, Johnson AC, Pepper TJ, Pottinger TG, Pulman KGT. 2006. Contamination of headwater streams in the United Kingdom by oestrognic hormones from livestock farms. Sci Total Environ 367:616-630.
63. Durhan EJ, Lambright CS, Makynen EA, Lazorchak J, Hartig PG, Wilson VS, Gray LE, Ankley GT. 2006. Identification of metabolites of tenbolone acetate in androgenic runoff from a beef feedlot. Environ Health Perspect 114:65-68.
64. Shore LS, Reichmann O, Shemesh M, Wenzel A, Litaor M. 2004. Washout of accumulated testosterone in a watershed. Sci Total Environ 332:193-202.

Chapter 11

Characterization of Chlortetracycline-induced Glutathione *S*-Transferase to Conjugate Chloroacetanilide and Chlorotriazine Herbicides

Michael H. Farkas[1], James O. Berry[1], and Diana S. Aga[2]

[1]Department of Biological Sciences, University at Buffalo
[2] Department of Chemistry, University at Buffalo

Glutathione S-transferases (GST) induced in maize by chlortetracycline (CTC) appear to inhibit the ability to detoxify chloroacetanilide herbicides, as determined by liquid chromatography/mass spectrometry. Furthermore, the same CTC-induced GSTs were able to conjugate chlorotriazine herbicides at a rate undistinguishable from GSTs isolated from untreated maize plants. This data suggests that CTC, while not toxic to maize, may have indirect effects on herbicide dtoxification in a class-specific manner. Chlortetracycline (CTC) is a commonly used antibiotic in animal husbandry. A majority of the antibiotic passes through the animal non-metabolized and the manure is applied to crop fields as fertilizer where it has been shown to have phytotoxic tendencies in certain crops, as well as accumulate in others. Soil amended with CTC has been shown to induce expression of glutathione *S*-transferases (GST) in maize and subsequent analysis by liquid chromatography/ion-trap mass spectrometry (LC/IT-MS) identified stable CTC products conjugated with glutathione (GSH). Purified GSTs isolated from maize treated with CTC were able to produce nearly twice as much conjugated product relative to the GSTs isolated from non-treated control plants. Due to the prevalence of antibiotics in the environment, this work has raised concerns with regards to inducing herbicide resistance among target weeds or susceptibility among non-target crop plants. Herbicide conjugation by GSTs has been well characterized and is one of the main determinants of plant susceptibility. When GSTs isolated from maize control and CTC-treated plants were added to separate *in vitro* reactions containing three chloroacetanilide herbicides (metolachlor, propachlor, and alachlor) and one chlorotriazine herbicide (atrazine), the GSTs from the CTC-treated maize showed a reduced ability to conjugate the herbicides. In addition, analysis via LC/IT-MS has made it possible, for the first time, to detect chloroacetanilide herbicides that have been conjugated with two GSH molecules, in addition to the single GSH conjugate.

Introduction

Pharmaceutical Usage in Agriculture

In the United States, veterinary pharmaceuticals are used to both treat illnesses and promote growth in every type of livestock. There is very little data as to the quantity of antibiotics used for animal husbandry purposes. The estimates that do exist are conflicting and most likely inaccurate. The Animal Health Institute has estimated that antibiotic use in agriculture is 10% of the total antibiotic consumption, of which 87% is used for therapeutic purposes (1). On the other hand, in a book sponsored by the Union of Concerned Scientists estimates that antibiotics used for growth promotion is 8 times higher than what is used in all of human medicine (2).

Regardless, of the disparity in usage estimates, it is known with great certainty that in both humans and animals a majority of the pharmaceuticals pass through the organism unaltered, in their active form. In some cases, 70% of the pharmaceutical is not metabolized and it is excreted in both urine and feces (3). While humans are a source of antibiotics in the environment, the focus this book is with regards to agricultural sources. Excrement from animals, largely in confined animal feeding operations (CAFOs), is collected and spread as a slurry onto crop fields where the pharmaceuticals are able to leach into the soil and into surrounding waters.

Tetracyclines are the most commonly used antibiotics for animal husbandry. For example, surveyed dairy farmers in Pennsylvania reported using tetracyclines 70% relative to other antibiotics (4). In Denmark, where actual usage numbers do exist, tetracyclines again accounted for a minimum of 25% of the total usage of antibiotics in all food animals (5). Tetracyclines raise greater concerns because they are also used in human therapeutics and multiple forms of resistant bacteria have been identified. Tetracyclines in the environment are probably the most studied of all the antibiotics, largely for the reasons stated.

Pharmaceuticals in the Environment

An important factor to consider when studying antibiotics in the environment is their interactions in soil. In other words, will they be tightly adsorbed to the soil or will they move freely through the soil and into surrounding waters? In either case, it is necessary to consider both the physical properties of the particular antibiotic, as well as the properties of the soil. Sulfonamides and tetracyclines are two examples that exhibit different distribution coefficients (K_d) in soil. Sulfonamides have a relatively low coefficients ranging from 0.6-4.9, whereas tetracyclines are more adsorbed to soil with coefficients varying from 290-1620(6,7). Therefore, it would be expected that sulfonamides are more likely to enter groundwater, and ultimately be more problematic. However, there are other factors to consider such as soil characteristics, i.e. soil composition, pH, soil minerals, and soluble organic matter (SOM)(6). For example, the relative composition of clay in the soil will increase the adsorption of tetracyclines. In addition, tetracyclines are well-known chelators of divalent cations. Soil composed of higher concentrations of Ca^{2+} will lead to increased adsorption. Kulshrestha, et al, determined that oxytetracycline adsorption to montmorillonite, in fact, decreases as the pH of the soil increases(8). This is likely due to a number of factors that account for both the antibiotic properties (it is positively charged at low pH and repulsed by soil cations) and soil properties (at higher soil pH, hydrophobic interactions are predominant).

As the understanding of antibiotics in the environment increases, there are a growing number of concerns that are being raised. The most prevalent and likely the most controversial concern is that sub-therapeutic levels of antibiotics in the environment will lead to an increase in drug-resistant pathogens. This is an area that requires careful consideration and investigation since many antibiotics are used for both veterinary purposes as well as for human therapeutics. Chee-Sanford and colleagues (2001) found that hog lagoons contained tetracycline resistance genes and these same genes could be found in groundwater up to 250 m downstream(9). More specifically, a similar study found that Tet M (a ribosome protection protein)(10) is the most prevalent resistance gene in hog lagoons and its prevalence is season-dependent(11).

Typically, when the effects of pharmaceuticals in the environment are studied, a single compound is used for the study. A more recent investigation looked at the effects of a mixture of pharmaceuticals at sub-therapeutic levels (ng/L) on human embryonic cells, *in vivo*(12). The investigators found that cell proliferation was significantly decreased and morphological changes were observed. This suggests potential adverse human health effects can occur from low levels of pharmaceuticals that can leach into groundwater and a potential adverse effect on aquatic organisms.

An area of concern that is beginning to be investigated more fully is that of plant uptake and accumulation of antibiotics. Uptake by plants leads to two different concerns: 1) accumulation of antibiotics in edible portions of plants will expose consumers to the antibiotic and 2) uptake will cause phytotoxicity by the antibiotic, thus resulting in lower crop yields or death of the crop. Many modeling studies have shown that antibiotics at environmentally-relevant concentrations will not kill plants, however others have shown decreased growth and crop production (13,14). A factor that has not been considered, and is the basis of this chapter investigates the indirect effects that may be caused by uptake of antibiotics into plants. All of these areas of plant uptake will be discussed in more detail in the subsequent sections.

Antibiotic Uptake into Plants

Relatively few studies have investigated uptake of pharmaceuticals into plants. A few studies have reported that some crop plants do take up commonly used veterinary pharmaceuticals into the edible portions of plants. The most recent studies have shown that chlortetracycline (CTC) is taken up by maize, green onion, and cabbage at ppb (ng g^{-1}) levels (15). However, this same study reported that tylosin, an antibiotic in the macrolide family used mainly for swine production, was not detected in any of the three plants. On the other hand, a study conducted by Boxall and colleagues (2006) observed uptake of a wider range of veterinary pharmaceuticals including: oxytetracycline, enrofloxacin, and tylosin to name a few into lettuce and carrots(14). Interestingly, uptake of compounds that are of xenobiotic nature into a plant is commonly believed to occur via absorption of the compound into the root of the plant. So, the octanol-water partition coefficient of the compound is a key factor in uptake (16). In other words, lipophilic compounds are expected to be taken up, and accumulated to a greater extent than would be more polar compounds. However, oxytetracycline uptake into alfalfa was shown to be facilitated in an energy dependent manner (17). Although the authors of this work were unable to deduce a specific channel responsible for uptake of oxytetracycline, they were able to show that uptake did follow Michaelis-Menten kinetics.

The first study to report antibiotic phytotoxicity was that of Batchelder in 1982 (13). The author showed that pinto bean development and nutrient uptake were stunted in the presence of CTC and oxytetracycline. However, the concentration of the antibiotics to observe a significant decrease in development and nutrient uptake was at a concentration of 160 mg kg^{-1}, a concentration 8 times higher than observed in the environment (18). Even though the concentration was so much greater than environmental concentrations, the study also showed that radish, wheat, and maize were not affected by the antibiotics in the soil (13). Since this first demonstration of antibiotic phytotoxicity, others have been published using environmentally relevant concentrations. For example, sulfadimethoxine is toxic to maize at a concentration of 1 mg kg^{-1} (19). Enrofloxacin, is toxic to radish, lettuce, pinto beans, and cucumber (20). In fact, enrofloxacin is efficiently accumulated in these plants in the range of µg g^{-1} and approximately 25% of this is converted to another active antibiotic, ciprofloxacin. Antibiotics are capable of leaching into surrounding waters, but in other cases such as flumequine (a quinolone derivative) it is directly introduced into waters used for aquaculture. This antibiotic is accumulated and phytotoxic to aquatic weeds, which are consumed by various organisms and possibly promoting dissipation throughout the food chain (21).

Plants are not totally defenseless against contaminants in the environment, including antibiotics. An aquatic fern, *Azolla filiculoides* Lam., efficiently accumulates sulfadimethoxine and it is believed to be degraded by symbiotic microorganisms (22). Degradation by microorganisms reduced uptake of active sulfadimethoxine into the plant, although metabolites were not identified. Hairy root cultures of sunflowers also showed rhizosphere degradation of antibiotics, however degradation was caused by recognition of a stressor and a defense response by the plant (23). In this study, oxytetracycline was applied to the hairy root cultures of sunflowers and a general defense response was activated where the roots produce Reactive Oxygen Species (ROS). The ROS consists of free radical oxygen species and are produced by enzymes such as NADPH oxidase and peroxidases, which in this case are exuded from the roots and into the media. This leads to oxidation of oxytetracycline, likely producing quinone derivatives (24).

Glutathione *S*-Transferase-mediated Antibiotic Detoxification in Plants

Our early research has focused on the observations of Batchelder (1982) where pinto bean development and nutrient uptake are directly inhibited in the presence of CTC, while maize plants are unaffected (13). We exposed maize and pinto bean seedlings to soil treated with 20 mg kg^{-1} of CTC and sampled the roots, shoots, and leaves of the plants for total protein content at 1, 2, and 3 days (specific methods can be found in (25)). Using SDS-PAGE as an initial screening tool, in maize we were able to determine that a difference in banding pattern existed between the control (untreated) and CTC-treated roots in the range of 20-30 kDa. Although this increased band could be any number of proteins, under the experimental conditions it was suggestive of an increase in glutathione s-transferases (GSTs). No significant changes were noticed in the shoots or leaves, but all were tested for GST activity, as were the pinto bean samples. The shoots and stems did not show any significant increase in GST activity, but the maize roots showed a significant ($p<0.001$) increase of the CTC-treated samples relative to the control samples (Figure 1). Pinto beans, on the other hand, showed no difference in GST activity and the activity of those samples were much lower compared to the maize.

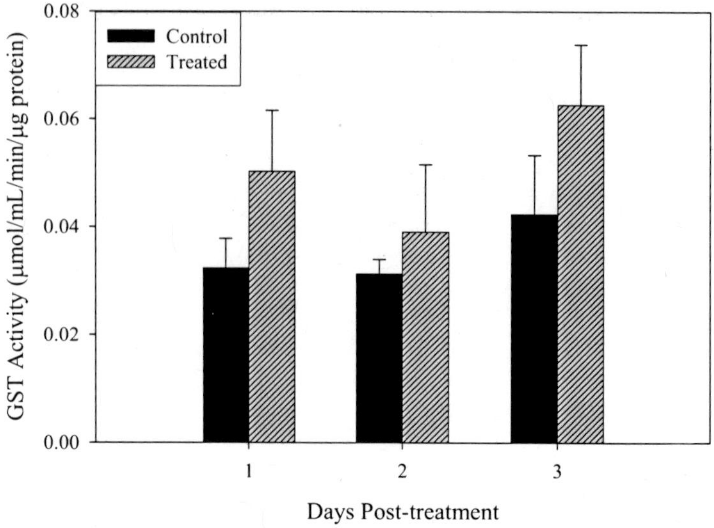

***Figure 1**. GST activity assay of total protein extracted from maize roots 1, 2, and 3 days after exposure to 20 mg kg^{-1} of chlortetracycline. Solid bars represent control samples and hatched bars are samples from plants treated with CTC. Adapted from (25).*

The early studies showing CTC phytotoxicity to pinto beans suggested that toxicity was caused by a reduced number of rhizobium, therefore, preventing sufficient symbiotic interactions and leading to stunted growth and deficient nutrient uptake (13). However, our group performed hydroponic uptake studies using ^{3}H-labeled tetracycline and found that in both maize and pinto beans, tetracycline is taken up by the root and approximately 50% of that is transported to the leaves of the plants (unpublished results). This observation allows for additional hypotheses to be made about the causes of phytotoxicity of tetracyclines in pinto beans. For example, the inability for pinto beans to recognize CTC as a xenobiotic and induce a stress response may allow for the antibiotic to act on some intracellular component in the plant. This is an interesting concept because it may be applicable to other plants and pharmaceuticals.

As mentioned earlier, maize is susceptible to sulfadimethoxine at concentrations of 1 mg kg^{-1}, which are environmentally-relevant (19). We also performed a laboratory experiment where we exposed maize to sulfadimethoxine in a manner similar to our earlier CTC study. Maize seedlings were transplanted into soil that had been separately treated with 1 mg kg^{-1} sulfadimethoxine, erythromycin, and tylosin. The plants were harvested at 1, 2, and 3 days post-treatment and assayed for GST activity (Figure 2). Interestingly, all antibiotics, regardless of known phytotoxicity, induced a significant ($p<0.001$) increase in GST activity relative to the untreated control.

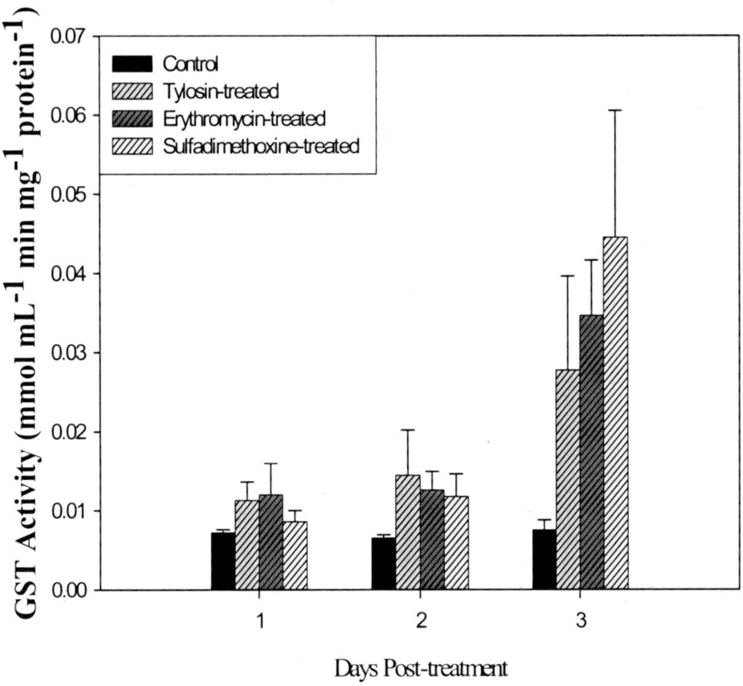

Figure 2. *GST activity assay for maize roots treated separately with 1 mg kg-1 of tylosin, erythromycin, and sulfadimethoxine.*

To really understand whether or not GSTs play a role in the ability of a plant to reduce the phytotoxicity of an antibiotic, an assay must be developed to determine if the induced GSTs can bind to and conjugate the antibiotic with GSH. To do just this, we developed *in vitro* assays (as described in (25)) where we compared the ability of GSTs purified from control and antibiotic-treated plants to conjugate a particular antibiotic.

While GST activity assays certainly sufficient for determining whether or not GSTs are induced in response to a stressor, they do not provide data to suggest if the GSTs are in fact conjugating the xenobioitc. The most common technique for determining GST conjugation is through the use of High Performance Liquid Chromatography. This is a useful technique; however it does not provide conclusive evidence that a GST conjugate does it exist, nor does it provide evidence into the site of conjugation. To identify whether or not the GSTs that we had isolated were capable of conjugating the antibiotics that we had previously used to treat the plants, we performed the *in vitro* assays and analyzed the reactions using liquid chromatography/mass spectrometry (LC/MS). Using both the chromatography and mass spectrometry data, we were able to identify putative GST conjugates. These conjugates were then confirmed using LC/MS/MS.

Using this assay allowed for positive identification of GST conjugation of some of the antibiotics, but not all. For example, CTC was conjugated by GSTs from both maize and pinto beans. However, the GSTs isolated from pinto beans conjugated about 2% as much as GSTs isolated form maize. A likely explanation for the data observed, i.e. negative results for an increase in GST activity, but positive conjugate data, is that the pinto beans are incapable of recognizing and inducing a defense response to CTC.

Glutathione *S*-Transferases

GSTs are part of a three phase detoxification system in plants. The first phase is typically used to introduce functional groups onto xenobiotic compounds. Cytochrome P450 monooxygenases are a class of enzymes that are most commonly implicated for Phase I detoxification. As their name suggests, these enzymes typically act by the addition of an oxygen. Phase II primarily employs GSTs as the main class of enzymes for detoxification. Here, xenobiotics are conjugated to glutathione to increase the polarity of the xenobiotic and prepare it for Phase III. Interestingly, Phase I and II are independent of one another such that an incoming xenobiotic can bypass Phase I processing and be acted upon by GSTs, as is the case in herbicide detoxification. Additionally, a xenobiotic may be acted upon only by Phase I enzymes and bypass Phase II processing. In Phase III, a conjugated xenobiotic is transported to the central vacuole via an ATP-dependent GS-X pump (26). Once inside the vacuole, the conjugated xenobiotic can be further processed which may include removing glutathione and glucosylation. The fate of the xenobiotic is relatively uncertain after processing within the vacuole. It is believed that the xenobiotic can be stored within the vacuole, transported to the apoplast, or transported back to the root and excreted. Aga and coworkers (1996), have provided evidence for the latter, where they identified metabolites of metolachlor in the soil of crop fields planted with maize. The metolachlor metabolite contained a sulfate group in place of chlorine, suggesting that this is a product of GST conjugation. Since GSTs are such an important class of enzymes for xenobiotic detoxification and the primary focus of this chapter, they will be discussed in more detail.

GSTs are dimeric enzymes where each monomer has a mass in the range of 20-30 kDa. GSTs can be found as homo- or heterodimers and the formation of a specific type of dimer is based on the needs of the plant. GSTI and GSTIII are homodimers that consist of subunits with molecular weights of 29 and26 kDa, respectively (27). These enzymes are constitutively expressed in maize. GSTII is a heterodimer and consists of subunits with molecular weights of 27 and 29 kDa. There are two sites in each monomer of GST that is responsible for binding. These sites are located in a cleft within the enzyme. The G site binds GSH and it is located deep within the cleft. The H site is responsible for binding xenobiotic compounds and can be found at the outer edge of the cleft.

GST-mediated conjugation to xenobiotics can occur due to the nucleophilic properties of GSH. However, this requires an electrophilic site on the xenobiotic. There are two types of electrophiles that can be conjugated by GSH. The first is a hard electrophile, which is typically an electrophilic atom such as chlorine. Conversely, a soft electrophile can be found in carbon-carbon double bonds. Figure 3 is a schematic diagram showing the conjugation of alachlor, a chloroacetanilide herbicide. Chlorinated herbicides such as those that belong to the chloroacetanilide and chlorotriazine classes are compounds that contain both hard and soft electrophiles. Interestingly, these compounds can be conjugated by GSTs at both electrophiles producing an herbicide with two GSHs attached (28). However, there appears to be preference for a single conjugate via removal of the chlorine atom.

Herbicide use for crop production is widespread and it is of the utmost importance that herbicides target only non-crop plants. Production of new herbicides is based on this relatively simple premise. However, in most cases, susceptibility is based on how well the plants can utilize detoxification pathways, mainly GSTs. In the following sections, we discuss how non-target plants like maize can show increased susceptibility when they are exposed to a secondary xenobiotic.

GSH conjugation to herbicides is one of the most well known detoxification reactions of GST. Until recently, studies investigating the conjugation of herbicides have used high performance liquid chromatography (HPLC) as the primary method for detecting these conjugates. While HPLC is a highly useful method, it does not provide sufficient data to 1) distinguish conjugates in complex matrices or 2) give specific details as to where GSH conjugates on the xenobiotic (chloroacetanilides have both hard and soft electrophilic sites). We developed a method to perform both of these functions using liquid chromatography-tandem mass spectrometry (LC-MS/MS). The specific details of our methodology can be found in reference (28).

The herbicides that we chose to analyze were those of the chloroacetanilide (alachlor, propachlor, and metolachlor) and chlorotriazine (atrazine) classes. These herbicides were chosen based on their high usage rates for crop production. In all cases, the LC data provided the expected results of a shift in retention time where the more polar conjugate eluted sooner than the parent herbicide. The MS data showed that the conjugated compound had lost the isotopic signature (M+2) of the chlorine atom and the m/z corresponded to GSH conjugation via removal of the chlorine atom. This data is significant because it corroborates early herbicide conjugation studies. To provide further verification that GSH is in fact conjugated to the herbicide, MS/MS fragmentation was performed on the putative conjugates. Figure 4 shows the MS/MS ion spectrum for the alachlor-GSH conjugate. The parent conjugate has a m/z of 539.6, which is representative of conjugation via removal of the chlorine atom. Xenobiotics conjugated with GSH have a relatively consistent fragmentation pattern: a loss of a water molecule (m/z 18) and glutamic acid (m/z 129). This has been observed with CTC (25), the chloroacetanilide herbicides (Figure 4) and atrazine (28),as well as other compounds. Interestingly, the chloroacetanilide herbicide conjugates provided extra fragmentation data that further confirmed GSH conjugation. The alachlor conjugate in Figure 4 shows ions at m/z 272.5, 254.3, and 179.3, which are not observed when the samples are analyzed using positive mode ionization. In addition to lower background, negative mode ionization is useful for identifying GSH and its conjugates. The ions at m/z 272.5, 254.3, and 179.3 correlate well with GSH fragmentation and the hypothesized fragments are shown in Figure 5.

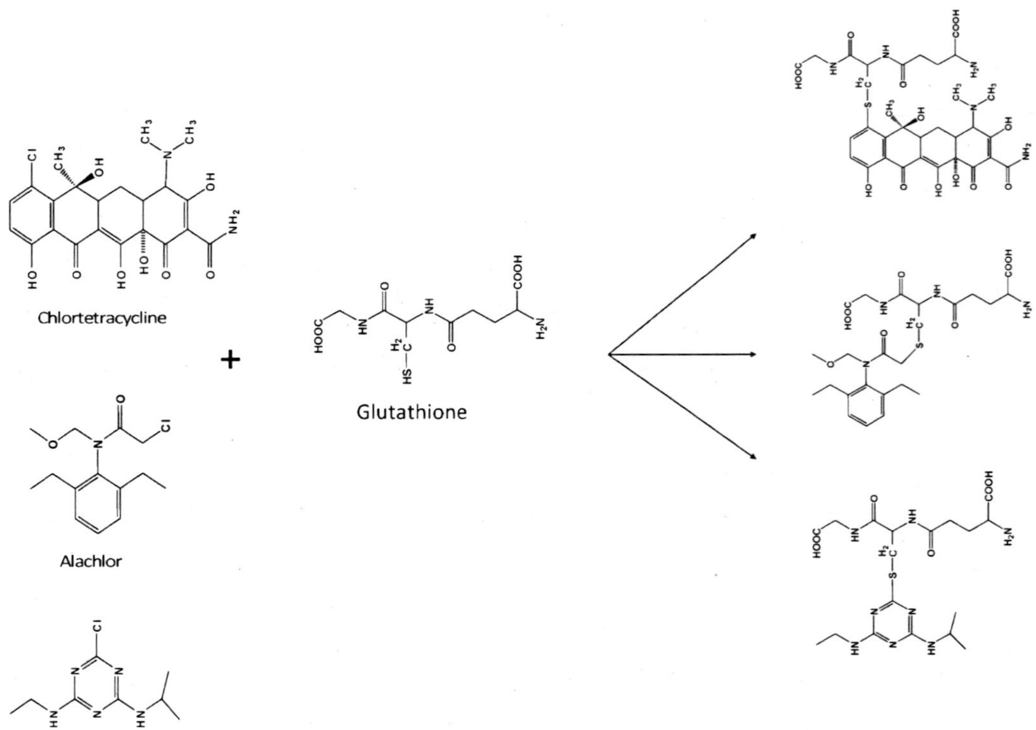

Figure 3. Schematic diagram depicting the conjugation of alachlor to glutathione by GST. The primary site of conjugation occurs at the site of the chlorine atom. A less prevalent conjugation can also occur via nucleophilic attack by GSH on the electrophilic aromatic ring of the herbicide.

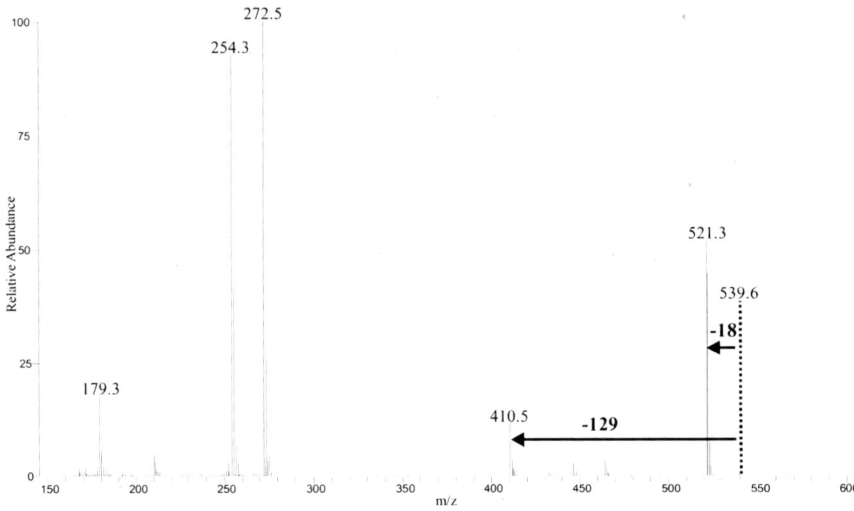

***Figure 4**. LC-MS/MS ion spectrum for the GSH conjugation product of alachlor produced in vitro. The parent conjugate at m/z 539.6 correlates to conjugation at the site of the chlorine atom. Characteristic losses of m/z 18 and 129 are representative of a water molecule and glutamic acid, respectively.*

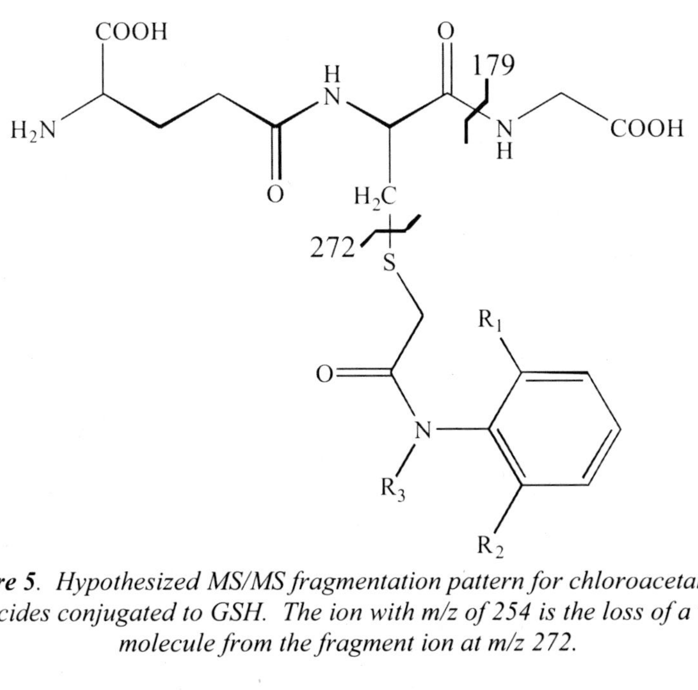

***Figure 5**. Hypothesized MS/MS fragmentation pattern for chloroacetanilide herbicides conjugated to GSH. The ion with m/z of 254 is the loss of a water molecule from the fragment ion at m/z 272.*

With the herbicide conjugation products sufficiently characterized, our next aim is to characterize, *in vitro*, the ability of GSTs isolated from CTC-treated maize to conjugate the chloroacetanilide herbicides and atrazine relative to GSTs isolated from untreated control maize. The precise methods can be found in reference (28). Individual *in vitro* reactions were setup containing one of the herbicides as a substrate and the reactions were stopped at 5, 10, 15, 30, and 60 minutes. Analysis was performed using LC/MS and peak areas for each time point were determined. Using propachlor as the representative chloroacetanilide substrate, Figure 6A shows that not only do the control GSTs conjugate more propachlor, but also the slope of the line is greater than that of the GSTs isolated from CTC-treated maize. However, Figure 6B shows the peak areas determined for atrazine conjugation and it appears that no difference exists between the GSTs isolated from CTC-treated and control maize. So, this data suggests that exposing maize plants to CTC may affect chloroacetanilide herbicide conjugation, whereas no observable effect can be determined with respect to atrazine conjugation.

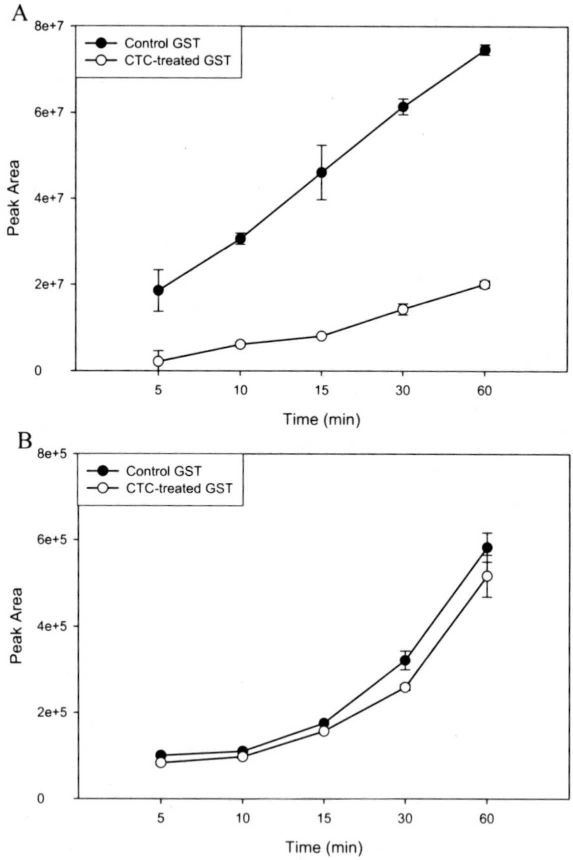

***Figure 6**. Time course measurement of in vitro conjugate formation by LC peak area. A) propachlor conjugate formation and B) atrazine conjugate formation mediated by untreated control GSTs (closed circles) and GSTs isolated from CTC-treated maize (open circles).*

Conclusions

We have found evidence that exposure to CTC, while not directly toxic to maize, may have indirect effects with regards to herbicide detoxification. However, other evidence suggests that this indirect effect is herbicide specific, rather than specific to all classes of herbicides. The *in vitro* data suggests that GSTs play a key role for both the detoxification of CTC and the herbicides when taken up separately. It appears that when maize is pre-exposed to CTC that a specific class of GSTs are induced in response to this particular stressor. And it is the induction of this specific class of GSTs that apparently disrupts the balance of constitutively expressed GSTs and results in the observed differences in herbicide conjugation to GSH with respect to the chloroacetanilide class of herbicides. The lack of significant change in atrazine conjugation mediated by GSTs isolated from CTC-treated and untreated control maize suggests that CTC induces a class of GSTs that are capable of binding and conjugating this class of herbicides. In fact, it is well documented that chlorotriazine and chloroacetanilide herbicides are detoxified by different classes of GSTs (27,29).

This research has far reaching implications for agriculture, especially where antibiotic-laden manure is used as fertilizer for crop production. Concerns can be raised regarding the possibility of reducing crop yields. While the overall affect may be minor, it still warrants judicious use of antibiotics and herbicides in agriculture.

There are at least two questions that must be answered to increase our understanding of these implications. First, we must specifically determine which GSTs are induced under CTC exposure. This will involve both a genomics and proteomics approach. If there are not any differences in the specific GSTs induced relative to controls, then we will determine the quantity and rate of induction. Second, our current research is a tightly controlled greenhouse study and all conjugation reactions are performed *in vitro*. This work will have to be performed in the field once we have identified the key GSTs induced in our laboratory experiments. This work will lead to a better understanding of the implications and remedies for applying herbicides in combination with antibiotics to crop fields.

Literature Cited

(1) www.ahi.org

(2) Mellon, M. 2001 Hogging It: Estimates of Antimicrobial Abuse in Livestock. *Union of Concerned Scientists.*

(3) D. L. Sedlak and K. E. Pinkston, Factors affecting the concentrations of pharmaceuticals released to the aquatic environment., *Journal of contemporary Water Research and Education*, 1, 56-65, 2001.

(4) A. A. Sawant, L. M. Sordillo and B. M. Jayarao, A survey on antibiotic usage in dairy herds in pennsylvania, *J. Dairy Sci.*, 88, 2991-2999, 2005.

(5) J. J. Dibner and J. D. Richards, Antibiotic growth promoters in agriculture: History and mode of action, *Poult Sci*, 84, 634-643, 2005.

(6) S. Thiele-Bruhn, Pharmaceutical antibiotic compounds in soils - a review, *Journal of Plant Nutrition and Soil Science*, 166, 145-167, 2003.
(7) A. B. A. Boxall, P. Blackwell, R. Cavallo, P. Kay and J. Tolls, The sorption and transport of a sulphonamide antibiotic in soil systems, *Toxicology Letters*, 131, 19-28, 2002.
(8) P. Kulshrestha, R. F. Giese and D. S. Aga, Investigating the molecular interactions of oxytetracycline in clay and organic matter: Insights on factors affecting its mobility in soil, 38, 4097-4105, 2004.
(9) J. C. Chee-Sanford, R. I. Aminov, I. J. Krapac, N. Garrigues-Jeanjean and R. I. Mackie, Occurrence and diversity of tetracycline resistance genes in lagoons and groundwater underlying two swine production facilities, *Appl. Environ. Microbiol.*, 67, 1494-1502, 2001.
(10) M. Roberts, Tetracycline resistance determinants: Mechanisms of action, regulation of expression, genetic mobility, and distribution, *FEMS Microbiology Reviews*, 19, 1-24, 1996.
(11) N. Peak, Knapp, C.W., Yang, R.K., Hanfelt, M.M., Smith, M.S., Aga, D.S., and Graham, D.W., Abundance of six tetracycline resistance genes in wastewater lagoons at cattle feedlots with different antibiotic use strategies, *Environmental Microbiology*, In Press, 2006.
(12) F. Pomati, S. Castiglioni, E. Zuccato, R. Fanelli, D. Vigetti, C. Rossetti and D. Calamari, Effects of a complex mixture of therapeutic drugs at environmental levels on human embryonic cells, 40, 2442-2447, 2006.
(13) A. R. Batchelder, Chlortetracycline and oxytetracylcine effects on plant growth and development in soil systems., *Journal of Environmental Quality*, 11, 675-678, 1982.
(14) A. B. A. Boxall, P. Johnson, E. J. Smith, C. J. Sinclair, E. Stutt and L. S. Levy, Uptake of veterinary medicines from soils into plants., *J. Agric. Food Chem.*, 54, 2288-2297, 2006.
(15) K. Kumar, S. C. Gupta, S. K. Baidoo, Y. Chander and C. J. Rosen, Antibiotic uptake by plants from soil fertilized with animal manure., *J Environ Qual*, 34, 2082-2085, 2005.
(16) J. G. Burken and J. L. Schnoor, Predictive relationships for uptake of organic contaminants by hybrid poplar trees, *Environ. Sci. Technol.*, 32, 3379-3385, 1998.
(17) W. D. Kong, Y. G. Zhu, Y. C. Liang, J. Zhang, F. A. Smith and M. Yang, Uptake of oxytetracycline and its phytotoxicity to alfalfa (medicago sativa l.), *Environmental Pollution*, 147, 187-193, 2007.
(18) D. S. Aga, S. O'Connor, S. Ensley, J. O. Payero, D. Snow and D. Tarkalson, Determination of the persistence of tetracycline antibiotics and their degradates in manure-amended soil using enzyme-linked immunosorbent assay and liquid chromatography-mass spectrometry, *J. Agric. Food Chem.*, 53, 7165-7171, 2005.
(19) L. Migliore, C. Civitareale, S. Cozzolino, P. Casoria, G. Brambilla and L. Gaudio, Laboratory models to evaluate phytotoxicity of sulphadimethoxine on terrestrial plants., *Chemosphere*, 37, 2957-2961, 1998.
(20) L. Migliore, S. Cozzolino and M. Fiori, Phytotoxicity to and uptake of enrofloxacin in crop plants., *Chemosphere*, 52, 1233-1244, 2003.

(21) L. Migliore, S. Cozzolino and M. Fiori, Phytotoxicity to and uptake of flumequine used in intensive aquaculture on the aquatic weed, lythrum salicaria l, *Chemosphere*, 40, 741-750, 2000.

(22) C. Forni, A. Cascone, M. Fiori and L. Migliore, Sulphadimethoxine and azolla filiculoides lam.: A model for drug remediation, *Water Research*, 36, 3398-3403, 2002.

(23) N. P. Gujarathi, Linden, J.C., Oxytetracycline inactivation by putative reactive oxygen species released to nutrient medium of helianthus annuus hairy root cultures., *Biotechology and Bioengineering*, 92, 393-402, 2005.

(24) C. Lau, J. Lu and M. Kai, Chemiluminescence determination of tetracycline based on radical production in a basic acetonitrile-hydrogen peroxide reaction, *Analytica Chimica Acta*, 503, 235-239, 2004.

(25) M. H. Farkas, J. O. Berry and D. S. Aga, Chlortetracycline detoxification in maize via induction of glutathione s-transferases after antibiotic exposure, *Environ. Sci. Technol.*, 41, 1450-1456, 2007.

(26) D. Sheehan, Meade, G., Foley, V.A., and Dowd, C.A., Structure, function and evolution of glutathione transferases: Implications for classification of non-mammalian members of an ancient enzyme superfamily, *Biochemical Journal*, 360, 1-16, 2001.

(27) D. C. Holt, V. J. Lay, E. D. Clarke, A. Dinsmore, I. Jepson, S. W. J. Bright and A. J. Greenland, Characterization of the safener-induced glutathione s-transferase isoform ii from maize, *Planta*, 196, 295-302, 1995.

(28) M. H. Farkas, J. O. Berry and D. S. Aga, Determination of enzyme kinetics and glutathione conjugates of chlortetracycline and chloroacetanilides using liquid chromatography/mass spectrometry, *The Analyst*, *In Press*.

(29) L. Prade, R. Huber and B. Bieseler, Structures of herbicides in complex with their detoxifying enzyme glutathione s-transferase - explanations for the selectivity of the enzyme in plants, *Structure*, 6, 1445-1452, 1998.

Chapter 12

Targets, Effects and Risks in Aquatic Plants Exposed to Veterinary Antibiotics

Richard A. Brain[1], Keith R. Solomon[2], and Bryan W. Brooks[1]

[1]Center for Reservoir and Aquatic Systems Research, Department of Environmental Studies, Baylor University, Waco, Texas, USA, 76798-7388.
[2]Department of Environmental Biology, University of Guelph, Guelph, Ontario, Canada N1G 2W1

Veterinary antibiotics are biologically active ubiquitous low-level contaminants which are continually introduced into the aquatic environment from a variety of sources ranging from therapeutic to prophylactic applications in livestock agriculture and aquaculture. The pharmacological modes of action of these compounds are generally well established and understood, though the toxicological modes of action in non-target organisms are comparatively less characterized. As a result, it is important to consider evolutionary conservation of metabolic pathways and receptors across multiple levels of biological organization, particularly in the case of plants, which present a number of targets for antibiotics due largely to the bacterial ancestry of plastid and mitochondrial organelles. Homologous targets for fluoroquinolone (chloroplast replication), tetracycline macrolide, lincosamide, β-aminoglycoside, and pleuromutilin (transcription and translation), sulfonamide (folate biosynthesis) and chlorophenol (fatty acid biosynthesis) antibiotics have been identified in plants. Applying the US FDA risk assessment paradigm to the available toxicity data, potential hazards are evident for blue-green algae exposed to several classes of antibiotics, particularly triclosan, which consistently posed the greatest hazard to both blue-green and green algae. Generally, however, antibiotics pose comparatively little hazard to green algae and higher plants, based on available information in the peer-reviewed literature.

Introduction

The first antibiotic discovered and applied for therapeutic purpose was penicillin, isolated accidentally from the mold *Penicillium notatum* (*1*) with no preconceived knowledge of it's bactericidal effect. Subsequently, a variety of antibiotic classes have been isolated, developed, and/or synthesized. However, the original biological purpose, as it pertains to the organisms in which they were discovered appears to have been forgotten, though this consideration is critically important when considering the ecotoxicology of these compounds. Antibiotics are biologically active and purposefully produced to affect other organisms. Therefore, given the numerous examples of evolutionary conservation of pathways and receptor targets within and among levels of biological organization, it stands to reason that antibiotics derived, synthesized, and inspired from these naturally produced compounds could ultimately induce toxic effects in non-target organisms (*2*). In the late 1990s the first comprehensive ecotoxicological reviews of pharmaceuticals were published conferring the general conclusion that the potential effects of these compounds on non-target aquatic organisms, particularly aquatic plants, were mostly unknown (*3,4*). Although little effect data for plants were available when these reviews were published, the potential for antibiotics, to adversely affect plants was certainly not a new concept. Reviews documenting the effects of antibiotics on plants have been compiled as far back as the 1950s (*5*). Although a number of compounds were identified as phytotoxic, these early studies focused largely on the applications of antibiotics for disease control in plants. The ecotoxicology of antibiotics with respect to aquatic plants is a much more recent area of research; this has been facilitated by an in depth understanding of the evolutionary conservation of pathways and targets between different levels of biological organization, particularly in the case of plants, between bacteria and plastids. This chapter is derived from Brain et al (*2*) with a special focus on veterinary antibiotics .

Estimates, Sources and Concentrations of Antibiotics in the Environment

Reports concerning the presence of antimicrobials and veterinary medicines in the environment date as far back as 1969 (*6*) but most are recent. Although there are two major sources of antibiotics in the environment; human and agricultural only veterinary sources will be considered here. Veterinary medicines provide the greatest array of possibilities for exposure and can be subdivided into substances used as growth promoters for pigs, as therapeutics in livestock production, cocciodiostatics used for poultry production, therapeutics for treatment of livestock on fields (e.g., antiparasitic agents), or into feed additives in aquaculture (*7*). In agricultural or veterinary applications, the parent compounds and/or metabolites are discharged into the environment through liquid manure and (waste) water or with storm water run-off from fields after manure application (*3,4*). In aquaculture antibiotics are directly added to receiving waters, formulated as feed additives (*4*), with 70 to 80 percent of administered drug entering the environment (*8*).

In the United States, roughly 40% of the 23,000 metric tons of antibiotics produced annually are used in agriculture, with more than 80 percent by weight used sub-therapeutically for growth promotion; the remainder is for therapy (*9*). The European Federation of Animal Health estimates that, of the 13,200 tons of antibiotics consumed in the EU in 1999, farm animals accounted for 35%, 29% for therapeutic purposes, and 6% for antimicrobial growth promoters (AGPs) which has fallen by 50% since 1997 (*10*). Although the use of antibiotics as growth promoters has significantly decreased in certain countries which have banned this application, such as Denmark, the total consumption of antibiotics has increased steadily by as much as 17% from 1996 to 2001 (*11*). The continual release of these compounds into the environment allows many to take on a pseudo-persistence, displaying similar exposure characteristics as many truly persistent compounds, since the rate of replacement can balance the rate of transformation or removal in the environment (*12*). This phenomena is particularly observed most commonly in effluent-dominated systems (*13*). In surface waters antibiotics have been detected at concentrations generally ranging from ng/L to µg/L in the U.S. (*14*), Canada, (*15-20*) and Europe (*3,4,21-24*), suggesting that these compounds are present ubiquitously in the aquatic environment.

Evolutionary Conservation of Receptors and Pathways

Antibiotics are biologically active compounds, designed to elicit a specific effect in microorganisms, therefore it may be expected that any effect could also occur in aquatic organisms with similar biological receptors. In this context it is necessary to consider evolutionary conservation of receptors across levels of biological organization. In order to obtain a respective response from a non-target organism, the organism must express the respective receptor or use the respective biosynthetic pathway, with enzymes exhibiting structures similar to the target counterpart (*25*). Plants provide a number of evolutionary conserved target sites for antibiotics as a result of the bacterial ancestry of plastid organelles and conservation of certain metabolic pathways (*26-28*). All plastid organelles arose from the engulfment of an endosymbiotic cyanobacterial-like prokaryotic cell by a larger eukaryotic cell, referred to as the primary endosymbiotic event, generating a plastid characterized by two membranes, such as those found in red and green algae, higher plants, and glaucophytes (*28*). Although plastids have undergone substantial modification during an estimated 500 million years of intracellular survival within their eukaryotic hosts, the genome, transcription, and translation equipment have remained fundamentally bacterial in nature (*28*). The bacterial nature of plastids therefore makes them susceptible as potential drug targets (*28*). The following sections discuss class-specific phytotoxicty data and risks of veterinary antibiotics in aquatic plants with specific focus on mechanism of action.

Plastid Replication and DNA Gyrases in Plants as Targets for Fluoroquinolone Antibiotics

Fluoroquinolone antibiotics act by inhibiting the activity of DNA gyrases, the bacterial equivalent of DNA topoisomerase (topo) which supercoils DNA by using the free energy of ATP hydrolysis (*29,30*). There are two types of topoisomerase enzymes, Types I and II, distinguished by the ability to transiently break one or both strands of the DNA (*29*). DNA gyrase is a Type-II enzyme required for replication and transcription (Figure 1) in prokaryotes, and is the only enzyme of this type that is able to catalyze ATP-dependent DNA supercoiling (*29,31*). The apparent absence of gyrases in eukaryotes and critical importance in prokaryotes has facilitated the development of a number of gyrase-specific antibacterial agents, including quinolones, coumarins, and cyclothialidines (*32*). Recent evidence, however, has demonstrated that plants contain DNA gyrases (*31,33*). The weak activity of an ATP-dependent topoisomerase in the green algae *Chlamydomonas reinhardtii*, which could supercoil DNA in vitro was demonstrated in the mid 1980s (*33*). Although the enzyme was not purified it was demonstrated that gyrase-specific drugs

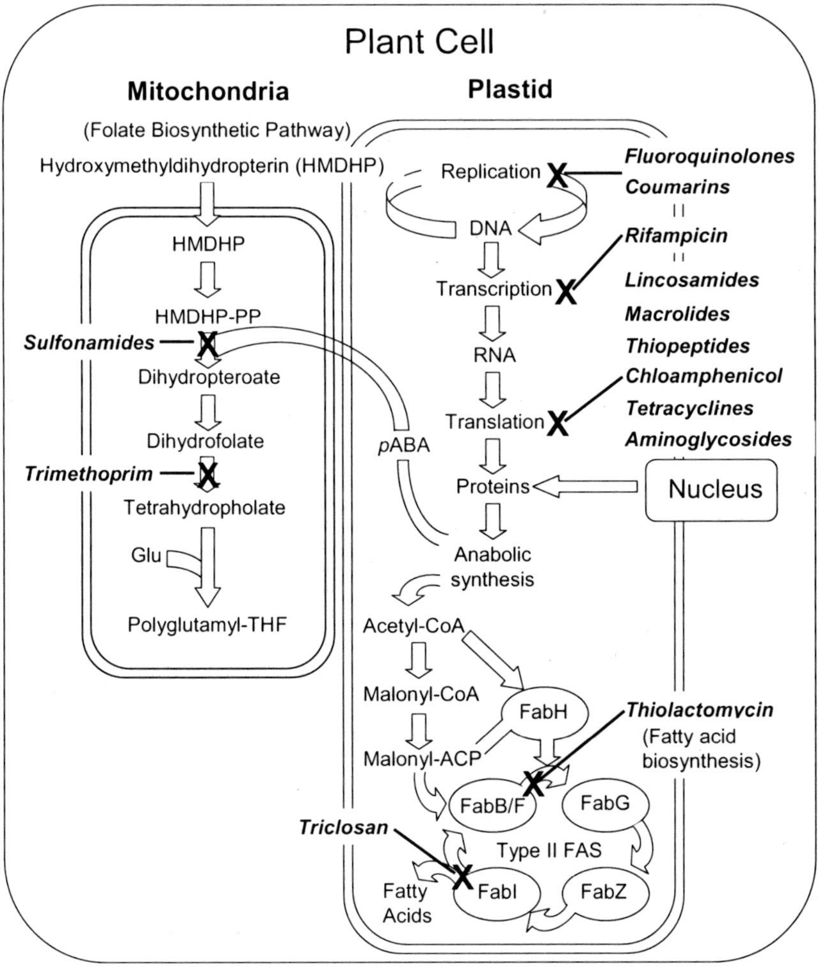

Figure 1. Select metabolic pathways within the mitochondria and plastid of a plant cell and the major classes of antibiotics believed to target these pathways or processes. A variety of antibiotic classes are suspected to inhibit specific processes within the chloroplast, such as replication, transcription, and translation. Triclosan and thiolactomycin are thought to disrupt fatty acid biosynthesis by inhibiting enoyl-ACP reductase (FabI) and β -ketoacyl-ACP synthase I and II (FabB and/or FabF), respectively. Sulfonamide antibiotics are thought to target dihydropteroate synthase (DHPS) in the folate biosynthetic pathway. Redrawn with permission from from (1). Copyright 2008 Springer Science+Buisiness Media, LLC.

(novobiocin a coumarin) inhibited chloroplast transcription. Recently, more definitive evidence has been published by Wall et al. (*31*) identifying four putative gyrase genes in the model plant *A. thaliana*; one gyrA and three gyrB homologues. Knockouts of *A. thaliana* gyrA were found to be embryo-lethal, whereas knockouts in the gyrB genes led to seedling-lethal phenotypes or severely stunted growth and development; treatment of seedlings and cultured cells with ciprofloxacin caused growth inhibition (*31*).

Shortly after its discovery, nalidixic acid was screened for antichloplastic activity in the facultative chloroplasts of *Euglena gracilis* causing a loss of green colony forming ability at 50 μg/L, and was suspected to have a mode of action analogous to that in bacteria (*34*). Among the different classes of antibiotics tested for phytotoxicity, quinolones have generally proven to be the most potent class. Ciprofloxacin, lomefloxacin, ofloxacin, levofloxacin, and norfloxacin were all found to induced phytotoxicity in *Lemna gibba* with 7-d fresh weight EC50 values ranging from 97-913 μg/L (*35*). Similarly ciprofloxacin, lomefloxacin, ofloxacin, levofloxacin, clinafloxacin, enrofloxacin, and flumequine induced phytotoxicity in *Lemna minor*, with 7-d frond number EC50 values ranging from 53 to 2,470 μg/L (*36*), showing good agreement with *L. gibba*. Cyanobacteria such as *Microcystis aeruginosa*, show comparable sensitivity to *Lemna* sp. when exposed to fluoroquinolones with EC50 values for growth ranging from 7.9-1,960 μg/L (*36*). Green algae such as *Selenastrum capricornutum*(recently renamed *Pseudokirchneriella subcapitata*) are up to 2 orders of magnitude less sensitive than cyanobacteria and *Lemna* sp. with EC50 values ranging from 1,100 to 22,700 μg/L (*36*). Since topoisomerase have been identified in green algae (*Chlamydomonas. reinhardtii*) (*33*) and higher plants (*31*), the mechanism(s) for the differential sensitivity are unknown. Chloroplastic type II topoisomerases and DNA replication in *C. reinhardtii* are inhibited with treatment of the coumarin novobiocin; however, after a delay of several hours, chloroplast chromosomes were found to initiate a novobiocin insensitive mode of DNA replication resulting in partial replication of the chloroplast chromosome (*37*). It is uncertain whether this phenomenon exists in the chloroplasts of cyanobacteria and higher plants, though differences within compensatory replication strategies could potentially explain this differential sensitivity (*2*). *Lythrum salicaria*, an aquatic higher plant was also found to be insensitive to flumequine at concentrations between 50-5,000 μg/L causing stimulation, only becoming inhibitory at 100 mg/L (*38*). Many parasites of the phylum *Apicomplexa*, which contain facultative chloroplasts suspected to be acquired through secondary endosymbiosis of a green alga, have similar sensitivity to green algae (*30,34,39-41*). Symptomology for all photosynthetic organisms exposed to DNA gyrase inhibitors is markedly similar however, typified by a pronounced bleaching of tissues or cultures (*34-36,40*).

Notable effects of ciprofloxacin on environmental samples taken upstream of a wastewater treatment plant (WWTP) include a significant reduction of the common diatom *Navicula* at 0.12 μg /L, reductions in the green algae *Chlamydomonas* at 0.12 μg /L and 1.2 μg /L, and *Sphaerocystis* at 1.2 μg/L ($P < 0.05$) (*42*). Consistent and dramatic declines in genus richness were found for increasing ciprofloxacin treatment. In comparisons of certain sampling dates, ciprofloxacin was found to cause statistically significant effects for *Synedra* and *Chlamydomonas* between upstream and downstream sampling sites (*42*).

Fluoroquinolones have also been implicated as active phytotoxic agents in 2 microcosm evaluations. Ciprofloxacin was found to be phytotoxic to *L. gibba*, *Myriophyllum sibiricum* and *Myriophyllum spicatum* at high (600, 1000, 1000 μg/L) treatment concentrations and cause growth inhibition in *L. gibba* at medium (6, 10, and 10 μg/L) treatment concentrations when treated in combination with ibuprofen (anti-inflammatory) and fluoxetine (selective serotonin reuptake inhibitor) (*43*). In the same systems, levoloxacin was found to be a phytotoxic agent in addition to sulfamethoxazole (sulfonamide antibiotic) and atorvostatin (blood lipid regulator) when treated in combination with acetaminophen (analgesic), caffeine (stimulant), carbamazepine (anti-epileptic), sertraline (SSRI), and trimethoprim (diaminopyrimidine antibiotic). Phytotoxicity was demonstrated in both rooted and floating macrophytes with total mixture molar EC50s ranging from 4.6-66.7 μmol/L for 35-day exposures in *M. sibiricum* and 1.15-12.4 μmol/L for 7-day exposures with *L. gibba* (*44*).

Translation and Transcription Processes in the Plastid as Targets for Tetracycline, Macrolide, Lincosamide, Aminoglycoside, and Pleuromutilin Antibiotics

The translational apparatus and mRNAs of plastids are similar to those in bacteria. Plastids are responsible for the autonomous transcription and translation of the genes encoded in their genome, and the machinery used to accomplish this shares a high degree of homology with that found in bacteria (*45*). Molecular phylogenetic analysis of c-type cytochrome and rRNA sequences have established a relationship between cyanobacteria and the green (euglenoids, green algae, and higher plants) and red (rhodophyte) chloroplasts, supporting the prokaryotic origins of chloroplasts (*46*). Plastid genes contain consensus prokaryotic promoters and bacterial-like sigma factors, also polymerases, and ribosome-binding sequences are found in many chloroplast mRNAs as a requirement for translation, presumably through a similar binding between mRNA and the ribosome as found in bacterial translation initiation (*45*). A high degree of homology between the chloroplast and bacteria with regard to general translation factors and a majority of the ribosomal proteins implies that many of the basic processes of translation are conserved between bacteria and the chloroplast (*2*).

Similar to prokaryotes, plastids and mitochondria both have 70S ribosomes composed of two subunits, a 50S and a 30S. Proteomic analysis of the 70S ribosome from the chloroplast of the green algae *C. reinhardtii* showed twenty-seven orthologs of *E. coli* large subunit proteins in the 50S subunit, as well as an ortholog of the spinach plastid-specific ribosomal protein (PSRP)-6, indicating that the large subunit proteins are very similar to those of spinach chloroplasts and *E. coli* (*47*). Overall, the chloroplast ribosome proteome of higher plants is composed of 59 distinct proteins: six PSRPs, a bacterial-type plastid ribosome recycling factor (pRRF), and 52 orthologues of eubacterial ribosomal proteins (*48*). Six ribosomal proteins are specific to higher plant chloroplast ribosomes (*49*).

Streptomycin, spectinomycin, neamine/kanamycin, and erythromycin resistance has been documented in the chloroplasts of *C. reinhardtii* and is conferred by base substitutions in conserved regions of the genes encoding the 16S and 23S chloroplast ribosomal RNAs (*50*). As macrolide and lincosamide antibiotics block protein synthesis by interacting with the peptidyl-tranferase domain of bacteria 23S rRNA, and aminoglycosides interfere with bacterial protein synthesis by irreversibly binding to 30S and 50S subunits of ribosomes, these compounds have the potential to disrupt transcription/translation in the chloroplasts of photosynthetic organisms (Figure 1).

Recently, through analyses of the plastidial transcription apparatus at least two distinct DNA-dependent RNA polymerases have been identified; one of these is a multi-subunit enzyme similar to those in bacteria and in the nuclei of eukaryotic cells (*51-53*). This enzyme is referred to as plastid-encoded polymerase (PEP) due to the intraorganellar coding sites of its core subunits (*51-53*). Plastid transcription utilizes an RNA polymerase (α_2, β, β'), which is homologous to that of cyanobacteria and other eubacteria (*51*). Two multi-subunit plastid RNA polymerases termed A and B have been identified in mustard (*53*). The B enzyme has a bacterial-type polypeptide composition and is sensitive to the prokaryotic transcription inhibitor rifampicin, whereas the A enzyme has a more complex subunit structure and is not sensitive (*53*). The bacterial nature of the plastidial RNA polymerase provides another potential target for antibiotics interfering with transcription/translation in photosynthetic organisms (*2*).

Tetracyclines

The phytotoxic effects of chlortetracycline and oxytetracycline have been reported in higher plants as far back as the 1950s (*5*). In bacteria, tetracyclines bind irreversibly to the 30S subunit of ribosomes, blocking the binding of aminoacyl transfer to DNA, inhibiting protein synthesis (*54*). Early studies with tetracyclines on several species of duckweeds, including *L. gibba*, and two species of ferns, indicated that oxytetracycline, tetracycline, and chlortetracycline were inhibitory at concentrations of 10 or 20 mg/L in 21-day experiments (*55*). However, under different exposure scenarios more recent studies have demonstrated toxicity to *Lemna* sp. at much lower concentrations of 219 to 4,920 μg/L, (*35,56*). The mechanism of phytotoxic action was suggested by Nickell and Gordon (*55*) to be largely through chelating of metal nutrients in growth solutions, and less so via direct interference with plastidial ribosomes. This hypothesis was supported by a recent investigation of the effects of a mixture of tetracyclines (tetracycline, chlortetracycline, oxytetracycline and doxycycline) in 12,000 L aquatic microcosms, where *L. gibba* showed no signs of phytotoxicity likely since nutrients were not limiting in these experimental units (*57*). The submerged macrophyte *Myriophyllum sibiricum* however experienced concentration-dependent decreases in dry mass to 69, 47, 30, and 7% of controls at treatment concentrations of 0.080, 0.218, 0.668, and 2.289 µmol/L, respectively, due to corresponding concentration-dependent attenuation of light penetration as high as 99.8% at a depth of 70 cm (*57*). In the same study, phytoplankton abundance showed an initial decrease in the 0.218, 0.668, and 2.289 µmol/L treatments after 7-days, however all treatments rebounded to levels above the controls after 35 days of exposure (*58*). Both blue-green and green algae show sensitivity to tetracyclines in laboratory studies with EC50 values ranging from 9-207 µg/L for *M. aeruginosa* and 170-4,500 µg/L *S. capricornutum*, respectively, though generally cyanobacteria are more sensitive (*59-61*). It is not known whether tetracyclines disrupt translation in algae, or whether toxicity is caused as a result of chelation of mineral nutrients as has been shown for *Lemna* sp. The mechanism is likely the latter since microcosm effect values are several orders of magnitude higher than the laboratory derived values (*62*). Inhibitory concentrations of tetracyclines in *E. gracilis* are also high, generally well above 1 mg/L, with the exception of doxycycline (*41*), suggesting that the mode of toxic action is likely not the pharmacological mode of action in photosynthetic organisms. Alternatively, tetracyclines may experience poor rates of uptake and/or translocation or high rates of metabolism, although this has not been investigated (*63*).

Macrolides and Lincosamides

The pharmacological mode of action for macrolide and lincosamide antibiotics is via interaction with the peptidyl-tranferase domain of bacterial 23S rRNA (*54*). Higher plants are generally insensitive to macrolides; for example, erythromycin is inhibitory to *Lemna* sp. only at concentrations >1 mg/L with an extrapolated EC50 of 5,620 µg/L (*64*). Tylosin causes stimulation in both laboratory tests with *L. gibba*, tylosin (25% increase in wet mass at 212 µg/L) and microcosm tests with *L. gibba* and *M. spicatum* (*65*). Macrolides also demonstrate little potency to protoplast-derived cells of tobacco (*Nicotiana plumbaginifolia*: erythromycin is toxic between 30 and 80 mg/L) (*66*). *Selenastrum capricornutum* displays variable sensitivity to macrolides, where the EC50s for tylosin and spiramycin are >1 mg/L (*59*), though erythromycin, lincomycin, and clarithromycin inhibit growth by 50% at 20, 70 and 2 µg/L, respectively (*61*). Compared to *S. capricornutum*, *M. aeruginosa* is considerably more sensitive to spiramycin and tylosin with EC50 values of 5 and 34 µg/L, respectively (*59*).

Aminoglycosides and Pleuromutilins

Aminoglycosides irreversibly bind to 30S and 50S subunits of bacterial ribosomes (*54*). The phytotoxicity profile for these compounds is similar to macrolides and lincosamides, with *M. aeruginosa* and *S. capricornutum* demonstrating sensitivity to streptomycin with EC50 values (7-133 μg/L) substantially lower than those found in *L. gibba* (>1,000 μg/L) (*35,59,66*). Pollock et al. (*66*) found that streptomycin, neomycin, kanamycin, gentamicin, G418, amikacin, and tobramycin, were all toxic to protoplast-derived cells of *Nicotiana plumbaginifolia* (Tex-Mex tobacco) to varying degrees. The suggested toxic mechanism was due to the action of the aminoglycosides on the 'prokaryotic-like' ribosomes of chloroplasts and mitochondria (*66*). Exposure of some strains of vigorously reproducing *E. gracilis* to high concentrations (100,000 μg/L) of streptomycin leads to fragmentation and loss of chloroplasts with consequent loss of chlorophyll and photosynthetic capabilities (*5,67*). Tiamulin is a pleuromutilin antibiotic that also binds to the 50S subunit of the bacterial ribosome and interacts at the peptidyl transferase center (*68*). Tiamulin is toxic to both green and blue-green algae with EC50 values of 3 and 165 μg/L for *M. aeruginosa* and *S. capricornutum*, respectively (*59*), though no data are available for sensitivity comparisons to higher plants.

Proteomic analysis of chloroplastic ribosomes from the algae *C. reinhardtii* indicate a high degree of homology with the bacterial ribosomes of *E. coli* (*47*), though similar to green algae, protein identification of the chloroplastic ribosomes of higher plants has also identified a full complement of *E. coli* ribosomal proteins (*48,49*). Therefore, since the appropriate receptor is present in both algae and higher plants, the differential sensitivity is again likely due to differences in the uptake, translocation and/or metabolism of these compounds, though this has not been investigated.

Mitochondrial Folate Biosynthesis in Plants as a Target for Sulfonamides

Pharmacologically, sulfonamide antibiotics disrupt folate biosynthesis via inhibition of the enzyme dihydropteroate synthase in the folate biosynthetic pathway acting as a structural analogue of the substrate *p*-aminobenzoic acid (*p*ABA) as shown in Figure 1 (*54*). Similarly, diaminopyrimidine antibiotics such as trimethoprim inhibit dihydrofolate reductase, two steps down from dihydropteroate synthase in the folate biosynthetic pathway also shown in Figure 1. Folates are essential cofactors for one-carbon transfer reactions in most living organisms and are required for the biosynthesis of nucleic acids, amino acids, pantothenate, and specifically in plants, lignin formation and photorespiration (*26,69*). Recent evidence has demonstrated that folate synthesis pathway in plants is essentially the same as in bacteria (*26*). There are currently no data available for the corresponding proteins in algae or fungi (*70*). However, the folate biosynthetic pathway has been identified in apicoplasts of *Plasmodium falciparum* and the efficacy of sulfonamide drugs on dihydropteroate synthase and genetic sequence alterations leading to sulfonamide resistance have been investigated (*71-73*). Resistance to trimethoprim has also been documented in *P falciparum* through alterations in dihydrofolate reductsase alleles (*73,74*).

Sensitivity to sulfonamide antibiotics is remarkably similar between *Lemna* sp. (EC50: 81-2,330 μg/L) (*56,57*) green algae (EC50: 146-7,800 μg/L) (*60,75*) and blue-green algae (EC50: 26.8-135 μg/L) (*60,61*). Although sulfamethoxazole was consistently the most toxic sulfonamide to aquatic plants, the protozoan *Plasmodium falciparum* is considerably less sensitive (Mean IC50: 89 mg/L), though sulfadoxine (Mean IC50: 63 mg/L) and trimethoprim (Mean IC50: 1.5 mg/L) are slightly more potent (*73*). In contrast to sulfonamides, green algae, blue-green algae and higher plants are comparatively insensitive to trimethoprim with corresponding EC50 values of 130,000, 112,000, and >1000 μg/L, respectively. Experiments with sulfamethoxazole and trimethoprim in *L. gibba* showed no significanct difference between exposures to sulfamethoxazole alone and combinatorial exposures to sulfamethoxazole and trimethoprim at equimolar concentrations (Brain, 2006 unpublished data). This discrepancy suggests another instance of differential uptake, translocation and/or metabolism since receptors exist for both sulfamethoxazole and trimethoprim in plants.

The Chlorplastic Fatty Acid Biosynthetic Pathway as a Target for Triclosan

In bacteria, fatty acid synthesis is regulated largely via fatty acid synthase (FAS), the principal enzymatic unit, and occurs as iterative elongations of acyl chains utilizing the 2-carbon donor malonyl coenzyme A (CoA) (*76*). FAS catalyzes essentially the same reactions for all organisms, though two different types are found in nature; Type I FAS, found in animals and yeast, consists of a single large multifunctional enzyme complex and Type II FAS, found in plants and most bacteria, consisting of multiple proteins with separate enzymatic functions (*27*). Fatty acid biosynthesis in plants takes place within plastids and utilizes the bacterial type II system (*27,77*). Long-chain fatty acids are assembled 2 carbons at a time from malonyl-ACP (acyl carrier protein) units (*27*). Condensation and reduction reactions in the FAS complex are catalyzed by a number of different enzymes, where the final reduction is catalyzed by enoyl-ACP reductase (FabI) (*76*). Pharmacologically, FabI is a major point of regulation for bacterial and plastidial fatty acid synthesis as well as the drug target for triclosan (Figure 1) (*77*). Triclosan is a chlorophenol antibiotic that inhibits fatty acid synthesis in bacteria by dramatically increasing the affinity of FabI for NAD+ creating an FabI-NAD+-triclosan complex resulting in a stable ternary complex, with the drug binding at the enoyl substrate site (*78*). The high degree of similarity between bacterial and plant (plastid) fatty acid biosynthesis suggests evolutionary conservation of this pathway and a likely drug target in plants (*2*).

Green algae, blue-green algae and diatoms are highly sensitive to triclosan with EC50s ranging from 0.7-19.1 μg/L, though *L. gibba* is less sensitive with EC50 values of >62.5 μg/L (the highest concentrations tested) (*79*). However, Orvos et al. (*79*) noted that studies with all plant species tested, including *Lemna*, confirmed the hypothesis that plants are indeed sensitive to the effects of triclosan. Although inhibition of FabI has not been demonstrated specifically in plants, triclosan has been shown to inhibit a plant-like FabI in *Plasmodium falciparum* at IC50 values ranging from approximately 150-2,000 μg/L (*80*). Triclosan has also been shown to cause a significant reduction in numbers of

Chlamydomonas at concentrations of 0.015, 0.15, and 1.5 µg/ L, in *Sphaerocystis* at 1.5 µg/L ($P < 0.05$), and in chroococcalian cyanobacteria at the concentration of 0.15 µg/L ($P<0.05$) in environmental samples taken upstream of a WWTP (*42*). Wilson (*42*) further suggested that increasing the environmental concentration of triclosan resulted in a consistent and dramatic decline in genus richness.

Thiolactomycin, a naturally produced antibiotic from *Nocardia sp*., is a potent inhibitor of FabB (β -ketoacyl-ACP synthase I and II) in *E. coli* (*81*) that has also been shown to be highly toxic to *P. falciparum* (*76*) and been shown to inhibit fatty acid synthesis in the leaves of the higher plant, *Avena* (oat) by 50% at 380 µg/L (*82*). This further suggest a high degree of conservation of enzymes involved in type II FAS between bacteria and plants, and suggests a receptor mediated response for plants exposed to FAS inhibiting antibiotics.

Other Antibiotics

ß-lactam antibiotics, including penicillin, inhibit the formation of peptidoglycan cross links in the bacterial cell wall by binding to the enzyme (transpeptidase) that links the peptidoglycan molecules (*54*). Blue-green algae, such as *M. aeruginosa,* are highly sensitive to both benzylpenicillin and amoxicillin with respective EC50 values of 6 and 3.7 µg/L, though green algae and higher plants are comparatively insensitive to β-lactams, with EC50 values generally well above 1 mg/L (*35,59,60*). Penicillin-binding proteins (PBPs) have recently been identified in 12 cyanobacterial genomes that are involved in the synthesis of the peptidoglycan layer of the cell wall during cyanobacterial cellular differentiation (*83*). Among plants however, only glaucophytes (mosses) have retained peptidoglycans, red algae have no genes for peptidoglycan synthesis, and green plants (*A. thaliana*) have retained some peptidoglycan genes, though their functions have not been determined (*84*). Furthermore, no wall-like structures have been detected in the plastids of green plants and knockouts of these genes have shown no effect on chloroplast division (*84*). Cephalexin, a carbacephem antibiotic with a similar mode of action, also showed little potency in *L. gibba* <1 mg/L (*35*). Monensin, an ionophore antibiotic that disrupts bacterial mitochondria cell membranes allowing excess water into organelles (*54*) is moderately toxic to *L. gibba* with an EC50 of 998 µg/L (*35*). No data are currently available for cephalexin and monensin toxicity to algae but microcosm studies suggest that algae are more sensitive than zooplankton (*85*).

Comparative Sensitivity of Blue-Green algae, Green Algal and Higher Plants

The phytotoxicity data for blue-green algae, green algae and *Lemna* sp. demonstrate a striking degree of class-specific differential sensitivity. Since potential antibiotic receptors have been demonstrated in each of these taxonomic groups it is reasonable to postulate that differences in sensitivity lie largely in the differential metabolic potentials of these organisms (*2*). Similar arrays in sensitivity have also been demonstrated for these groups of plants treated with a number of herbicides, where receptors are known in each species (*86-88*) though susceptibility varies markedly and suggested to be due to differential metabolic capacities (*88*). Metabolism in plants typically involves oxidation, reduction and hydrolysis reactions by enzymes such as cytochrome p-450s (Phase I) followed by conjugation and secondary conjugation to glucose, amino acids and glutathione (Phase I and II) (*89*).

Cytochrome P450s are present in all kingdoms; though in bacteria, they are soluble, whereas in eukaryotes, they are membrane anchored (*90*). In plants the role of these endogenous enzymes is generally poorly understood; and only a few dozen P450 enzymes have been characterized to any extent (*91*). As of 2004, 1,098 plant P450s have been named, and projections suggest 10,500 P450 sequences are needed to completely characterize the compliment of P450 proteins in higher plants (*92*), indicating incredible diversity among the plant kingdom. Species specific variations therefore likely account for the differential potential for metabolizing xenobiotics. Similarly, differences in the properties of GST isoenzymes also account for conspicuous species-specific differences in metabolism, and hence for the vast differences in the susceptibility of different plant species toward certain herbicides (*93*) and likely antibiotics.

Risk

The current prospective risk assessment process for antibiotics relies on hazard or risk quotient (HQ or RQ) approaches, where a PEC (predicted environmental concentration) or MEC (measured environmental concentration) is divided by a statistically derived toxicological benchmark concentration (TBC: NOEC or ECx value), usually under worst-case circumstances (*94*). If the HQ or RQ is >1 then a potential hazard or 'risk' may exist, however values <1 indicate low potential hazard or 'risk'. The US FDA (*94*) guidance for the environmental assessments of pharmaceuticals suggests a PEC value of 0.10 μg/L and a TBC derived from Tier 2 effects assessments, where the EC50 for the most sensitive organism is divided by a safety or assessment factor of 100 resulting in the following HQ equation: (0.1 μg/L)/(EC50/100). Applying this equation to the the available phytotoxicity data indicates no instances of significant hazard or 'risk' for aquatic higher plants, however 8 of 23 antibiotics indicated potential hazards to blue green algae and 2 of 26 for green algae. Figure 2 shows the hazard quotients by class for blue-green algae, green algae and higher plants exposed to antibiotics. Blue-green algae were consistently the most sensitive group accounting for 80% of the HQ exceedence; specifically *M. aeruginosa* exposed to aminoglycoside, pleuromutilin, macrolide, β-lactam, quinolone, and tetracycline antibiotics and *Anabaena flos-aquae* exposed to triclosan (*2*). For green algae, *S. capricornutum* and *Scendesmus subspicatus* species showed potential risks from exposure to triclosan, as well as clarithromycin, a macrolide, in the case of *S. capricornutum*. Therefore, under the US FDA paradigm several classes of antibiotics do pose a potential hazard to algal species, particularly triclosan.

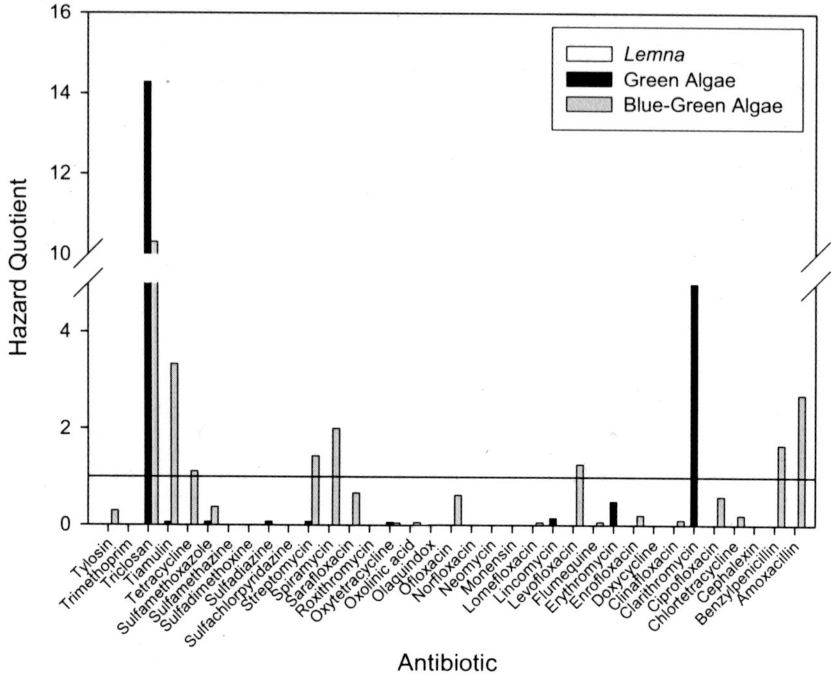

__Figure 2__. Hazard quotients (HQs) calculated for blue-green algae, green algae, and higher plants exposed to antibiotics. The HQs were calculated as the ratio of the predicted environmental concentration (PEC: 0.10 μg/L) and the lowest EC50 value or lowest toxicity value (divided by a factor of 100) available for each compound according to the US FDA framework. The critical HQ of 1 is indicated by a reference line. In some cases toxicity data was not available for one or more of the blue-green algae, green algae or Lemna groupings in which case zero values denote missing data. The data used to generate this figure were originally compiled in (2).

Conclusions

Coupling the evolutionary conservation of metabolic pathways and receptors, particularly between plastids and bacteria, with the recent widespread detection of antibiotics in the environment raises potential concerns regarding toxicity in non-target aquatic plants. Receptors or pathways with bacterial homology have been identified in plants for a number of antibiotics including fluoroquinolones (chloroplast replication), tetracyclines macrolides, lincosamides, β-aminoglycosides, and pleuromutilins (transcription and translation), sulfonamides (folate biosynthesis) as well as triclosan and thiolactomycin (fatty acid biosynthesis). Toxicological investigations indicate susceptibility across multiple plant species, though sensitivity to these compounds varies widely between blue-green algae, green algae and higher plants in a rather inconsistent manner, except that cyanobacteria are largely the most sensitive group (*2*). The differential susceptibility among aquatic plants is likely a consequence of differential metabolic potentials as well as uptake kinetics. According to the US FDA risk assessment paradigm potential hazards are evident for blue-green algae exposed to several classes of antibiotics, particularly triclosan, which consistently posed the greatest hazard to both blue-green and green algae.

References

1. Fleming, A. On the antibacterial action of cultures of a *penicillium*, with special reference to their use in the isolation of *B. influenzæ British Journal of Experimental Pathology* **1929**, *10*, 226–236.
2. Brain, R. A.; Hanson, M. L.; Solomon, K. R.; Brooks, B. W. Aquatic Plants Exposed to Pharmaceuticals: Effects and Risks *Rev. Environ. Contam. Toxicol.* **2007**, *Accepted 27/11/2006*.
3. Daughton, C. G.; Ternes, T. A. Pharmaceuticals and personal care products in the environment: Agents of subtle change? *Environ. Hlth. Perspect.* **1999**, *107*, 907-938.
4. Halling-Sørensen, B.; Nors Nielsen, S.; Lanzky, P. F.; Ingerslev, F.; Holten Lutzhoft, H. C.; Jørgensen, S. E. Occurrence, fate and effects of pharmaceutical substances in the environment - a review *Chemosphere* **1998**, *36*, 357-393.
5. Brian, P. W. Effects of antibiotics on plants *Annu. Rev. Plant. Physiol.* **1957**, *8*, 413-426.
6. Swann, M. M.; Blaxter, K. L.; Field, H. I.; Howie, J. W.; Lucas, I. A. M.; Millar, E. L. M.; Murdoch, J. C.; Parsons, J. H.; White, E. G. "Report of the joint committee on the use of antimicrobials in animal husbandry and veterinary medicine," Her Majesty's Stationery Office, **1969**.
7. Halling-Sørensen, B.; Jørgensen, S. E. Drugs in the environment *Chemosphere* **2000**, *40*, 691-699.
8. Samuelsen, O. B.; Lunestad, B. T.; Husevag, B.; Holleland, T.; Ervik, A. Residues of oxolinic acid in wild fauna following medication in fish farms *Dis. Aquat. Org.* **1992**, *12*, 111-119.

9. Levy, S. B. Multidrug resistance - a sign of the times *N. Engl. J. Med.* **1998**, *338*, 1333-1338.
10. EUC European Union Commission Press Release (IP/02/1891) Food Law News., **2002**; Vol. 2003.
11. Bager, F.; Emborg, H.-D.; Heuer, O. E. "DANMAP 2001-Use of antimicrobial agents and occurrence of antimicrobial resistance in bacteria from food animals, foods and humans in Denmark," Danish Veterinary Institute, **2001**.
12. Daughton, C. G. Environmental stewardship and drugs as pollutants *The Lancet* **2002**, *360*, 1035-1036.
13. Brooks, B.; Riley, T.; Taylor, R. Water quality of effluent-dominated stream ecosystems: ecotoxicological, hydrological, and management considerations *Hydrobiol.* **2006**, *556*, 365-379.
14. Kolpin, D. K.; Furlong, E. T.; Meyer, M. T.; Thurman, M. E.; Zaugg, S. D.; Barber, L. B.; Buxton, H. T. Pharmaceuticals, hormones, and other organic wastewater contaminants in U.S. streams, 1999-2000: A national reconnaissance *Environ. Sci. Technol.* **2002**, *36*, 1202-1211.
15. Lissemore, L.; Hao, C.; Nguyen, B.; Kleywegt, S.; Yang, P.; Solomon, K. R. Determination of pharmaceuticals in environmental waters by liquid chromatography/tandem mass spectrometry *Anal. Chem.* **2004**, *Submitted 28/07/04*.
16. Lissemore, L.; Hao, C.; Yang, P.; Sibley, P. K.; Mabury, S. A.; Solomon, K. R. An exposure assessment for selected pharmaceuticals within a model watershed in Southern Ontario *Chemosphere* **2006**, *in press*.
17. Metcalfe, C. D.; Koenig, B. G.; Bennie, D. T.; Servos, M.; Ternes, T. A.; Hirsch, R. Occurrence of neutral and acidic drugs in the effluents of canadian sewage treatment plants *Environ. Toxicol. Chem.* **2003**, *22*, 2872-2880.
18. Miao, X.-S.; Metcalfe, C. D. Determination of pharmaceuticals in aqueous samples using positive and negative voltage switching microbore liquid chromatography/electrospray ionization tandem mass spectrometry *Journal of Mass Spectrometry* **2003**, *38*, 27-34.
19. Miao, X.-S.; Bishay, F.; Chen, M.; Metcalfe, C. Occurrence of antimicrobials in the final effluents of wastewater treatment plants in Canada *Environ. Sci. Technol.* **2004**, *38*, 3533-3541.
20. Sosiak, A.; Hebben, T. "A Preliminary Survey of Pharmaceuticals and Endocrine Disrupting Compounds in Treated Municipal Wastewaters and Receiving Rivers of Alberta," Alberta Environment, **2005**.
21. Heberer, T. Occurrence, fate, and removal of pharmaceutical residues in the aquatic environment: a review of recent research data *Toxicol. Lett.* **2002**, *131*, 5-17.
22. Hirsch, R.; Ternes, T.; Herberer, K.; Kratz, K.-L. Occurrence of antibiotics in the aquatic environment *Sci, Tot. Environ.* **1999**, *225*, 109-118.
23. Jones, O. A. H.; Voulvoulis, N.; Lester, J. N. Human pharmaceuticals in the aquatic environment a review *Environmental Technology* **2001**, *22*, 1383-1394.

24. Kümmerer, K. Drugs in the environment: emission of drugs, diagnoistic aids and disinfectants into wastewater by hospitals in relation to other sources - a review *Chemosphere* **2001**, *45*, 957-969.
25. Seiler, J. P. Pharmacodynamic activity of drugs and ecotoxicology-can the two be connected? *Toxicol. Lett.* **2002**, *131*, 105-115.
26. Basset, G. J. C.; Quinlivam, E. P.; Gregory, J. F.; Hanson, A. D. Folate synthesis and metabolism in plants and prospects for biofortification *Crop. Sci.* **2005**, *45*, 449-453.
27. Harwood, J. B. Recent advances in the biosynthesis of plant fatty-acids *Biochim. Biophys. Acta.* **1996**, *1301*, 7-56.
28. McFadden, G. I.; Roos, D. S. Apicomplexan plastids as drug targets *Trends in Microbiology* **1999**, *7*, 328-333.
29. Champoux, J. J. DNA topoisomerases: structure, function, and mechanism *Annu. Rev. Biochem.* **2001**, *70*, 369-413.
30. Divo, A. A.; Sartorelly, A. C.; Patton, C. L.; Frank, J. B. Activity of fluoroquinolone antibiotics against *Plasmodium falciparum in vitro Antibacterial Agents and Chemotherapy* **1988**, *32*, 1182-1186.
31. Wall, M. K.; Mitchenall, L. A.; Maxwell, A. Arabidopsis thaliana DNA gyrase is targeted to chloroplasts and mitochondria *Proc. Nat. Acad. Sci. U.S.* **2004**, *101*, 7821-7826.
32. Maxwell, A. DNA gyrase as a drug target *Trends in Microbiology* **1997**, *5*, 102-109.
33. Thompson, R. J.; Mosig, G. An ATP-dependent supercoiling topoisomerase of Chlamydomonas reinhardtii affects accumulation of specific chloroplast transcripts *Nucleic Acids Research* **1985**, *13*, 873–891.
34. Lyman, H.; Jupp, A. S.; Larrinua, I. Specific inhibition of chloroplast replication in *Euglen gracilis* by naladixic acid *J. Cell. Biol.* **1967**, *35*, 726-730.
35. Brain, R. A.; Johnson, D. J.; Richards, S. M.; Sanderson, H.; Sibley, P. K.; Solomon, K. R. Effects of 25 pharmaceutical compounds to *Lemna gibba* using a seven-day static renewal test *Environ. Toxicol. Chem.* **2004**, *23*, 371-382.
36. Robinson, A. A.; Belden, J. B.; Lydy, M. J. Toxicity of fluoroquinolone antibiotics to aquatic organisms *Environ. Toxicol. Chem.* **2005**, *24*, 423-430.
37. Woelfle, M. A.; Thompson, R. J.; Mosig, G. Roles of novobiocin-sensitive topoisomerases in chloroplast DNA replication in *Chlamydomonas reinhardtii Nucleic Acids Research* **1993**, *21*, 4231-4238.
38. Migliore, L.; Cozzolino, S.; Fiori, M. Phytotoxicity to and uptake of flumequine used in intensive aquaculture on the aquatic weed, *Lythrum salicaria* L. *Chemosphere* **2000**, *40*, 741-750.
39. Hashimoto, H.; Murakami, S. Chloroplast replication and loss of chloroplast DNA induced by nalidixic acid in *Euglena gracilis Cell Structure and Function* **1982**, *7*, 111-120.
40. Krajcovic, J.; Ebringer, L.; Polonyi, J. Quinolones and coumarins eliminate chloroplasts from Euglena gracilis *Antibacterial Agents and Chemotherapy* **1989**, *33*, 1883-1889.

41. Woods, K. M.; Nesterenko, M. V.; Upton, S. J. Efficacy of 101 antimicrobials and other agents on the development of *Cryptosporidium parvum in vitro Ann. Trop. Med. Parasitol.* **1996**, *90*, 603-615.
42. Wilson, B. A.; Smith, V. H.; Denoyelles, F.; Larive, C. K. Effects of three pharmaceutical and personal care products on natural freshwater algal assemblages *Environ. Sci. Technol.* **2003**, *37*, 1713-1719.
43. Richards, R. A.; Wilson, C. J.; Johnson, D. J.; Castle, D. M.; Lam, M. W.; Mabury, S. A.; Sibley, P. K.; Solomon, K. R. Effects of pharmaceutical mixtures in aquatic microcosms *Environ. Toxicol. Chem.* **2003**, *23*, 1035-1042.
44. Brain, R. A.; Johnson, D. J.; Richards, R. A.; Hanson, M. L.; Sanderson, H.; Lam, M. W.; Young, C.; Mabury, S. A.; Sibley, P. K.; Solomon, K. R. Microcosm evaluation of the effects of an eight pharmaceutical mixture to the aquatic macrophytes *Lemna gibba* and *Myriophyllum sibiricum Aquat. Toxicol.* **2004**, *70*, 23-40.
45. Manuell, A.; Beligni, M. V.; Yamaguchi, K.; Mayfield, S. P. Regulation of chloroplast translation: interactions of RNA elements, RNA-binding proteins and the plastid ribosome *Biochem. Soc. Trans.* **2004**, *23*, 601-605.
46. Giovannoni, S. J.; Turner, S.; Olsen, G. J.; Barns, S.; Lane, D. J.; Pace, N. R. Evolutionary relationships among cyanobacteria and green chloroplasts *Journal of Bacteriology* **1988**, *170*, 3584-3592.
47. Yamaguchi, K.; Beligni, M. V.; Prieto, S.; Haynes, P. A.; McDonald, W. H.; Yates, J. R.; Mayfield, S. P. Proteomic characterization of the *Chlamydomonas reinhardtii* chloroplast ribosome: Identification of proteins unique to the 70S ribosome *J. Biol. Chem.* **2003**, *278*, 33774-33785.
48. Yamaguchi, K.; Subramanian, A. R. The plastid ribosomal proteins: Identification of all the proteins in the 50S subunit of an organelle ribosome (chloroplast) *J. Biol. Chem.* **2000**, *275*, 28466–28482.
49. Yamaguchi, K.; Subramanian, A. R. Proteomic identification of all plastid-specific ribosomal proteins in higher plant chloroplast 30S ribosomal subunit PSRP-2 (U1A-type domains), PSRP-3a/b (ycf65 homologue) and PSRP-4 (Thx homologue) *Eur. J. Biochem.* **2003**, *270*, 190-205.
50. Harris, E. H.; Burkhart, B. D.; Gillham, N. W.; Boynton, J. E. Antibiotic resistance mutations in the chloroplast 16s and 23s rRNA genes of *Chlamydomonas reinhardtii*: Correlation of genetic and physical maps of the chloroplast genome *Genetics.* **1989**, *123*, 281-292.
51. Gray, M. W.; Lang, B. F. Transcription in chloroplasts and mitochondria: a tale of two polymerases *Trends in Microbiology* **1998**, *6*, 1-3.
52. Maliga, P. Two plastid RNA polymerases of higher plants: an evolving story *Trends in Plant Science* **1998**, *3*, 4-6.
53. Pfannschmidt, T.; Ogrzewalla, K.; Baginsky, S.; Sickmann, A.; Meyer, H. E.; Link, G. The multisubunit chloroplast RNA polymerase A from mustard (Sinapis alba L.) Integration of a prokaryotic core into a larger complex with organelle-specific functions *Eur. J. Biochem.* **2000**, *267*, 253-261.
54. Sweetman, S. C., Ed. *Martindale The complete drug reference*; 33 ed.; Pharmaceutical Press: London, UK, **2002**.

55. Nickell, L. G.; Gordon, P. N. Effects of the tetracycline antibiotics, their degradation products, derivatives and some synthetic analogs on the growth of certain plant systems *Antibacterial Agents Ann.* **1961**, *1960*, 588-593.
56. Pro, J.; Oritz, J. A.; Boleas, S.; Fernández, C.; Carbonell, G.; Tarazona, J. V. Effect assessment of Antimicrobial pharmaceuticals on the aquatic plant *Lemna minor Bull. Environ. Contam. Toxicol.* **2003**, *70*, 290-295.
57. Brain, R. A.; Wilson, C. J.; Johnson, D. J.; Sanderson, H.; Bestari, B.-J.; Hanson, M. L.; Sibley, P. K.; Solomon, K. R. Toxicity of a mixture of tetracyclines to *Lemna gibba* and *Myriophyllum sibiricum* evaluated in aquatic microcosms. *Environ. Pollut.* **2005**, *138*, 426-443.
58. Wilson, K. E.; Thompson, J. E.; Huner, N. P. A.; Greenberg, B. M. Effects of ultraviolet-A exposure on ultraviolet-B–induced accumulation of specific flavonoids in *Brassica napus Photochem. Photobiol.* **2001**, *73*, 678-684.
59. Halling-Sørensen, B. Algal toxicity of antibacterial agents used in intensive farming *Chemosphere* **2000**, *40*, 731-739.
60. Holten Lützhøft, H. C.; Halling-Sørensen, B.; Jørgensen, S. E. Algal toxicity of antibacterial agents applied in Danish fish farming *Arch. Environ. Contam. Toxicol.* **1999**, *36*, 1-6.
61. Isidori, M.; Lavorgna, M.; Nardelli, A.; Pascarella, L.; Parrella, A. Toxic and genotoxic evaluation of six antibiotics on non-target organisms *Sci. Tot. Environ.* **2005**, *346*, 87-98.
62. Wilson, C. J.; Brain, R. A.; Sanderson, H.; Johnson, D. J.; Bestari, K. T.; Sibley, P. K.; Solomon, K. R. Structural and functional responses of plankton to a mixture of four tetracyclines in aquatic microcosms *Environ. Sci. Technol.* **2004**, *38*, 6430-6439.
63. Brain, R. A. Phytotoxicity and hazard assessment of pharmaceuticals in aquatic higher plants In *Environmental Biology/Toxicology*; University of Guelph: Guelph, **2006**; Vol. Doctor of Philosophy, p 273.
64. Pomati, F.; Netting, A. G.; Calamari, D.; Neilan, B. A. Effects of erythromycin, tetracycline and ibuprofen on the growth of Synechocystis sp. and *Lemna minor Aquat. Toxicol.* **2004**, *67*, 387-396.
65. Brain, R. A.; Bestari, K. T.; Sanderson, H.; Hanson, M. L.; Wilson, C. J.; Johnson, D. J.; Sibley, P. K.; Solomon, K. R. Aquatic microcosm assessment of the effects of tylosin on *Lemna gibba* and *Myriophyllum spicatum Environ. Pollut.* **2005**, *133*, 389-401.
66. Pollock, K.; Barfield, D. G.; Shields, R. The toxicity of antibiotics to plant cell cultures *Plant Cell Reports* **1983**, *2*, 36-39.
67. Montandon, P.-E.; Nicolas, P.; Schürmann, P.; Stutz, E. Streptomycin-resistance of *Euglena gracilis* chloroplasts: identificatio of a point mutation in the 16S rRNA gen in an invariant position *Nucleic Acids Research* **1985**, *13*, 4299-4310.
68. Högenauer, G.; Ruf, C. Ribosomal binding region for the antibiotic tiamulin: stoichiometry, subunit location, and affinity for various analogs *Antimicrob. Agents. Chemother.* **1981**, *19*, 260-265.
69. Hanson, A. D.; Roje, S. One-carbon metabolism in higher plants *Annu. Rev. Plant. Physiol.* **2001**, *52*, 119–137.
70. Rébeillé, F. D.; Macherel, D.; Mouillon, J. M.; Garin, J.; Douce, R. Folate biosynthesis in higher plants: Purification and molecular cloning of a

bifunctional 6-hydroxymethyl-7,8-dihydropterin pyrophosphokinase/7,8-dihydropteroate synthase localized in mitochondria *EMBO reports* **1997**, *16*, 947-957.

71. Brooks, D. R.; Wang, P.; Read, M.; Watkins, W. M.; Sims, P. F. G.; Hyde, J. E. Sequence variation of the hydroxymethyldihydropterin pyrophosphkinase; dihydropteroate synthase gene in lines of the human malaria parasite, *Plasmodium falciparum*, with differing resistance to sulfadoxine *Eur. J. Biochem.* **1994**, *224*, 397-405.
72. Wang, P.; Read, M.; Sims, P. F. G.; Hyde, J. E. Sulfadoxine resistance in the human malaria parasite *Plasmodium falciparum* is determined by mutations in dihydropteroate synthetase and an additional factor associated with folate utilization *Molecular Microbiology* **1997**, *23*, 979-986.
73. Khalil, I.; Rønn, A. M.; Alifrangis, M.; Gabar, H. A.; Satti, G. M. H.; Bygbjerg, I. C. Dihydrofolate reductase and dihydropteroate synthase genotypes associated with in vitro resistance of Plasmodium falciparum to pyrimethamine, trimethoprim, sulfadoxine, and sulfamethoxazole *Am. J. Trop. Med. Hyg.* **2003**, *68*, 586-589.
74. Iyer, J. K.; Milhous, W. K.; Cortese, J. F.; Kublin, J. G.; Plowe, C. V. Plasmodium falciparum crossresistance between trimethoprim and pyrimethamine *The Lancet* **2001**, *358*, 1066-1067.
75. Ferrari, B.; Mons, R.; Vollat, B.; Fraysse, B.; Paxéus, N.; Giudice, R. L.; Pollio, A.; Garric, J. Environmental risk assessment of six human pharmaceuticals: Are the current environmental risk assessment procedures sufficient for the protection of the aquatic environment? *Environ. Toxicol. Chem.* **2004**, *23*, 1344-1354.
76. Waller, R. F.; Ralph, S. A.; Reed, M. B.; Su, V.; Douglas, J. D.; Minnikin, D. E.; Cowman, A. F.; Besra, G. S.; McFadden, G. I. A type II pathway for fatty acid biosynthesis presents drug targets in *Plasmodium falciparum* *Antimicrob. Agents. Chemother.* **2003**, *47*, 297-301.
77. Ryall, K.; Harper, J. T.; Keeling, P. J. Plastid-derived Type II fatty acid biosynthetic enzymes in chromists *Gene.* **2003**, *313*, 139-148.
78. Heath, R. J.; Rubin, J. R.; Holland, D. R.; Zhang, E.; Snow, M. E.; Rock, C. O. Mechanism of triclosan inhibition of bacterial fatty acid synthesis *J. Biol. Chem.* **1999**, *274*, 11110-11114.
79. Orvos, D. R.; Versteeg, D. J.; Inauen, J.; Capdevielle, M.; Rothenstein, A.; Cunningham, V. Aquatic toxicity of triclosan *Environ. Toxicol. Chem.* **2002**, *21*, 1338-1349.
80. McLeod, R.; Muench, S. P.; Rafferty, J. B.; Kyle, D. E.; Mui, E. J.; Kirisits, M. J.; Mack, D. G.; Roberts, C. W.; Samuel, B. U.; Lyons, R. E.; Dorris, M.; Milhous, M. K.; Rice, D. W. Triclosan inhibits the growth of Plasmodium falciparum and Toxoplasma gondii by inhibition of Apicomplexan Fab I *Int. J. Parasitol.* **2001**, *31*, 109-113.
81. Furukawa, H.; Tsay, J.-T.; Jackowski, S.; Takamura, Y.; Rock, C. O. Thiolactomycin resistance in *Escherichia coli* is associated with the multidrug resistance efflux pump Encoded by *emrAB* *Journal of Bacteriology* **1993**, *175*, 3723-3729.

82. Kato, M.; Nishida, I.; Kawaguchi, A.; Yamada, M. Effect of thiolactomycin on fatty acid synthesis in higher plants *Plant. Cell. Physiol.* **1987**, *28*, 851-855.
83. Leganés, F.; Blanco-Rivero, A.; Fernández-Pinas, F.; Redondo, M.; Fernández-Valiente, E.; Fan, Q.; Lechno-Yossef, S.; Wolk, C. P. Wide variation in the cyanobacterial complement of presumptive penicillin-binding proteins *Archives of Microbiology* **2005**, *184*, 234-248.
84. Machida, M.; Takechi, K.; Sato, H.; Chung, S. J.; Kuroiwa, H.; Takio, S.; Seki, M.; Shinozaki, K.; Fujita, T.; Hasebe, M.; Takano, H. Genes for the peptidoglycan synthesis pathway are essential for chloroplast division in moss *Proc. Nat. Acad. Sci. U.S.* **2006**, *103*, 6753-6758.
85. Hillis, D. G.; Lissemore, L.; Sibley, P. K.; Solomon, K. R. Effects of monensin on zooplankton communities in aquatic microcosms *Environ. Sci. Technol.* **2007**, *Submitted 2007 03 10.*
86. Fairchild, J. F.; Ruessler, D. S.; Haverland, P. S.; Carlson, A. R. Comparative sensitivity of *Selenastrum capricornutum* and *Lemna minor* to sixteen herbicides *Arch. Environ. Contam. Toxicol.* **1997**, *32*, 353-357.
87. Fairchild, J. F.; Ruessler, D. S.; Carlson, A. R. Comparative sensitivity of five species of macrophytes and six species of algae to atrazine, metribuzin, alachlor, and metolachlor *Environ. Toxicol. Chem.* **1998**, *17*, 1830-1834.
88. Mohammad, M.; Kishimoto, T.; Itoh, K.; Suyama, K.; Yamamoto, H. Comparative sensitivity of *Pseudokirchnetiella subcapitata* vs. *Lemna* sp. to eight sulfonylurea herbicides *Bull. Environ. Contam. Toxicol.* **2005**, *75*, 866-872.
89. Shimabukuro, R. H. Herbicide Physiology In *Detoxification of herbicides, in Weed Physiology*; CRC Press: Boca Raton, **1985**; Vol. 2, pp 215-240.
90. Paquette, S. M.; Bak, S.; Feyereisen, R. Intron–exon organization and phylogeny in a large superfamily, the paralogous cytochrome P450 genes of *Arabidopsis thaliana DNA and Cell Biology* **2000**, *19*, 307-317.
91. The-Arabidopsis-Genome-Initiative Analysis of the genome sequence of the fowering plant *Arabidopsis thaliana Nature* **2000**, *408*, 796-815.
92. Nelson, D. R.; Schuler, M. A.; Paquette, S. M.; Werck-Reichhart, D.; Bak, S. Comparative genomics of rice and *Arabidopsis*. Analysis of 727 cytochrome P450 genes and pseudogenes from a monocot and a dicot *Plant. Physiol.* **2004**, *135*, 756-772.
93. Kreuz, K.; Tommasini, R.; Martinoia, E. Old enzymes for a new job *Plant. Physiol.* **1996**, *111*, 349-353.
94. USFDA "Guidance for industry Environmental impact assessments (EIA's) for veterinary Medicinal products (VMP's)-Phase 1," U.S. Department of Health and Health Services, Food and Drug Administration, Center for Veterinary Medicine, **2001**.

Chapter 13

Risks of agricultural pharmaceuticals in surface water systems and soils

Keith R. Solomon[1], Derek G. Hillis[1], Linda Lissemore[2], Paul K. Sibley[1]

[1]Centre for Toxicology and Department of Environmental Biology, University of Guelph, Guelph, ON, Canada.
[2] Laboratory Services Division, University of Guelph, Guelph, ON, Canada.

Geographic and temporal distributions of agricultural pharmaceuticals measured in surface water at several locations in Grand River watershed in Southern Ontario, Canada showed 5 pharmaceuticals with pulses occurring between May and November at similar but varying times. Compared to species sensitivity distributions of acute toxicity values, pharmaceuticals detected in the surface waters presented small risks to aquatic organisms from acute effects. Effects on population and community responses in microcosms support the observation of low risk to the environment. In vitro bioassays of twelve pharmaceuticals on the arbuscular mycorrhizal fungi, *Glomus intraradices* grown on *Daucus carota* root organ cultures showed that, doxycycline, 17-α-ethinyl estradiol, and carbamazepine were selectively toxic to *G. intraradices* with 28-d EC50s less than 100 μg/L. Risks to plants and mycorrhizal fungi from estimated environmental concentrations were small.

Introduction

Increasing research and attention to human- and agricultural-use pharmaceuticals in the last decade have highlighted the extensive use of these products and their presence in the environment, particularly in surface waters, sediments and soils. Estimates of total pharmaceutical use are not readily available in many countries, especially North America and, as a result, it can be difficult to determine exactly what is being used and in what quantities. Some usage data for human prescription pharmaceuticals are available as sales information; however, this does not include livestock production or veterinary use. The United States and individual EU member states report primarily on the sales of antibacterial agents. Similar information is not readily available in Canada. Data on the sale and usage of veterinary medicines from Europe and the United States suggest that antimicrobial substances represent the highest amounts of pharmaceuticals used in livestock production with approximately half of the 22,700 tonnes of antibiotics produced annually in the US used in agriculture (*1,2*). The estimate for agricultural use in the United Kingdom for 2000 was 897 tonnes (*3*).

Since early reports drew attention to the issue of pharmaceuticals in the environment (*4*), a number of studies have reported concentrations in effluents from sewage treatment plants (STPs) (*5,6*) and surface waters (*7-9*). The application of manure and/or biosolids to agricultural fields, a recommended component of sustainable agriculture today, represents a potential non-point source for inputs of pharmaceuticals into aquatic environments (*10*). Although pharmaceuticals have been observed to degrade in manure (*11*), movement into the soil is also reported (*12*). Residues of chlortetracycline, tetracycline, oxytetracycline, and tylosin decreased to approx 1 μg/kg at 60 cm depth from 9-12 μg/kg in the upper 10 cm of soil in fields amended with manure slurry (*11*). Chlortetracycline, oxytetracycline, tetracycline, and tylosin were also reported at concentrations at the limit of detection (LOD) in the water samples near fields where manure had been applied (*13*). Agricultural-use antibiotics were reported from watersheds in Ontario (*14*) and temporal sampling over one season revealed the presence of several antibiotics and pharmaceuticals with large frequencies of detection and relatively large concentration ranges (Figure 1) in a subwatershed of the Grand River where intensive production of animals is practiced (data from *10*). While livestock production clearly represents a primary input to agricultural areas, contributions of human pharmaceuticals from application of biosolids to agricultural lands and seepage from septic systems cannot be disregarded. Leaching from biosolids spread on agricultural lands was suggested as the most likely source of carbamazepine in a portion of the Grand River watershed that received no inputs from STPs (*10*).

This paper presents an overview of environmental effects of pharmaceuticals in water systems and soil. Although the primary focus is on pharmaceuticals used in agriculture, studies on human-use products are also included. While we concentrate mainly on approaches and methods for assessing environmental effects and risks from agricultural pharmaceuticals, these principles apply to pharmaceuticals in general.

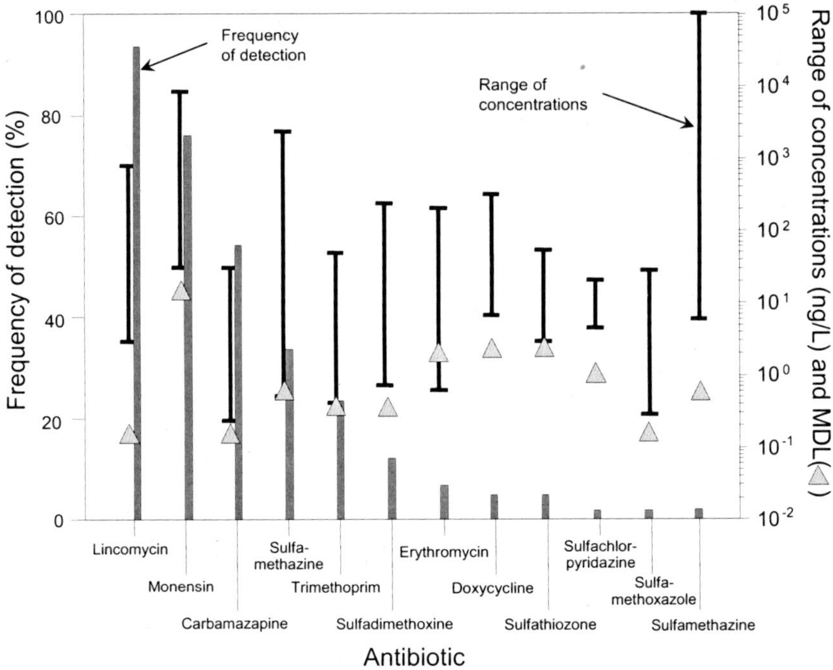

Figure 1. Frequency and range of concentrations of pharmaceuticals detected in surface waters a subwatershed of the Grand River, Ontario, Canada April 2003 and 2004 (data from 10).

Problem formulation and special considerations for pharmaceuticals

Surface water systems

As with all substances in the environment (*15,16*), assessing hazards and risks from pharmaceuticals requires knowledge of exposure as well as effects. Characterization of exposure is made more complex because pharmaceuticals from human sources (STPs) are frequently present as mixtures. These mixtures may change seasonally in response to pharmaceuticals used to treat seasonal infections but in the shorter term, consistent mixtures of components are a reality. For substances with a common mechanism of action, additivity is likely (*17,18*) but synergism and antagonism cannot be ruled out. Furthermore, continuous releases of pharmaceuticals result in chronic exposures, even for those substances that are not highly persistent in the environment. This pseudopersistence must be considered in developing tools to characterize effects.

Another issue is that, with a few exceptions such as parasiticides (*19*), pharmaceuticals are not designed to be highly acutely toxic to the organism in which they are used and therefore are usually not acutely lethal to non-target animals, although plants may be affected (*20*). The effects of pharmaceuticals may not be easily observed unless they produce clear biomarker responses, such as have been observed in fish exposed to estradiol, estrone, and ethinyl estradiol from STPs (*21*). In this latter case, lethality was not a suitable endpoint – reproduction and population structure were the most likely to be affected. Similarly, effects of diclofenac on vultures were not acutely lethal (*22*), although serious in the end. As a result, the environmental effects and subtle changes associated with pharmaceuticals in aquatic ecosystems remain relatively unknown.

In an attempt to address the issue of effects of mixtures (and single substances), we have used aquatic microcosms (12,000 L outdoor pools) as a surrogate for an ecosystem(*23-26*). Because microcosms contain many species in several trophic levels, most of which interact in a food web, subtle and non-lethal responses in one or more groups of organisms may be observed as changes in the structure (diversity and abundance of organisms) of the community. If a range of concentrations is tested in these systems, it is possible to use concentration-dependent changes in structure to identify causal relationships for further investigation. In a sense, these systems are hypothesis generating in terms of subtle effects (acute or chronic) but also have the additional advantage that they are also exposed to environmental factors, such as sunlight, thus providing the tools to assess fate processes, such as photolysis under more realistic conditions.

Chronic exposures are difficult to attain in large-volume microcosms because of the amounts of water that would be required to mimic a flowing system such as a river or stream. Chronic exposures may, however, be achieved through regular addition of the test substance as long as these amounts are small in relation to the total volume of the system (*23-27*).

If the proportions of the components of environmentally-relevant mixtures are known from measurements and the probability of occurrence can be estimated, these values may be used to extrapolate to greater concentrations than occur in the environment; responses to which may be used to determine margins of exposure and to identify possible biomarkers for use in field assessments (Figure 2). When these concentrations are based on distributions of values, probability of co-occurrence can be estimated. For example, if a mixture of chemicals is highly correlated and they always occur together in a fixed proportion, then a combination of these at the 99th centile will have a 1% probability of exceedence. If, at the other extreme, the components of the mixture are not correlated at all, the probability of co-occurrence is the product of the individual probabilities. Thus, for a three-component mixture at the 99th centile, the product of the individual exceedences (1% x 1% x 1%) is 0.0001%. In reality, as no mixture will always be completely correlated and complete independence also will not occur, the exceedence probability will be between these two values with the greatest (1% in this example) being the worst case.

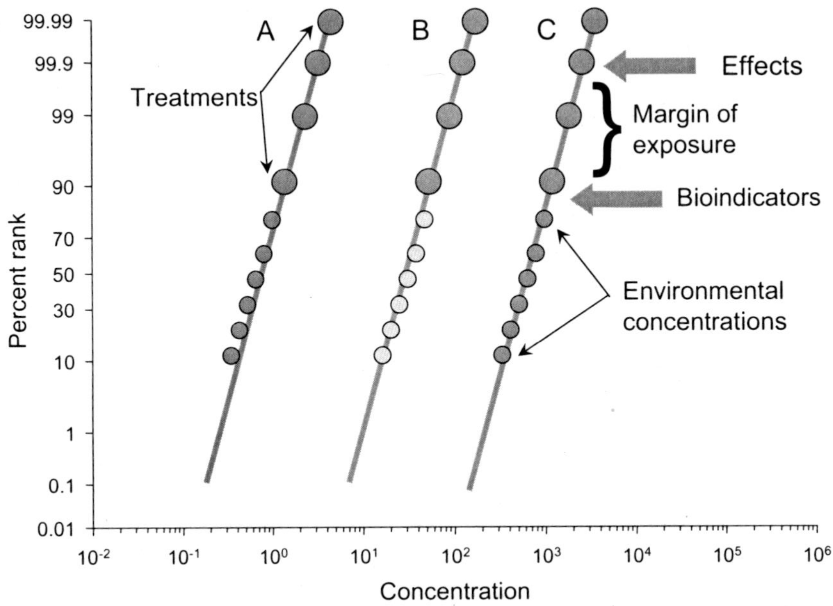

Figure 2. Illustration of the concept of testing mixtures of pharmaceuticals (A, B, and C) combined on the basis of cumulative distributions of measured or modeled values to allow the selection of treatment concentrations for microcosm studies.

Soils

Veterinary pharmaceuticals enter terrestrial systems through the common practice of amending soils with manure from intensive livestock operations or directly from grazing and free-ranging livestock. Human pharmaceuticals and personal care products may enter soils through the application of biosolids collected from the grit chamber, settling tanks, and waste activated sludge of municipal wastewater treatment systems. The application of pharmaceutical-containing manure and biosolids may lead to direct phytotoxic responses, resulting in a decrease in agricultural crop protection. Of equal importance is the indirect effect pharmaceuticals may have on soil health and productivity through effects on beneficial microbial and symbiotic processes that contribute to plant growth and yield. For example, most plants, including the majority of crop species, form a beneficial symbiosis with arbuscular mycorrhizal fungi (AMF), which improves plant nutrient uptake, tolerance to drought, pathogenic and toxic stressors as well as improved seedling establishment and soil stability (*28*). In addition, the below-ground mycorrhizal community has been shown to influence the plant community structure in terms of dominance and growth (*29*). Preferential feeding on AMF by soil invertebrates, such as Collembola, also highlights this plant symbiont as a key trophic level in soil food webs (*30*). Despite the fact that these specialized fungi provide ecological functions that are not duplicated by any other organism, within the fungal or any other kingdom, they have been largely ignored in the evaluation of effects caused by the introduction of a chemical stressor to an agro-ecosystem.

Standardized plant growth and emergence experiments (*31,32*) were used to evaluate the impact of ten common pharmaceuticals and personal care products (PCPPs) directly on growth of three economically important agricultural crop species. A new technique using root organ cultures to host AMF in culture (*33*) was also used to evaluate the impact of 12 different PCPPs on root growth of carrot (*Daucus carota*) as well as growth and propagule endpoints for the AMF, *Glomus intraradices*.

Responses to pharmaceuticals in the environment

Surface water systems

A number of studies have reported on the acute toxicity of agricultural antibiotics and pharmaceuticals to aquatic organisms. In laboratory studies, the most sensitive non-target organisms to antibiotics are generally plants (*34*). Of the antibiotic classes tested, only members of the fluoroquinolone, sulfonamide, and tetracycline classes of antibiotics displayed significant phytotoxicity. The most toxic members of each of these classes tested were lomefloxacin, sulfamethoxazole, and chlortetracycline, with wet weight EC25 values of 38, 37, and 114 mg/L, respectively (*34*). These responses are most likely as a result of the evolutionary relationship between the chloroplasts and other key functions found in modern plants (the chloroplast, DNA gyrases, etc.) and similar processes in bacteria and the blue-green algae (*20,34,35*). Similar observations have been made in microcosms. For example, the antibiotic growth promoter, tylosin, promoted growth of the aquatic macrophyte, *Lemna gibba* in laboratory bioassays with a lowest observed effect concentration (LOEC) of 300 μg/L whereas, in the field, no effects were observed in this species at concentrations up to 3000 μg/L from a single treatment. However, exposures of the aquatic macrophyte *Myriophyllum spicatum* demonstrated a LOEC of 3000 μg/L in the microcosms. No effects were observed at higher trophic levels in microcosms treated with tylosin at concentrations up to 3,000 μg/L. Monensin, an ionophore antibiotic widely used in the poultry and beef industry was of low toxicity to plants (*34*). Effects on zooplankton community structure and population dynamics were evaluated in microcosms after a single treatment at concentrations ranging from 0.5 to 500 μg/L. Monensin did not significantly affect community structure within trophospecies (Rotifera, Cladocera, Copepoda adults, Copepoda nauplii, Ostracoda, and macroinvertebrates). However, significant changes within trophospecies groups were observed with decreases in the abundance of Rotifera and Copepoda nauplii and in the richness of Rotifera and Cladocera (*36*). A concentration-dependent increase in Ostracoda abundance was also observed.

Studies with mixtures of tetracyclines (oxytetracycline, chlortetracycline, tetracycline, and doxycycline) at total concentrations of 0.08, 0.22, 0.67, and 2.29 μM over a period of 35 days in microcosms showed no direct effects on macrophyte growth, although they were toxic to *L. gibba* at EC50 concentrations ranging from 219 to 1010 µg/L when tested individually under laboratory conditions (because of differences in molecular weight, concentrations of mixtures of different pharmaceuticals have been expressed in molar units). In microcosms, a concentration-dependent decrease in the growth of *M. spicatum,* but not *L. gibba,* was observed at all treatment concentrations tested. This was ascribed to the formation of colored breakdown products in the microcosms which then reduced the penetration of light into the water column, thus reducing rates of photosynthesis and, as a result, growth (*25*). As *L. gibba* is a floating macrophyte with leaves above the surface, it was unaffected. Depths and volumes of water used in test systems with *M. spicatum* in the laboratory were such that any formation of colored compounds would not interfere significantly with availability of light (*25*). Effects on some phytoplankton endpoints were observed in the 0.22 µM and greater treatment concentrations. The largest responses were concentration-dependent reductions in total phytoplankton abundance and species richness (*26*). Abundance of phytoplankton recovered to control levels in all microcosms after treatment was terminated, and resilience (return to normal operating range before the removal of the stressor) was observed with respect to species richness of phytoplankton. Despite the effects on phytoplankton and rooted macrophytes and the potential for foodchain-driven interactions, zooplankton were generally unaffected by treatment with the tetracyclines. No-observed-effect-concentrations observed in microcosms were 40 to 100-fold greater than maximum concentrations measured in the environment (*26*).

Treatments of microcosms with more complex mixtures of pharmaceuticals have not shown effects that could not be explained by the action of one or more components of the mixture. Treatment of microcosms with a mixture of ibuprofen (a nonsteroidal anti-inflammatory drug), fluoxetine (a selective serotonin reuptake inhibitor), and ciprofloxacin (a DNA gyrase–inhibiting antibiotic) in the ratio of 6:10:10 at total concentrations of 0.09, 0.88, and 8.8 µM for 35 days resulted in few responses at 0.09 µM (*23*). However, mortality of fish occurred within the 35 d period at 0.88 µM and within 4 d at 8.8 µM. Fish mortality was attributed to fluoxetine and occurred at concentrations below therapeutic values for humans, suggesting enhanced sensitivity in aquatic vertebrates. Phytoplankton increased in abundance and decreased in diversity (number of taxa) at 8.8 µM, with consistent trends at 0.88 and 0.09 µM. Zooplankton increased in abundance but decreased in diversity at 8.8 µM with a similar trend at 0.99 µM, suggesting an interaction between phyto- and zoo-plankton. *L. gibba* and *Myriophyllum spp*. showed mortality at 8.8 µM with reduced growth of *L. gibba* at 0.88 µM. Effects on these macrophytes were attributed to sensitivity to ciprofloxacin, a response confirmed in laboratory studies (*34*). Treatment of microcosms for 35 d with a mixture of atorvastatin, acetaminophen, caffeine, sulfamethoxazole, carbamazepine, levofloxacin, sertraline, and trimethoprim) at total molar concentrations of 0, 0.044, 0.608, 2.66, and 24.5 µmol/L showed concentration-dependent effects on growth of *L. gibba* and *M. spicatum*, with EC25s of 0.5 and 0.6 µmol/L, respectively. Effects on plants were attributed to the combined response to levofloxacin (a DNA-gyrase inhibitor) and atorvastatin a lipid-reducing drug used in humans that also inhibits an analogous target in plants, 3-hydroxy-3-methylglutaryl coenzyme-A reductase (HMGR) (*37*). These effects were observed at concentrations 10 to 100-fold greater than measured or predicted environmental concentrations.

A commonality to all of the laboratory and microcosm studies with agricultural antibiotics was that effects of individual compounds and mixtures of these were only observed at concentrations 10 to 100-fold larger than the greatest observed or estimated environmental concentration. Based on laboratory test data with individual pharmaceuticals, synergistic or antagonistic interactions were not observed with mixtures in the microcosms; however, the studies were not designed to specifically address this and low levels of interaction would not have been noticed.

Soils

Exposure of carrot (*D. carota*), alfalfa (*Medicago sativa*) and lettuce (*Lactuca sativa*) to ten antibiotics were found to elicit phytotoxic responses to seed germination, root, shoot and total growth, in vitro, typically at concentrations greater than 1,000 µg/L. Exceptions were the sulfonamide, fluoroquinolone and tetracycline classes of antibiotics where the results varied considerably between 10-1,000 µg/L dependent upon the tested plant species (*38*). These results are not unexpected as germination and radical growth are a highly conserved mechanism, with much of the carbohydrates, lipids, and nutrients needed for initial growth provided by seed reserves and mechanisms of action related to processes such as photosynthesis would have little effect. For example, the sulfonamide class of pharmaceuticals inhibits plant folate synthesis in a mechanism similar to that which causes its antibacterial activity (*39,40*). Folates are essential cofactors in one-carbon transfer reactions for all organisms. Initial folate concentrations in seeds have been shown to support root elongation for the initial growth period (*41*), indicating that longer duration plant-based experiments are required to evaluate effects. Plant emergence experiments in soil and root organ experiments in culture were conducted extending from 4 to 8 weeks. The majority of the tested compounds did not result in a decrease of any measured plant growth endpoint at exposure concentrations less than 1,000 µg/L. The most phytotoxic pharmaceutical observed was the sulfonamide antibacterial, sulfamethoxazole, with significant decreases in root growth observed at 10 µg/L. Exposure to atorvastatin, levofloxacin and chlortetracycline also resulted in significant decreases in plant growth. Measured environmental concentrations of these compounds in biosolids or manure, however, are typically an order of magnitude less than concentrations where phytotoxicity was observed. A lack of plant response to pharmaceuticals at environmentally relevant concentrations is consistent with the literature (*42-44*).

Concurrent to the root organ culture phytotoxicity tests, the mycorrhizal endpoints of hyphal growth and spore production were measured using the AMF, *G. intraradices,* in root-organ culture (*45*). In general, if effects were observed on the plant root organ culture-AMF system, three characteristic plant-microbe responses were noted. Some pharmaceuticals resulted in primarily phytotoxic responses such as the previously described sulfamethoxazole and atorvastatin. Mycorrhizal fungi are obligate symbionts with the plant host and consequently decreases in hyphal growth and spore production were observed following the phytotoxic response. Other pharmaceuticals elicited a narcotic or generalized toxicity where the plant and mycorrhizal fungi were impacted at a similar rate of toxicity. This response type, which is characterized by parallel reductions in the measured endpoints, occurred with the fluoroquinolone, levofloxacin, and chlortetracycline. Of greater interest, were the few pharmaceutical compounds which resulted in little or no phytotoxicity but resulted in significant reductions in mycorrhizal growth. These compounds were the antibacterial, doxycycline, the synthetic estrogen, 17-α-ethinyl estradiol and the anti-epileptic, carbamazepine (Figure 3). As was reported with the evaluation of the potential for phytotoxicity due to pharmaceuticals in soil, the concentrations required to elicit a negative response in AMF were in general, much greater than that measured in manure and biosolids. The importance of the observed negative effects to beneficial microbes without showing evidence of effect through an immediate phytotoxic response should, however, be further considered. The vast majority of pharmaceuticals are antibiotics and effects of these compounds on bacterially regulated soil mineralization processes as well as the symbiotic relationships between legumes and rhizobacteria still require examination.

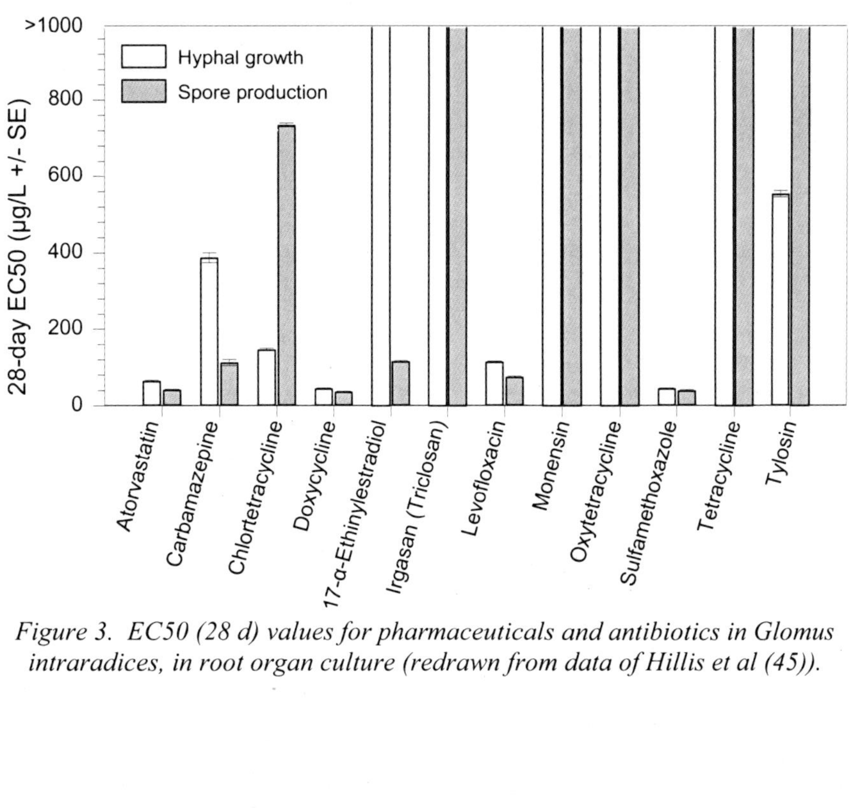

Figure 3. EC50 (28 d) values for pharmaceuticals and antibiotics in Glomus intraradices, in root organ culture (redrawn from data of Hillis et al (45)).

General conclusions

Some pharmaceuticals have been observed to have adverse effects in the environment; however, these are few in number. Natural and synthetic estrogens have been observed to have effects in surface waters with a high proportion of inputs from STPs (*21*) and the veterinary drug diclofenac has been shown to be the cause of decline of Indian vultures (*22*). Other pharmaceuticals, such as the selective serotonin reuptake inhibitors have been shown to be toxic to algae (*17*), affect behavior in pelagic animals (*46*), and reproduction (*18*) but the relevance of this at the population level or in the field is still uncertain. The antiparisiticide, avermectin, (also used an insecticide) has been demonstrated to cause changes in ecosystem structure and function at concentrations ≥30 ng/L, below the predicted environmental concentration (*19*), suggesting the potential for adverse effects, however, the current paucity of measures of environmental concentrations prevents the use of a full environmental risk assessment.

As discussed above, although there are more data on ecological effects for antibiotics used in agriculture, not all antibiotics have been tested at this time; only those used in large amounts or detected frequently in the environment have been characterized. Based on responses to acute bioassays in the laboratory and currently reported concentrations from surface waters, there is a low acute risk of adverse environmental effects for individual antibiotics. At large concentrations of mixtures of antibiotics, significant acute and chronic effects have been observed in microcosms at all levels of biological organization. At this time, these responses were observed at concentrations 10 to 100-fold above currently measured environmental concentrations and risks from exposures to individual substances and mixtures of these appear to be low (*10*). Although there are few data for sediment-dwelling organisms, responses appear to occur at concentrations greater than those measured in the environment (*47*), consistent with observations for water-dwelling organisms. Overall, risks from agricultural use of antibiotics and pharmaceuticals appear to be small.

Other considerations which influence the potential risk of PCPPs to soil systems is that reported biosolid and manure concentrations are often reported before aging or anaerobic digestion of human and animal waste, which typically results in further reductions of the active pharmaceutical parent compound. In addition, the dilution of the pharmaceutical concentration which occurs from the application of manure and biosolids across an agricultural landscape should also be considered. There is a paucity of measured concentration data in the literature for pharmaceuticals in soil following amendment with biosolids or manure.

Acknowledgements

The authors gratefully acknowledge funding from Livestock Environment Initiative - Agriculture & Agri-Food Canada, Canada Canadian Pork Producers, Canadian Network of Toxicology Centres, RX&D, NSERC, Beef Cattle Research Council, the Ontario Ministry of the Environment (BIS Program), and Elanco. Graduate students Richard Brain, David Johnson, Amanda Warne,

Christian Wilson, Eve Dussault, Andrea Wojtyniak, and Erin McGreggor also contributed to aspects of the work as did Drs. Sean Richards, Hans Sanderson, Jim Bestari, Mark Hanson, Scott Mabury, and John Klironomos.

References

1. Montforts, M. H. M. M.; Kalf, D. F.; van Vlaardingen, P. L. A.; Linders, J. B. H. J. The exposure assessment for veterinary medicinal products *Sci. Tot. Environ.* **1999**, *225*, 119-133.
2. Meyer, M. T.; Bumgarner, J. E.; Daughtridge, C.; Kolpin, D.; Thurman, E. M.; Hostetler, K. A. Occurrence of antibiotics in liquid waste at confined animal feeding operations and in surface and groundwater In *AFO Technical Meeting*: Fort Collins, CO, USA, **1999**.
3. Thiele-Bruhn, S.; Aust, M. O. Effects of pig slurry on the sorption of sulfonamide antibiotics in soil *Arch. Environ. Contam. Toxicol.* **2004**, *47*, 31-39.
4. Daughton, C. G.; Ternes, T. A. Pharmaceuticals and personal care products in the environment: agents of subtle change? *Environ. Health Perspect.* **1999**, *107*, 907-938.
5. Metcalfe, C. D.; Koenig, B. G.; Bennie, D. T.; Servos, M.; Ternes, T. A.; Hirsch, R. Occurrence of neutral and acidic drugs in the effluents of Canadian sewage treatment plants *Environ. Toxicol. Chem.* **2003**, *22*, 2872–2880.
6. Miao, X.-S. Occurrence of antimicrobials in the final effluents of wastewater treatment plants in Canada *Environ. Sci. Technol.* **2004**, *38*, 3533-3541.
7. Kolpin, D. W.; Furlong, E. T.; Meyer, M. T.; Thurman, E. M.; Zaugg, S. D.; Barber, L. B.; Buxton, H. T. Pharmaceuticals, hormones, and other organic wastewater contaminants in United States streams, 1999-2000: A national reconnaissance *Environ. Sci. Technol.* **2002**, *36*, 1202-1211.
8. Metcalfe, C.; Koenig, B. "A survey of levels of ibuprofen in surface water in the lower Great Lakes," Water Quality Centre, Trent University, **2001**.
9. Metcalfe, C. D.; Miao, X.-S.; Koenig, B. G.; Struger, J. Distribution of acidic and neutral drugs in surface waters near sewage treatment plants in the lower Great Lakes, Canada *Environ. Toxicol. Chem.* **2003**, *22*, 2881-2889.
10. Lissemore, L.; Hao, C.; Yang, P.; Sibley, P. K.; Mabury, S. A.; Solomon, K. R. An exposure assessment for selected pharmaceuticals within a model watershed in Southern Ontario *Chemosphere* **2006**, *64*, 717-729.
11. Ingerslev, F.; Halling-Sorensen, B. Biodegradability of metronidazole, olaquindox and tylosin and formation of tylosin degradation products in aerobic soil-manure slurries *Ecotoxicol. Environ. Safety* **2001**, *48*, 311-320.
12. Carlson, J. C.; Mabury, S. A. Dissipation kinetics and mobility of chlortetracycline, tylosin, and monensin in an agricultural soil in Northumberland county, Ontario, Canada *Environ. Toxicol. Chem.* **2006**, *25*, 1-10.

13. Hamscher, G.; Sczesny, S.; Hoper, H.; Nau, H. Determination of persistant tetracycline residues in soil fertilized with liquid manure by high-performance liquid chromatography with electrospray ionization tandem mass spectrometry *Anal. Chem.* **2002**, *74*, 1509-1518.
14. Hao, C.; Lissemore, L.; Nguyen, B.; Yang, P.; Solomon, K. R. Detection and identification of pharmaceuticals in agricultural surface waters by liquid chromatography coupled to electrospray ionization tandem mass spectrometry *Anal. Bioanal. Chem.* **2006**, *384*, 505–513.
15. USEPA "Framework for ecological risk assessment," United States Environmental Protection Agency, **1992**.
16. USEPA "Guidelines for Ecological Risk Assessment," United States Environmental Protection Agency, **1998**.
17. Johnson, D. J.; Sanderson, H.; Brain, R. A.; Wilson, C. J.; Solomon, K. R. Toxicity and hazard of selective serotonin reuptake inhibitor antidepressants fluoxetine, fluvoxamine, and sertraline to algae *Ecotoxicol. Environ. Safety* **2007**, *67*, 128-139.
18. Brooks, B. W.; Foran, C. M.; Richards, S. M.; Weston, J.; Turner, P. K.; Stanley, J. K.; Solomon, K. R.; Slattery, M.; La Point, T. W. Aquatic toxicity of fluoxitine *Toxicol. Lett.* **2003**, *142*, 169-183.
19. Sanderson, H.; Laird, B.; Pope, L.; Brain, R. A.; Wilson, C.; Johnson, D.; Bryning, G.; Peregrine, A. S.; Boxall, A.; Solomon, K. R. Assessment of the environmental fate and effects of ivermectin in aquatic mesocosms *Aquat. Toxicol.* **2007**, *85*, 229-240.
20. Brain, R. A.; Hanson, M. L.; Solomon, K. R.; Brooks, B. W. Aquatic plants exposed to pharmaceuticals: Effects and risks *Rev. Environ. Contam. Toxicol.* **2007**, *192*, 67–115.
21. Sumpter, J. P.; Johnson, A. C. Lessons from endocrine disruption and their application to other issues concerning trace organics in the aquatic environment *Environ. Sci. Technol.* **2005**, *39*, 4321-4332.
22. Oaks, J. L.; Gilbert, M.; Virani, M. Z.; Watson, R. T.; Meteyer, C. U.; Rideout, B. A.; Shivaprasad, H. L.; Ahmed, S.; Chaudhry, M. J. I.; Arshad, M.; Mahmood, S.; Ali, A.; Khan, A. A. Diclofenac residues as the cause of vulture population decline in Pakistan *Nature* **2004**, *427*, 630-633.
23. Richards, S. M.; Wilson, C. J.; Johnson, D. J.; Castle, D. M.; Lam, M.; Mabury, S. A.; Sibley, P. K.; Solomon, K. R. Effects of pharmaceutical mixtures in aquatic microcosms *Environ. Toxicol. Chem.* **2004**, *23*, 1035-1042.
24. Brain, R. A.; Johnson, D. J.; Richards, R. A.; Hanson, M. L.; Sanderson, H.; Lam, M. W.; Mabury, S. A.; Sibley, P. K.; Solomon, K. R. Ecotoxicological evaluation of the effects of an eight pharmaceutical mixture to the aquatic macrophytes *Lemna gibba* and *Myriophyllum sibiricum* exposed in aquatic microcosms *Aquat. Toxicol.* **2004**, *70*, 23–40.
25. Brain, R. A.; Wilson, C. J.; Johnson, D. J.; Sanderson, H.; Bestari, K.; Hanson, M., L; Sibley, P. K.; Solomon, K. R. Effects of a mixture of tetracyclines to *Lemna gibba* and *Myriophyllum sibiricum* evaluated in aquatic microcosms *Environ. Pollut.* **2005**, *138*, 426-443.
26. Wilson, C. J.; Brain, R. A.; Sanderson, H.; Johnson, D. J.; Bestari, K. T.; Sibley, P. K.; Solomon, K. R. Structural and functional responses of

plankton to a mixture of four tetracyclines in aquatic microcosms *Environ. Sci. Technol.* **2004**, *38*, 6430-6439.

27. van den Brink, P. J.; Tarazona, J. V.; Solomon, K. R.; Knacker, T.; van den Brink, N. W.; Brock, T. C. M.; Hoogland, J. P. H. The use of terrestrial and aquatic microcosms and mesocosms for the ecological risk assessment of veterinary medicinal products *Environ. Toxicol. Chem.* **2005**, *24*, 820–829.
28. Hart, M. M.; Trevors, J. T. Microbe management: Application of mycorrhizal fungi in sustainable agriculture *Frontiers in Ecology and the Environment* **2005**, *3*, 533-539.
29. Klironomos, J. N. Feedback with soil biota contributes to plant rarity and invasiveness in communities *Nature* **2002**, *417*, 67-70.
30. Klironomos, J. N.; Ursic, M. Density-dependent grazing on the extraradical hyphal network of the arbuscular mycorrhizal fungus, *Glomus intraradices*, by the collembolan, *Folsomia candida Biology and Fertility of Soils* **1998**, *26*, 250-253.
31. Environment Canada "Biological Test Method: Toxicity Test for Measuring Emergence and Growth of Terrestrial Plants Exposed to Contaminants in Soil," Environment Canada, **2005**.
32. ASTM International Standard Guide for Conducting Terrestrial Plant Toxicity Tests In; ASTM International: West Conshohocken, PA, USA, **2003**; Vol. E 1963-02, pp 1534-1554.
33. Bécard, G.; Fortin, J. A. Early events of vesicular-arbuscular mycorrhiza formation on Ri T-DNA transformed roots *New. Phytol.* **1988**, *108*, 211-218.
34. Brain, R. A.; Johnson, D. J.; Richards, S. M.; Sanderson, H.; Sibley, P. K.; Solomon, K. R. Effects of 25 pharmaceutical compounds to *Lemna gibba* using a 7-day static renewal test *Environ. Toxicol. Chem.* **2004**, *23*, 371-382.
35. Brain, R. A.; Solomon, K. R.; Brooks, B. W. Targets, Effects and Risks in Aquatic Plants Exposed to Veterinary Antibiotics. In *Veterinary Pharmaceuticals in the Environment*; Henderson, K., Coats, J. R., Eds.; ACS Symposium Series 1018; American Chemical Society: Washington, DC, **2009**; pp169-189.
36. Hillis, D. G.; Lissemore, L.; Sibley, P. K.; Solomon, K. R. Effects of monensin on zooplankton communities in aquatic microcosms *Environ. Sci. Technol.* **2007**, *41*, 6620-6626.
37. Brain, R. A.; Rietsma, T. S.; Lissemore, L. I.; Bestari, K. T.; Sibley, P. K.; Solomon, K. R. Herbicidal effects of statin pharmaceuticals in *Lemna gibba Environ. Sci. Technol.* **2006**, *40*, 5116-5123.
38. Hillis, D. G.; L, B.; Solomon, K. R.; Sibley, P. K. Effects of antibiotics on the seed germination and root elongation of *Daucus carota*, *Lactuca sativa* and *Medicago sativa*. *Environ. Pollut.* **2008**, *Submitted May 2008*.
39. Basset, G. J. C.; Quinlivam, E. P.; Gregory, J. F.; Hanson, A. D. Folate synthesis and metabolism in plants and prospects for biofortification *Crop. Sci.* **2005**, *45*, 449-453.
40. Basset, G.; Quinlivan, E. P.; Ziemak, M. J.; Díaz de la Garza, R.; Fischer, M.; Schiffmann, S.; Bacher, A.; Gregory, I.; Jesse, F.; Hanson, A. D. Folate synthesis in plants: The first step of the pterin branch is mediated by a

unique bimodular GTP cyclohydrolase I *Proc. Nat. Acad. Sci. USA.* **2002**, *99*, 12489-12494.

41. Gambonnet, B.; Jabrin, S.; Ravanel, S.; Karan, M.; Douce, R.; Rébeillé, F. Folate distribution during higher plant development *J. Sci. Food. Agric.* **2001**, *81*, 835-841.
42. Boxall, A. B. A.; Johnson, P.; Smith, E. J.; Sinclair, C. J.; Stutt, E.; Levy, L. S. Uptake of veterinary medicines from soils into plants *J. Agric. Food. Chem.* **2006**, *54*, 2288-2297.
43. Batchelder, A. R. Chlortetracycline and oxytetracycline effects on plant growth and development in soil systems *J. Environ. Qual.* **1982**, *11*, 675-678.
44. Migliore, L.; Cozzolino, S.; Fiori, M. Phytotoxicity to and uptake of enrofloxacin in crop plants *Chemosphere* **2003**, *52*, 1233-1244.
45. Hillis, D. G.; Solomon, K. R.; Sibley, P. K. Structural responses of *Daucus carota* root organ cultures and the arbuscular mycorrhizal fungi, *Glomus intraradices*, to 12 pharmaceuticals *Chemosphere* **2008**, *Accepted May 2008.*
46. Fong, P. P. Antidepressants in aquatic organisms: A wide range of effects In *Pharmaceuticals and Personal Care Products in the Environment: Scientific and Regulatory Issues*; Daughton, C. G., Jones-Lepp, T. L., Eds.; American Chemical Society: Washington, DC, USA, **2001**; pp 264-281.
47. Dussault, È. B.; Balakrishnan, V. K.; Sverko, E.; Solomon, K. R.; Sibley, P. K. Toxicity of human pharmaceutical and personal care products to benthic invertebrates *Environ. Toxicol. Chem.* **2008**, *27*, 435-442.

Chapter 14

Risk assessment considerations for veterinary medicines in aquatic ecosystems

Bryan W. Brooks[1,2], Richard A. Brain[1,2], Duane B. Huggett[3], and Gerald T. Ankley[4]

[1]Department of Environmental Science, [2]Center for Reservoir and Aquatic Systems Research, Baylor University, Waco, Texas, USA, 76798.
[3]Department of Biological Sciences, Institute of Applied Sciences, University of North Texas, Denton, Texas, USA.
[4]U.S. Environmental Protection Agency, Duluth, Minnesota, USA .

In the United States, Canada, the European Union, Australia/New Zealand and Japan, prospective ecological risks of veterinary medicines are deterministically assessed following guidelines from the International Cooperation on Harmonization of Technical Requirements for Registration of Veterinary Medicinal Products. However, no regulatory framework is available for retrospective risk assessments of these compounds. This chapter provides a critical evaluation of prospective and retrospective risk assessment approaches for veterinary medicines in aquatic ecosystems and provides recommendations for alternative approaches for hazard characterization. Existing prospective risk assessment approaches include tiered testing, aquatic hazard information is generated from standardized toxicity tests and subsequent application of uncertainty factors. These approaches are not, however, adequate to characterize aquatic risks for select compounds (e.g., the synthetic anabolic steroid trenbolone). Because veterinary medicines often have specific biological properties that are identified during the drug development process, specific assay types and species chosen for hazard assessment should consider known pharmacology information and extrapolation to "read across" from target to non-target species for modes of action. In higher tiered risk assessment, probabilistic approaches and lentic and lotic mesocosm studies are useful for risk characterization. Finally, we propose a hazard assessment framework for veterinary medicines that incorporates alterative endpoints in ecological risk assessment.

Introduction

In the past decade, increasing information has become available for the environmental occurrence, fate and effects of select veterinary medicines (VMs). As introduced previously in this book, such information has led to unprecedented attention from the scientific, regulatory and public sectors to VMs in the environment (*1*). In particular, studies with the synthetic androgen 17β-trenbolone (*2*) and the parasiticide ivermectin (*3*) indicate that impacts from select VMs in aquatic ecosystems are possible under environmentally realistic scenarios. The purpose of this chapter is to provide our perspectives on several limitations of existing risk assessment approaches for VMs. We also propose some other considerations for future efforts to characterize risk of VMs in aquatic ecosystems.

Overview and Limitations of Current Regulatory Framework for Veterinary Medicines in the Environment

Although the recent interest in VMs has resulted from retrospective evaluations of the occurrence of VMs in aquatic ecosystems, regulatory efforts to assess VM risk to aquatic organisms have focused historically on prospective or product registration environmental assessments. For example, relatively recent regulatory guidelines were developed for the European Union, Canada, Australia/New Zealand, Japan and the United States (US) through the International Cooperation on Harmonization of Technical Requirements for Registration of Veterinary Medicinal Products (VICH) to deterministically assess ecological risks of VMs (*4,5*). This harmonized approach includes a two-phase process that was finalized in June 2000 (*4*) and October 2004 (*5*).

Under the first phase of VICH guidelines, most VMs applied in aquaculture settings with environmental concentrations ($EIC_{aquatic}$) of <1 μg L^{-1} and substances used as livestock treatments with predicted concentrations in soil (EIC_{soil}) of <100 μg kg^{-1} are not considered to be ecologically important, resulting in no testing of these substances (*4*). There are a few exceptions where VMs are not considered by VICH; for example, VMs given to pets are not considered important because loadings to aquatic ecosystems are believed to be too small to cause ecologically relevant effects. If the assessment of a VM does not stop in the first phase of VICH, the second phase is initiated in a two tier process.

During the first tier of phase two in VICH, standardized acute toxicity tests with algae, fish and cladocera are required (*5*). Similar to a screening-level ecological risk assessment, acute toxicity data with uncertainty factors (e.g., 100 for algal EC50 values, 1000 for cladocera, fish LC50 values) are utilized to generate predicted no effect concentrations, which together with the predicted exposure concentrations, are then compared to generate a deterministic risk quotient (RQ). If potential risk is identified in the first tier of VICH phase two, then additional standardized toxicity tests are performed with algae, cladocera, benthic invertebrate, and fish models to identify toxicological benchmark concentrations (e.g., no observed effect concentrations, NOECs) for sublethal responses such as growth or reproduction (*5*). An uncertainty factor of 10 is applied to these NOECs, and the RQ is recalculated. A third tier can be used in the US to further evaluate a specific VM if a RQ remains >1. Unfortunately, unlike the first two compulsatory tiers no guidelines are provided in the third tier of the US Food and Drug Administration (FDA) regulatory guidance documents. However, when a VM is known to act through a specific mode of action in the target animal (e.g., hormone modulation), it may be possible to assess more sensitive and ecologically relevant sublethal effects (e.g., fish reproduction) of VMs with alternative toxicity tests such as the 21 d *Pimephales promelas* (fathead minnow) reproduction protocol developed by Ankley et al. (*6*).

Although VICH provides an excellent example of a multinational effort to harmonize regulatory guidelines for a group of environmental contaminants, the biologically active nature of pharmaceuticals has led some to suggest that aquatic organisms may be impacted in ways not detected using current test methodologies and data evaluation strategies (*7,8,9*). For example, Ankley et al. (*2*) performed a 21 d fathead minnow reproduction study with the synthetic androgenic growth promoter trenbolone. In that study, the no observable effect concentration (NOEC) for reproduction (egg production) was about 3 ng L^{-1}, which is well within the range of environmentally relevant concentrations of the steroid (*10*). Such an observation with trenbolone is consistent with observations for 17β-ethinylestradiol (EE2), a potent synthetic estrogen agonist that affects fish at environmentally realistic concentrations (*11*). Specifically, Kidd et al (*11*) reported an elimination of *P. promelas* in an experimental lake treated with 5-6 ng L^{-1} EE2 for 3 years. However, Clubbs and Brooks (*12*) recently utilized a model organism (*Daphnia magna*) and endpoint (mortality) that is included in the VICH guidelines to evaluate aquatic hazard of EE2, and reported a mean LC_{50} value of 2.6 mg L^{-1}. Thus, if the current risk assessment framework for aquatic systems or approaches to develop a national ambient water quality criterion (NAWQC) in the US were to be applied to EE2, the toxicological benchmark concentration (e.g., EC50) identified by standardized toxicity tests with model organisms (e.g., *D. magna* 96 hr mortality) and the NAWQC would be orders of magnitude higher than levels known to adversely affect fish (*8*). Before examining additional considerations for the aquatic hazard assessment of VMs, it is instructive to evaluate several factors that influence the site specific exposure of aquatic organisms to veterinary medicines.

Site-Specific Exposure Considerations

Metcalfe et al (*13*) recently reviewed pertinent scientific issues associated with aquatic exposures to VMs. There are several specific items that should be considered during site-specific evaluations of aquatic exposure. Whereas investigators have examined occurrence and fate of veterinary pharmaceuticals in surface waters, therapeutics may partition to biota or sediments and persist in benthic communities much longer than pelagic counterparts (*14*). Subsequently, sediments, suspended sediments and organic material in aquatic ecosystems may present more continuous exposure at higher levels than associated surface water, highlighting the importance of understanding exposure contributions for VMs via aqueous and dietary routes (*8*). Building on our understanding of aquatic partitioning of non-ionized hydrophobic compounds, partitioning predictions based on log K_{ow} values and *f*oc may be appropriate for some pharmaceuticals: log K_{ow} values of > 4 have been reported for EE2 and trenbolone (*15*). However, many pharmaceuticals such as the fluoroquinilone antibiotic ciprofloxacin are charged (cations, anions, zwitterions) and can adsorb to clay and organic matter through ion exchange in a pH-dependent manner (*14*). Belden et al. (*17*) recently demonstrated higher K_D values for ciprofloxacin partitioning to fine particulate organic matter (FPOM) than for coarse particulate organic matter (CPOM), which may result in higher exposure to benthic macroinvertebrates utilizing FPOM (e.g., collector-gatherers). An understanding of fate and exposure pathways are much less developed for ionizable compounds than more historical approaches based on log K_{ow}.

Because the vast majority of pharmaceuticals are ionizable compounds, ambient pH may influence ionization state and aquatic toxicity of human pharmaceuticals and VMs. In fact, it has been long recognized that the un-ionized form of an ionizable compound is more lipophilic than the dissociated form of the molecule. This subsequently leads to greater bioavailability, uptake rates, and potential bioaccumulation of the nonionized compounds (*18*). For example, the herbicide 2:4-dichlorophenxyacetic acid demonstrated typical pH-dependent toxicity of a weak acid when a 10-fold difference was observed for *Lemna minor* LC_{50} values in tests performed between 4.6 and 6.1 (*19*). Recognizing such influences of pH on aquatic exposure and toxicity, the U.S. Environmental Protection Agency (U.S. EPA) developed NAWQC for the weak base ammonia (*21*) and the weak acid pentachlorophenol (*21*) that allow for site-specific modifications of allowable loadings from wastewater treatment plants based on ambient pH of a receiving system. Unfortunately, an understanding of the influence of ambient pH on the bioavailability and aquatic toxicity of human drugs and VMs is not developed (*22*); NAWQC do not exist for human and VMs in the US.

Site-specific factors may also influence the fate of VMs via photolysis. For example, photodegradation appears less likely to occur for a compound that is susceptible to photolysis once it is bound to sediment, thus increasing its persistence, particularly in turbid ecosystems (*14*). Belden et al. (*17*) recently demonstrated that the photodegradation half-life of ciprofloxacin was four-fold greater in the presence of FPOM, suggesting that increased persistence may be expected in systems with higher FPOM. Further, the photolabile selective serotonin reuptake inhibitor fluoxetine has been reported at higher levels in sediments downstream from effluent outfalls than in effluents or surface waters (*23*). Fluoxetine is also known to persist in sediments longer than in water (*24,25*), and was detected in piscivorous, omnivorous and benthivorous fish collected from a municipal effluent-dominated stream (*26*).

Previous studies have indicated that traditional contaminants such as nutrients and pathogens and "emerging" contaminants, such as VMs, may be transported from application fields and lagoons to adjacent aquatic ecosystems, particularly after rain events (*1,27,28,29,30*). As described below, watershed scale fate and transport models can be used to predict instream contaminant concentrations and potential exposures to VMs. In addition, these models allow for "what if" simulations of the influence of various management scenarios to reduce contaminant loadings to aquatic systems. Prochnow et al. (*31*) employed the state-of-the-art Soil Water and Assessment Tool (*32*), which includes various landscape attributes such as slope, soils and landcover in ArcGIS®, to partition the relative loadings of nutrients and suspended sediments to various agricultural practices in the North Bosque River watershed, located in the central Brazos River Basin, Texas, USA.

Environmental quality of the North Bosque River is severely impacted by non-point source nutrient pollution from dairy concentrated animal feeding operations (CAFOs; *28,31*), which represent the only permitted CAFOs in this watershed. At least 80,000 dairy cattle were present in Erath County, Texas during 2002, an area that encompasses the headwaters of the North Bosque watershed (*31*). Runoff from these CAFOs largely occurs during fall, winter and spring storm events, resulting in elevated nutrient (phosphorus, nitrogen) and pathogen loads to the river, periphytic and benthic macroinvertebrate impairments in subwatersheds with high CAFO densities, and hypereutrophication of Waco Lake, a reservoir formed by the confluence of the North, Middle and South Bosque Rivers (*31*). Subsequently, the Texas Commission on Environmental Quality (TCEQ) developed a Total Maximum Daily Load (TMDL) for soluble reactive phosphorus (SRP). The goal of this TMDL and the associated Implementation Plan is to reduce instream SRP by 50% (www.tceq.state.tx.us).

Because of the high density of confined animals and a semi-arid climate that likely provides a worst-case scenario for CAFO related impacts, the North Bosque River was recently identified a model watershed for retrospective assessment of VMs from dairy CAFO impacts on aquatic systems (*33*). In fact, arid and semi-arid headwater streams located adjacent to CAFOs may represent worse case scenarios for aquatic exposure and potential VM effects (*34*). Prochnow et al (*31*) simulated the effectiveness of several watershed management activities to reduce nutrient and sediment loadings, which provides useful information for multiple stakeholder groups, watershed managers and regulatory agencies. It may be that transport of other contaminants (e.g., nutrients, pathogens) that co-occur with VMs would be useful indicators when selecting watersheds for retrospective risk assessments of VMs.

It is clear that exposure to VMs in the environment may be influenced by compound- and site-specific partitioning behavior, bioavailability and density of CAFOs relative to receiving systems in watersheds. Subsequently, prospective and retrospective risk assessments of VMs should specifically consider relative partitioning and persistence in aquatic ecosystems strongly influenced by CAFOs as unique scenarios, likely contributing the upper centiles of environmental exposures in probabilistic exposure distributions (*14*), particularly for these sites with ambient pH that result in nonionized forms of ionizable compounds. The advantages of probabilistic techniques are described later in this chapter.

Importance of Considering Pharmacological Mechanism in the Ecological Risk Assessment of Veterinary Medicines

Under the Overview of the Ecological Risk Assessment (ERA) Process in the Office of Pesticide Programs of the U.S. EPA, the preliminary assessment of toxicity is facilitated by a complete understanding of the product chemistry and biochemical of the active ingredient, where mode of action is listed as a primary component of the risk assessment processes (*35*). While mortality, growth and reproductive toxicity endpoints are routinely used to calculate screening-level risk assessment risk quotients (RQs), they do not represent a limitation on the types of toxicity endpoints that may be considered in the risk assessment (*35*). Over the course of evaluation of available toxicity data, risk assessors may find other effects data that provide insight on endpoints not routinely considered for RQ calculation (*35*). However, professional judgment is necessary to determine whether and how available data on other toxicological endpoints are included in the risk assessment, including whether the information is applicable to the assessment endpoints established for the risk assessment (*35*). Data applicability is determined by utilizing available lines of evidence to ascertain if the toxicological endpoints can be linked to assessment endpoints in a reasonable and plausible manner (*35*). For example, the PMRA/EPA NAFTA project to update plant ecotoxicology testing for pesticide registration has identified significant gaps in the existing terrestrial plant tests and protocols, which result in a high level of uncertainty in terrestrial assessments, primarily because the measurement endpoints do not consider the mode of action (*36*).

The use and application of mechanistic endpoints in ERA requires several considerations with respect to selection criteria for investigations of pathway or target specific responses. The mechanistic endpoint should be: 1) specific to a particular pathway or process of interest, 2) sensitive to the contaminant of interest, 3) biologically relevant and 4) directly related to effects at the physiological level. Simply stated, the relationship must also be logical biochemically, physiologically, morphologically, and ecologically. This requires a fairly robust assessment and discussion of the biochemical pathway or target and how it is plausibly related to apical outcomes. If the mechanistic (pathway or target specific) endpoint is less sensitive and specific than the morphological endpoint, it has limited value as a measure of effect in the risk assessment process. However, if an endpoint provides some indicator of exposure then it may be suitable for inclusion in exposure assessment applications.

In the ERA decision making tree outlined in Figure 1, the first consideration is whether the compound, in this case a VM, causes adverse effects on the apical endpoint of concern. If no biologically significant effects are realized at exposure concentrations up to 1,000 μg L^{-1} with a battery of test organisms, there is no reasonable justification to continue the ERA process. However, if biologically significant effects are observed the ERA process should move either directly to assessing whether or not there is a homologous receptor present in the non-target test organism based upon weight-of-evidence, biological plausibility, evolutionary conservation of pathway or target receptors, experimental verification of the receptor, etc (*9*).

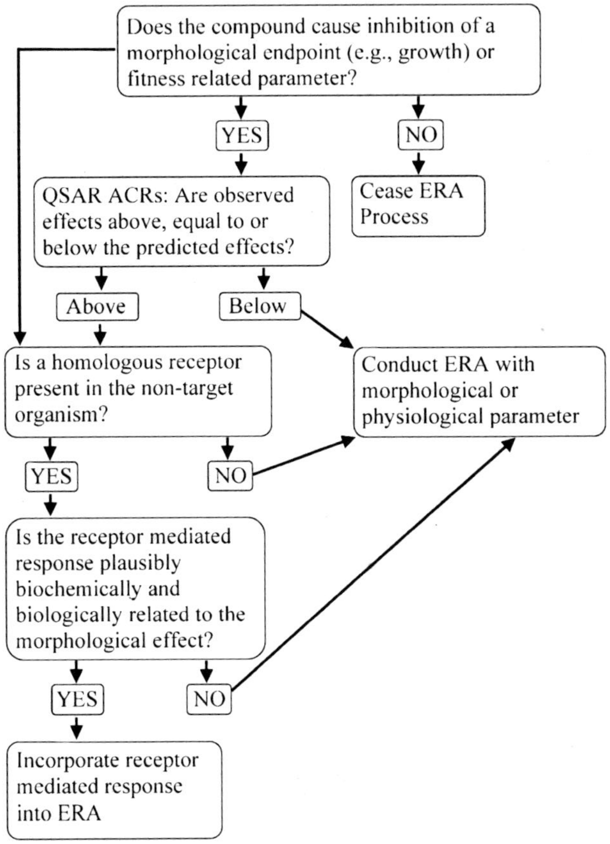

Figure 1. Proposed model for identifying whether a non-traditional response should be included in an effects analysis portion of an ecological risk assessment (ERA) of a veterinary medicine in aquatic ecosystems. Quantitative Structure Activity Relationship (QSAR) predictions and traditional acute to chronic ratios (ACRs; e.g., 10-20) may be used to screen compounds for non-traditional target mediated responses.

Indirectly, quantitive structure-activity relationship (QSARs) predictions of target related interactions for the various MOAs of VMs may be possible in the future. In the meantime, QSAR predictions of baseline (narcosis) acute toxicity and coupling such QSAR predictions with acute to chronic ratios (ACR; *36*) for chronic responses may be used to suggest whether the responses are target mediated or not. In the case of acute toxicity, if the toxicological benchmark value (e.g., LC50) obtained via experimentation is greater than a QSAR prediction, which might be expected for insecticides used as VMs (e.g., diazinon, ivermectin), then a target mediated response is likely. For chronic responses, if the acute QSAR generated value combined with a default ACR (e.g., 10-20) is similar (within an order of magnitude) to the experimental ACR value, the response is likely not target mediated and an ERA can be conducted with traditional morphological or physiological metrics. However, if the experimental ACR is much larger than the QSAR-predicted ACR value (e.g., trenbolone), the response may potentially be target mediated and mechanistic endpoints should be considered and investigated. Further, it may be possible to utilize such QSAR predictions and default ACRs as a diagnostic tool to screen compounds for target specific responses among species. Once a plausible homologous receptor has been identified the next critical step is to provide weight-of-evidence relating the target mediated response to effects at the physiological or morphological level. If no plausible relationship can be established we recommend that the ERA process could revert to traditional morphological or physiological metrics; however, if a strong biochemical and/or biological evidence is demonstrated then the ERA process should incorporate the target mediated metric as either a biomarker or exposure or a measure of effect (Figure 1). It is important to note that such non-traditional responses should not serve as regulatory "stop signs" during ERA (*38*); rather, biomarker responses can provide important mechanistic information to support a weight of evidence approach in the ERA of VMs.

Although the prospect of understanding target-specific responses to human drugs and VMs in non-target organisms may initially seem daunting, an understanding of pharmacological safety data from target organisms (e.g., mammals) may be coupled with comparative molecular genetics, biochemistry and physiology to identify non-target organisms that may be susceptible to compounds with specific MOAs. Pharmaceuticals typically have a known receptor or enzyme- based mechanism of action and represent some of the most widely studied compounds during their development process in terms of their pharmacological and toxicological properties. It has been recognized in the mammalian safety sciences that acute lethality studies may not be predictive of effects observed in sub-chronic or chronic studies. Therefore, additional studies and endpoints are evaluated during the drug development process to ensure human safety and efficacy. Studies investigating fertility, reproduction, mutagenicity, genotoxicity, carcinogenicity, as well as multiple tests to establish pharmacological activity and adsorption, distribution, metabolism and excretion characteristics (ADME) are conducted (*39*). In addition, the potency of the active pharmaceutical ingredient against a multitude of other pharmacological targets is studied. Therefore, it is known which, if any, other pharmacological targets and physiological systems may be altered in mammals when a pharmaceutical product is administered. The mammalian data generated during the drug development process, as well as the mammalian testing paradigm could be used as the conceptual backbone for the regulatory or prospective ecological risk assessment process for determining the chronic ecotoxicological impacts of VMs in the environment.

Since there is a considerable amount of target organism data generated in support of a drug registration, it would be beneficial to leverage these data for assessing the potential impact of pharmaceuticals on non-target species in the environment. Given the premise of conservation of enzymes and receptors across species, non-target species may be susceptible to the same pharmacological and toxicological effects as those observed in humans or laboratory animals. Review of all the pertinent pharmacology and safety data collected during the drug development process should be the first step in determining the need to perform chronic ecotoxicity tests. In terms of ecotoxicity, one of the most important pieces of information is determining the receptors/enzymes involved in a mammalian pharmacological/toxicological response, and assessing whether they are conserved in your species of interest. In addition, target organism safety data can highlight any toxicological concerns (reproduction, fertility, etc) that may be of interest. Figure 2 includes a conceptual model for informing environmental decision making for a VM from information obtained through the drug development efforts for a target organism. Employing such a "read across" approach can provide important weight-of-evidence information in ERAs of VMs.

TARGET ORGANISM DATA	*READ-ACROSS*	ENVIRONMENTAL DECISIONS
• Discovery Pharmacology – *In Vitro* Therapeutic Activity – Safety Pharmacology	→ Identify key receptor & enzyme interactions	• Mechanistic Ecotoxicology – Targeted Single Gene/Protein Assays – Target Related Responses
• Exposure – Internal Dose (C_{max}, etc) – Metabolism – Physio-chemical properties	→ Identify key metabolites, effective plasma concentrations, biotransformation routes, partitioning	• Exposure – Partitioning and Biotransformation • Parent vs. Metabolites – Extrapolation of Internal and Effective Doses – Physio-chemical properties
• Standard Testing – Reproduction, fertility, etc	→ Identify key therapeutic and toxic *in vivo* responses	• Standard Testing – Targeted endpoints and assays • Alternative Endpoints

Figure 2. Extrapolation of existing target organism data for environmental decision making.

The effective and non-effective target organism doses can be reviewed to determine if effects would be expected in non-target environmental organisms at environmentally relevant levels. Subsequent comparative pharmacological studies could be performed with the non-target organism and the compound under consideration. For example, Gould et al. (*40*) recently compared the binding affinities of several selective serotonin reuptake inhibitors (SSRIs) to serotonin reuptake transporters of different fish species and the Sprague-Dawley rat model. In the Gould et al. (*40*) study, similarities in K_D values were observed between models, demonstrating functional similarities in the SERT target for SSRIs among fish and rats. Depending on such mechanism or safety finding, researchers may be directed to investigate a particular environmental species or endpoint. For instance, researchers have highlighted that invertebrates, especially mollusks, are particularly sensitive to SSRIs (*41*). Therefore, compounds in drug development that target the serotonergic system in mammals may need to be screened against these non-target organisms.

A recent example of a mammalian to environmental "read across" effort is a conceptual model to bridge MOA across species using therapeutic or toxic No Observed Effect Level (NOEL derived on therapeutic or toxic plasma concentrations) and (predicted) environmental concentrations to calculate a Plasma Exposure Ratio (PER) (*42,43*). This PER is a modified hazard index that may be an additional methodology to determine which compound(s) need additional environmental toxicological follow-up in aquatic vertebrates. The general basis for this theoretical approach is the conservation of receptors and enzymes between mammals and aquatic vertebrates. In this approach, a therapeutic or toxic NOEL plasma concentration derived from traditional testing (e.g., with rats, dogs) of a given pharmaceutical is compared to a predicted steady state fish plasma concentration. This predicted fish steady state plasma concentration (Predicted F_{SS}PC) is obtained by considering absorption of a drug across the gills of an aquatic vertebrate. Specifically, the (predicted) environmental concentration of the pharmaceutical is multiplied by its theoretical blood:water partitioning coefficient ($\text{Log P}_{\text{Blood:Water}} = 0.73\text{x Log P} - 0.88$; *43*). The PER is then calculated by dividing the mammalian plasma effect concentration by the Predicted F_{SS}PC. A PER < 1 indicates that the predicted fish steady state concentration will be greater than the mammalian maximum therapeutic plasma concentration, thus indicating a high likelihood for pharmacological responses in aquatic vertebrates. A PER> 1 indicates that the Predicted F_{SS}PC is less than the mammalian plasma effect concentration, theoretically indicating that effects on aquatic vertebrates are less likely. Building on previous reports of select pharmaceutical bioaccumulation in fish tissue (*26, 45*), Brown et al. (*46*) successfully applied the Huggett model to fish exposed to effluent containing human pharmaceuticals. However, bioconcentration of specific therapeutics varied among their three study locations, indicating that site-specific differences influenced exposure (*46*). Future studies should utilize and expand the Huggett model to further evaluate the relative influence of water quality characteristics (e.g., pH, humic materials) and the dietary route of exposure on internal dosimetry of pharmaceuticals in the environment.

Ecotoxicogenomics in Ecological Risk Assessment of Veterinary Medicines

The comparative pharmacology or "read across" approach described above may provide a logical basis for toxicity testing and selection of standardized or alternative endpoints to subsequently contribute important weight-of-evidence information for ERAs of VMs. However, it is possible that VMs may have different or additional MOAs in non-target species relative to target organisms. Herein, toxicogenomic data may provide useful information during the testing of

VMs. Recent advances in analytical chemistry technologies and bioinformatics has allowed for collection of an unprecedented amount of molecular and biochemical data in many different organisms (*47,48*). These genomic techniques enable the collection of "global" information for an organism concerning gene or protein expression or endogenous metabolite profiles. In more recent years, genomic information has become available for several aquatic organisms (e.g., *Daphnia* sp., rainbow trout, fathead minnow), leading to the development of arrays that often focus on a number of gene products (*49*). Such genomic information in mammals and aquatic organisms can allow for comparative assessments of targets for VMs; for example, comparative genomic approaches have allowed for human–fish sequence evaluations (*50*).

Ankley et al. (*51*) identified a number of advantages that targeted genomic approaches can provide to the various tiers of ERA. For example, a database of gene expression profile data could be developed for algae, cladocerans, and fish (organisms used in Tier I ERA) exposed to reference VMs with defined MOAs for screening purposes (*8*). Other organisms could be selected for study based on a comparison of gene sequences among species. For example, when a target of interest is known, the degree of similarity of the gene or amino acid sequence could be evaluated in non-target species, which could provide an indication of whether the receptor is present and how similar it is to the target. It seems logical that reference chemicals and organisms selected for such "library" development would include toxicity pathways known to be influenced by the various classes of VMs or other common modes of toxicological action (*37*).

Development of a genomics database focused on targeted pathways presents advantages for both prospective and retrospective ERAs of VMs. After a gene expression profile database is assembled for common MOAs in non-target organisms, fingerprints for new VMs could be compared to their MOA profile (and other known MOAs) based on *a priori* knowledge of the VM's MOA (*8*). If the database profile data is consistent with profile data generated for the new compound, then the VM may not possess an unanticipated MOA, confirming the appropriateness of species/endpoints normally used for testing. If expression response profiles are different from those expected based on MOA, the chemical may exert its toxicity through other pathways, the most common of which (ideally) would be included in the database, thus identifying alternative species, endpoints and test designs for ERA. Such a database could also be very useful for retrospective assessments. A library of specific gene (or protein, metabolite) expression biomarkers appear to present excellent opportunities in biomonitoring exposure to individual or mixtures of VMs in the field, particularly in those watersheds that may have multiple loading scenarios (and, hence, chemical stressors) from agriculture or municipal entities.

Probabilistic Ecological Risk Assessment of Veterinary Medicines

There is no allowance for the application or incorporation of probabilistic techniques under the current VICH guidance for ERA of VMs. However, probabilistic methods incorporating species sensitivity distributions (SSDs) have been developed and refined over the last 20 years particularly in the risk assessment of agrochemicals (*52*). Furthermore, although the USEPA does not use the term SSD to set water quality criteria for aquatic life, guidelines regarding this concept have been derived and written (*53,54*). Probabilistic techniques have received comparatively limited application with respect to pharmaceuticals due largely to deficiencies in comprehensive exposure data. Thus, as previously introduced, current risk assessments of VMs rely on traditional highly conservative RQs and HQs, which are mandated for environmental impact assessments (EIAs) under the current European guidelines (*55,56*). However, as has been shown for agrochemicals, probabilistic techniques, which are based on a continuum of potential exposures and effects, allow the risk assessor to include estimates of uncertainty as well as stochastic properties of both exposures and effects (*57*). Furthermore, this type of assessment allows the risk assessor to conduct the assessment independent of most value judgments, and therefore offers more useful data for decision-making than hazard quotients.

In light of the exposure data deficiencies for many VMs, alternative probabilistic approaches have been developed and applied for veterinary antibiotics. For example, a probabilistic ecological hazard assessment (PEHA) methodology using intra-species endpoint sensitivity distributions (IESDs) and chemical toxicity distributions (CTDs) has been evaluated on pharmaceutical toxicity data sets for aquatic macrophytes using the same concepts outlined for SSDs (*58*). This methodology does not incorporate an exposure distribution, simply a point estimate or criterion concentration is used, and based on a distribution of effects the probability of finding a distributional value below a specified exposure concentration can be calculated (*58*). Using an SSD or IESD in a PEHA therefore allows the risk assessor to estimate the probability of encountering a species or effect measure at or below a particular concentration of interest. Alternatively, if the risk assessor wishes to consider the uncertainty that a species will have a more sensitive response, a criterion concentration corresponding to this level of uncertainty can be calculated directly from the SSD or IESD (*58*). In the case of CTDs the risk assessor can evaluate the probability of finding a chemical displaying potency below a particular threshold based on a distribution of toxic potencies for a particular group of chemicals to a common endpoint (*58*). CTDs appear to be particularly useful for hazard assessment with a variety of standardized or nontraditional endpoints for chemical classes for which limited environmental exposure data exists (*58*). Using distributions of species or endpoints allows for a more realistic, statistically derived threshold of response in place of a NOEC based on a single response (*59*). The advantage of using this type of approach over the traditional RQ approach is that the NOEC is usually calculated by applying a safety factor to the statistical summary of a single toxicity test, typically representing the most sensitive organism/response, with an applied safety factor that is assumed to be highly conservative but does not include any expression of uncertainty (*59*). Alternatively, an SSD incorporates available toxicity effects data for a range of different species and permits assemblage-level variability and uncertainty to be estimated and expressed quantitatively (*60*). Therefore toxicity thresholds derived from SSDs or IESDs provide a more realistic threshold of response than is possible with arbitrary uncertainty factors applied to a NOEC or ECx value in a deterministic HQ assessment (*59*).

As outlined in the VICH phase II guidance (*5*), it is not possible to evaluate the effects of VMPs on every species in the environment that may be exposed and, thus, the taxonomic groups tested are intended to serve as surrogates or indicators for the range of species present in the environment. Under the Tier A of VICH testing, only three taxonomic groups are recommended: at least one fish, one aquatic invertebrate and one algal species (*5*). For pesticide registration under the USEPA guidance it is recommended that toxicity data be generated for birds (mallard duck, bobwhite quail), mammals (laboratory rat), freshwater fish (bluegill sunfish, rainbow trout, fathead minnow), freshwater invertebrates (*Daphnia magna*), estuarine/marine fish (sheepshead minnow), estuarine/marine invertebrates (*Crassostrea virginica*, *Mysidopsis bahia*), terrestrial plants (corn, soybean, carrot, radish or sugar beet), oat (wheat or ryegrass), tomato, onion, cabbage (cauliflower or brussels sprout), lettuce, cucumber), and finally algae and aquatic plants (*Lemna gibba*, *Skeletonema costatum*, *Anabaena flos-aquae*, *Selenastrum capricornutum*, *Clorella vulgaris*, *Scenedesmus subspicatus*) (*35*). Thus, the number of species required for testing in the pesticide registration process under the USEPA is much more amenable to the application of probabilistic techniques using SSDs. Given that VICH maintains that impacts of greatest potential concern are usually those at the community and ecosystem function levels, with the aim being to protect most species (*5*), the demonstrated advantages of probabilistic techniques provides a much more data driven means of attaining these policy goals.

Future Perspectives: Integration of Target Related Endpoints and Higher Tier Approaches to the Ecological Risk Assessment of Veterinary Medicines

As discussed throughout this chapter and elsewhere (*8,9*), there would be numerous advantages to developing a more flexible ERA paradigm for VMs. Figure 3 outlines a new framework for ERA of VMs that specifically would allow comparative pharmacology information to shape the assessment *a priori* by including alternative endpoints in the problem formulation of a prospective ERA. Initially, knowledge of the MOA of a VM and an understanding of the presence of a pharmacological target would be determined for groups of non-target organisms (Figure 3). If a target is not present in a non-target organism (e.g., cladoceran), then it would be included in an initial Tier A battery of standardized acute toxicity tests similar to those currently employed in VICH. Alternatively, if a pharmacological target is present in a non-target organism, then an understanding of physiological responses to target interaction with a VM should be determined. When there does not appear to be physiologically relevant responses at environmentally relevant exposure levels, the approach presented in Figure 3 would include performing standardized sub-chronic toxicity tests as part of a Tier B phase.

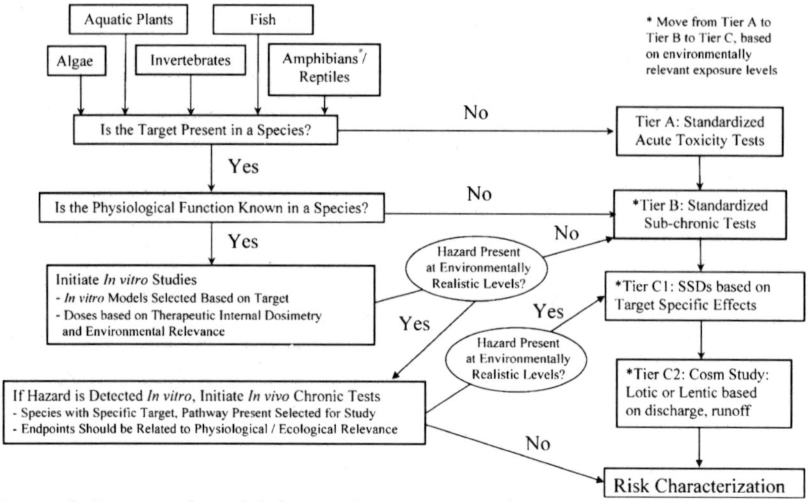

Figure 3. Proposed model for evaluating the ecological hazard of a veterinary medicine in aquatic ecosystems.

When the MOA of a VM can be linked to a physiologically relevant response, Figure 3 includes a two phased approach that would initiate *in vitro* studies specifically selected based on pharmacological targets in non-target organisms and MOA information for a VM. Experimental doses for these *in vitro* studies should be selected to encompass environmentally relevant exposures and internal doses associated with typical usage regimes in a target organism (e.g., therapeutic plasma levels). Unfortunately, environmental exposure data are limited for many VMs. As described above, herein CTDs present an excellent approach to characterize hazard of a compound or group of VMs that act through a specific MOA (Figure 1). Such hazard characterization may then be compared to predicted environmental concentrations of a VM, preferably generated from a probabilistic exposure distribution. If a hazard is identified by the *in vitro* studies at environmentally relevant exposures, then more in-depth chronic studies would be required to establish TBCs for biomarkers and physiologically and ecologically relevant endpoints. An example of such a study could be a 21 d fathead minnow reproduction test (see *6*) with additional endpoints related to the MOA of a VM.

Following the chronic studies depicted in Figure 3, when a VM is identified to present a hazard to aquatic life at environmentally realistic levels, a SSD would be developed in a Tier C1 process. This SSD would include a distribution of responses specifically related to the VM MOA in different species. If the 5th centile (or, HC5) of the SSD, for example, exceeds environmental exposures (preferably a probabilistic exposure distribution is available to allow for construction of a joint probability curve) Figure 3 includes a final hazard assessment approach – a Tier C2 –cosm study. Lentic –cosm experiments have been performed for select VMs, but lotic –cosm studies may be more ecologically relevant for many VMs because VM loadings may result primarily from runoff to lotic systems (*8,14*). To date, -cosm studies with human or VMs have not included MOA related endpoints; the whole lake study by Kidd et al

(*11*) with EE2 is the only experimental field study to have evaluated target related endpoints in fish and assessed fish population and community responses to environmentally relevant pharmaceutical treatment.

Ankley et al. (*9*) recently provided several recommendations for the assessment of human pharmaceuticals and VMs in the environmental. Ankley et al. (*9*) argued that building flexibility into the current ERA regulatory process for VMs is scientifically justified, particularly if *a priori* knowledge of MOA for a VM can allow for selection of target related endpoints in animal or plant models that are ecologically relevant (*9*). The model we propose in Figure 3 provides an example of such a model within the tiered ERA framework currently used for VMs. Although basic comparative biology, pharmacology and toxicology information is needed for both plant and animal ecological receptors, integration of target related, alternative endpoints in exposure and effects assessments can reduce ecotoxicological uncertainties and contribute to a weight-of-evidence approach in ERAs of VMs.

References

1. Boxall, A.B.A.; Fogg, L.A.; Blackwell, P.A.; Kay, P.; Pemberton, E.J.; Croxford, A. Veterinary medicines in the environment. *Rev. Environ. Contam. Toxicol.* **2004**, *180*, 1-91.
2. Ankley, G.T.; Jensen, K.M.; Makynen, E.A.; Kahl, M.D.; Korte, J.J.; Hornung, M.W.; Henry, T.R.; Denny, J.S.; Leino, R.L.; Wilson, V.S.; Cardon, M.C.; Hartig, P.C.; Gray, L.E. Effects of the androgenic growth promoter 17-beta-trenbolone on fecundity and reproductive endocrinology of the fathead minnow. *Environ. Toxicol. Chem.* **2003,** *22*, 1350-1360.
3. Sanderson, H.; Laird B.; Pope L.; Brain R.; Wilson C.; Johnson D.; Bryning G.; Peregrine, A.S.; Boxall, A.; Solomon, K.R. Assessment of the environmental fate and effects of ivermectin in aquatic mesocosms. *Aquat. Toxicol.* **2007**, *85*, 229-240.
4. VICH. Environmental impact assessment (EIAs) for veterinary medicinal products (Veterinary medicines) – Phase 1. International Cooperation on Harmonisation of Technical Requirements for Registration of Veterinary Medicinal Products. VICH-GL6 (Ecotoxicity Phase I). **2000**.
5. VICH. Environmental Impact Assessment for Veterinary Medicinal Products: Phase II Guidance. International Cooperation on Harmonisation of Technical Requirements for Registration of Veterinary Medicinal Products. VICH-GL38 (Ecotoxicity Phase II). **2004**.
6. Ankley, G.T.; Jensen, K.M.; Kahl, M.D.; Korte, J.J.; Makynen, E.A. Description and evaluation of a short-term reproduction test with the fathead minnow (*Pimephales promelas*). *Environ. Toxicol. Chem.* **2001**, *20*, 1276-1290.
7. Daughton, C.G.; Ternes, T.A. Pharmaceuticals and personal care products in the environment: agents of subtle change? *Environ. Health Perspect.* **1999**, *107*(Suppl 6), 907-938.
8. Brooks, B.W.; Ankley, G.; Hobson, J.; Meyerhoff, R.; Lazorchak, J.; Solomon, K. Aquatic Hazard Assessment of Veterinary Medicines. In

Veterinary Medicines in the Environment; Crane, M.; Barrett, K.; Boxall, A., Eds; SETAC Press: Pensacola, FL; **2008**; In press.
9. Ankley, G.L.; Brooks, B.W.; Huggett, D.B.; Sumpter J. Repeating history: pharmaceuticals in the environment. *Environ. Sci. Technol.* **2007**, *41*, 8211-8217..
10. Durhan, E.J.; Lambright, C.S.; Makynen, E.A.; Lazorchak, J.; Hartig, P.C.; Wilson, V.S.; Gray, L.E.; Ankley, G.T. Identification of metabolites of trenbolone acetate in androgenic runoff from a beef feedlot. *Environ. Health Perspect.* **2006,** *114*(Suppl. 1): 65-68.
11. Kidd, K.A.; Blanchfield, P.J.; Mills, K.H.; Palace, V.P.; Evans, R.E.; Lazorchak, J.M.; Flick, R.W. Collapse of a fish population after exposure to a synthetic estrogen. *Proc. Natl. Acad. Sci. USA* **2007**, *104*, 8897-8901.
12. Clubbs, R.L.; Brooks, B.W. *Daphnia magna* responses to a vertebrate estrogen receptor agonist and an antagonist: A multigenerational study. *Ecotox. Environ. Saf.* **2007**, *67*, 385-398.
13. Metcalfe, C.; Boxall, A.; Fenner, K.; Kolpin, D.; Servos, M.; Silberhorn, E.; Staveley, J. Exposure Assessment of Veterinary Medicines in Aquatic Systems. In *Veterinary Medicines in the Environment*; Crane, M.; Barrett, K.; Boxall, A.; Eds. SETAC Press: Pensacola, FL; **2008**, In Press.
14. Brooks, B.W.; Maul, J.; Belden, J.B. Emerging contaminants: antibiotics in aquatic and terrestrial ecosystems. In *Encyclopedia of Ecology*; Jorgensen, S.E., Ed; Elsevier: Oxford, UK; **2008**; In press.
15. Holthaus, K.I.E.; Johnson, A.C.; Jurgens, M.D.; Williams, R.J.; Carter, J.E. The potential for estradiol and ethinylestradiol to sorb to suspended and bed-sediments in some English rivers. *Environ. Toxicol. Chem.* **2002**, *21*, 2526-2535.
16. Kummerer, K.; Ericson, J.F.; Hannah, R.; Johnson, A.; Sedlak, D.L.; Weston, J.J. Environmental fate and transport of human pharmaceuticals. In *Human pharmaceuticals: Assessing the impacts on aquatic ecosystems*; Williams, R.T., Ed. SETAC Press: Pensacola, FL; **2005**; p. 111-148.
17. Belden, J.B.; Maul, J.D.; Lydy, M.J. Partitioning and photodegradation of ciprofloxacin in aqueous systems in the presence of organic matter. *Chemosphere* **2007**, *66*, 1390-1395.
18. Fent, K. Ecotoxicology of organotin compounds. *Crit. Rev. Toxicol.* **1995,** *26*, 1-117.
19. Blackman, G.E.; Robertson-Cunningham, R.C. The influence of pH on the phytotoxicity of 2:4-dichlorophenxyacetic acid to *Lemna minor. New Phytologist* **1953**, *52*, 71-75.
20. United States Environmental Protection Agency. *1999 update of ambient water quality criteria for ammonia.* Office of Water. United States Environmental Protection Agency; Washington, DC. EPA-822-R-99-014. **1999.**
21. United States Environmental Protection Agency. Ambient water quality criteria for pentachlorophenol- 1986. Office of Regulations and Standards, Criteria and Standards Division. United States Environmental Protection Agency; Washington, DC. EPA-440-5-86-009. **1986**.

22. Orvos, D.R.; Versteeg, D.J.; Inauen, J.; Capdevielle, M.; Rothenstein, A.; Cunningham, V. Aquatic toxicity of triclosan. *Environ. Toxicol. Chem.* **2002**, *21*, 1338-1349.
23. Furlong, E.T.; Kinney, C.A.; Burkhardt, M.R.; Zaugg, S.D.; Werner, S.L. Organic wastewater contaminants in biosolids and biosolid-derived products. In *Abstracts with Programs*; Geological Society of America Meeting, Seattle, WA; **2003**, *35*, 149.
24. Johnson, D.J.; Sanderson, H.; Brain, R.A.; Wilson, C.J.; Bestari, K.J.T.; Solomon, K.R. Exposure assessment and microcosm fate of selected selective serotonin reuptake inhibitors. *Regul. Toxicol. Pharmacol.* **2005**, *42*, 313-323.
25. Kwon, J.W.; Armbrust, K.L. Laboratory persistence and fate of fluoxetine in aquatic environments. *Environ. Toxicol. Chem.* **2006**, *25*, 2561-2568.
26. Brooks, B.W.; Chambliss, C.K.; Stanley, J.K.; Ramirez, A.J.; Banks, K.E.; Johnson, R.D.; Lewis, R.J. Determination of select antidepressants in fish from an effluent-dominated stream. *Environ. Toxicol. Chem.* **2005**, *24*, 464-469.
27. Geary, P.M.; Davies, C.M. Bacterial source tracking and shellfish contamination in a coastal catchment. *Water Sci. Technol.* **2003**, *47*, 95-100.
28. McFarland, A.M.S.; L. Hauck. Relating agricultural land uses to in-stream stormwater quality. *J. Environ. Qual.* **1999**, *28*, 836-844.
29. Orlando, E.F.; Kolok, A.S.; Binzcik, G.A.; Gates, J.L.; Horton, M.K.; Lambright, C.S.; Gray L.E.; Soto, A.M.; Guillette Jr, L. Endocrine-disrupting effects of cattle feedlot effluent on an aquatic sentinel species, the fathead minnow. *Environ. Health Perspect.* **2004**, *112*, 353-358.
30. Yang, S.W.; Carlson, K. Evolution of antibiotic occurrence in a river through pristine, urban and agricultural landscapes. *Water Res.* **2003**, *37*, 4645-4656.
31. Prochnow, S.J.; White, J.D.; Scott, T.;Filstrup, C. Small reservoir impact on watershed-scale simulated nutrient loads. **2008**, accepted.
32. Arnold, J.G.; Fohrer, N. SWAT2000: Current capabilities and research opportunities in applied watershed modeling. *Hydrological Processes* **2005**, *19*, 563-572.
33. de Knect, J.; Boucard, T.; Brooks, B.; Crane, M.; Gerould, S.; Koschorreck, J.; Scheef, G.; Solomon, K.; Zhixing, Y. Environmental risk assessment of veterinary medicines. In *Veterinary Medicines in the Environment*; Crane, M.; Barrett, K.; Boxall, A.; Eds. SETAC Press: Pensacola, FL; **2008**, In Press.
34. Brooks, B.W.; Riley, T.M.; Taylor, R.D. Water quality of effluent-dominated stream ecosystems: ecotoxicological, hydrological, and management considerations. *Hydrobiologia* **2006**, *556*, 365-379.
35. United States Environmental Protection Agency. *Overview of the Ecological Risk Assessment Processin the Office of Pesticide Programs, U.S. Environmental Protection Agency: Endangered and Threatened Species Effects Determinations*. Office of Prevention, Pesticides and Toxic Substances Office of Pesticide Programs. United States Environmental Protection Agency; Washington, DC. **2004**.

36. Davy, M.; Petrie, R.; Smrchek, J.; Kuchnicki, T.; Francois, D. Proposal to update non-target plant toxicity testing under NAFTA. US Environmental Protection Agency; Washington, DC; **2001**; http://www.epa.gov/scipoly/sap/. 158 p
37. Ankley, G.T.; Black, M.C.; Garric, J.; Hutchinson, T.H.; Iguchi, T. A framework for assessing the hazard of pharmaceutical materials to aquatic species. In *Human pharmaceuticals: Assessing the impacts on aquatic ecosystems*; Williams, R.T., Ed.; SETAC Press: Pensacola, FL; **2005**; pp 183-237.
38. Hutchinson, T. H.; Ankley, G. T.; Segner, H.; Tyler, C. R. Screening and testing for endocrine disruption in fish —biomarkers as "signposts," not "traffic lights" in risk assessment. *Environ. Health Perspect.* **2006**, *116*(Suppl. 1), 106-114.
39. United States Food and Drug Administration. Resources for drug registration studies. Available at http://www.fda.gov/ceder/guidance. **2002.**
40. Gould, G.G.; Brooks, B.W.; Frazer, A. [^{3}H] Citalopram binding to serotonin transporter sites in minnow brains. Basic Clin. Pharmacol. Toxicol. **2007**, *101*, 203-210.
41. Fong, P.; Huminski, P.; D'Urso, L. Induction and potentiation of parturition in fingernail clams by selective serotonin re-uptake inhibitors. *J. Exp. Zool.* **1998**, *280*, 260-264.
42. Huggett, D. B.; Cook, J.C.; Ericson J.E.; Williams, R.T. Theoretical model for prioritizing potential impacts of human pharmaceuticals to fish. *J. Human Eco. Risk Ass.* **2003**, 9, 1789-1799.
43. Huggett, D.B.; Benson, W.H.; Chipman, K.; Cook, J.C.; Gray, L.E.; Kinter, L.B.; Meyerhoff, R.D.; Trudeau, V.L. Role of mammalian data in determining pharmaceutical responses in aquatic species. In *Human pharmaceuticals: Assessing the impacts on aquatic ecosystems*; Williams, R.T., Ed. SETAC Press: Pensacola, FL; **2005**; p. 149-181.
44. Fitzsimmons, P.N.; Fernadez, J.D.; Hoffman, A.D.; et al. Branchial elimination of superhydrophobic organic compounds by rainbow trout (*Oncorhynchus mykiss*). *Aquat. Toxicol.* **2001**, *55*, 23-34.
45. Ramirez, A.J.; Mottaleb, M.A.; Brooks, B.W.; Chambliss, C.K. Analysis of pharmaceuticals in fish tissue using liquid chromatography - tandem mass spectrometry. *Anal. Chem.* **2007**, *79*, 3155-3163.
46. Brown, J.N.; Paxéus, N.; Förlin, L.; Larsson, D.G.J. Variations of bioconcentration of human pharmaceuticals from sewage effluents into fish blood plasma. *Environ. Toxicol. Pharmacol.* **2007**, *24*, 267-274.
47. MacGregor, J.T. The future of regulatory toxicology: impact of the biotechnology revolution. *Toxicol. Sci.* **2003**, 75, 236–248.
48. Waters, M.D.; Fostel, J.M. Toxicogenomics and systems toxicology: aims and prospects. *Nature Rev Genetics* **2004**, *5*, 936–948.
49. Lettieri, T. Recent applications of DNA microarray technology to toxicology and ecotoxicology. *Environ. Health Perspect.* **2006**, *114*, 4-9.
50. Ahituv, N.; Rubin, E.M.; Nobrega, M.A. Exploiting human–fish genome comparisons for deciphering gene regulation *Human. Mol. Genet.* **2004**, *13*, R261-R266.

51. Ankley, G. T.; Daston, G. P.; Degitz, S. J.; Denslow, N. D.; Hoke, R. A.; Kennedy, S. W.; Miracle, A. L.; Perkins, E. J.; Snape, J.; Tillitt, D. E.; Tyler, C. R.; Versteeg, D. Toxicogenomics in regulatory ecotoxicology. *Environ. Sci. Technol.* **2006**, *40*, 4055–4065.
52. Solomon, K.R.; Takacs, P. Probabilistic risk assessment using species sensitivity distributions. In *Species Sensitivity Distributions in Risk Assessment*; Posthuma, L.; Traas, T.; Suter, G.; Ed. CRC Press: Boca Raton, FL; **2002**.
53. Stephan, C.E.; Mount, D.I.; Hansen, D.J.; Gentile, J.H.; Chapman, G.A.; Brungs, W.A. Guidelines for deriving numerical national water quality criteria for the protection of aquatic organisms and their uses. US EPA ORD ERL; Duluth MN. PB 85-227049: **1985**. p 97.
54. Stephan, C.E. Use of species sensitivity distributions in the derivation of water quality criteria for aquatic life by the U.S. Environmental Protection Agency. In *Species Sensitivity Distributions in Risk Assessment*; Posthuma, L.; Traas, T.; Suter, G.; Eds. CRC Press: Boca Raton, FL. **2002**. p 564.
55. EMEA. Committee for Medicinal products for Veterinary Use (CVMP): Guideline on Environmental Impact Assessment for Veterinary Medicinal Products Phase II. 7 Westferry Circus, Canary Wharf, London, E14 4HB, UK: European Medicines Agency Veterinary Medicines and Inspections. CVMP/VICH/790/03-FINAL. **2004**.
56. EMEA. Committee for Medicinal Products for Human Use (CHMP): Guideline on the Environmental Risk Assessment of Medicinal Products for Human Use. 7 Westferry Circus, Canary Wharf, London E14 4HB, UK: European Medicines AF. **2004**.
57. Solomon, K.R.; Gipsy, J.P.; Jones, P. Probabilistic risk assessment of agrochemicals in the environment. *Crop Protect.* **2000**, *19*, 649-655.
58. Brain, R.A.; Sanderson, H.; Sibley, P.K.; Solomon, K.R. Probabilistic ecological hazard assessment: Evaluating pharmaceutical effects on aquatic higher plants as an example. *Ecotox. Environ. Saf.* **2006**, *64*, 128-135.
59. Hanson, M.L.; Solomon, K.R. New technique for estimating thresholds of toxicity in ecological risk assessment. *Environ. Sci. Technol.* **2002**, *36*, 3257-3264.
60. Grist, E.P.M.; Leung, K.M.Y.; Wheeler, J.R.; Crane, M. Better bootstrap estimation of hazardous concentration thresholds for aquatic assemblages. *Environ. Toxicol. Chem.* **2002**, *21*, 1515-1524.

Indexes

Author Index

Aga, D. S., 153–167
Anderson, R. C., 27–44
Ankley, G. T., 205–223

Bandeff, J. M., 135–152
Berry, J. O., 153–167
Brain, R. A., 169–189, 205–223
Brooks, B. W., 169–189, 205–223

Casey, F. X. M., 11–25
Cerniglia, C. E., 105–120
Coats, J. R., 1–2, 3–7, 93–104, 121–131

Fan, Z., 11–25
Farkas, M. H., 153–167

Guo, T., 45–65

Hakk, H., 11–25, 27–44
Hedman, C. J., 45–65
Henderson, K. L. D., 1–2, 3–7, 93–104, 121–131
Hillis, D. G., 191–204
Hu, D., 93–104
Huang, Y., 135–152
Huggett, D. B., 205–223

Karl, W., 67–91
Kim, Y.-H., 105–120

Larsen, G. L., 11–25, 27–44
Lissemore, L., 191–204

Moorman, T. B., 121–131

Nam, P. K., 135–152
Niyogi, D. K., 135–152

Oliver, C. E., 27–44

Pedersen, J. A., 45–65

Rupert, K. F., 45–65

Schneider, J., 67–91
Sibley, P. K., 191–204
Smith, D. J., 27–44
Solomon, K. R., 169–189, 191–204

Wetzstein, H.-G., 67–91

Subject Index

A

Abiotic process, tetracycline antibiotics, 47
Acid/base-catalyzed reactions, erythromycin A, 115
Active transport outside cell, antibiotic resistance, 111
Adsorption on clay
  ciprofloxacin, 208
  erythromycin A, 115*f*
Aerobic reduction, chlorate, 39
Aeromonas hydrophila
  indigenous bacteria, 108*t*, 109
  oxytetracycline- and tetracycline-resistant, 110
  tetracyclines against, 106*t*
Agricultural Research Service (ARS), pre-harvest safety tools, 29
Agricultural soils
  antibiotic exposure, 198–199
  biotransformation of fluoroquinolones, 76*t*
  fate of parent fluoroquinolones, 81–82
  veterinary pharmaceuticals, 195–196
  wheat straw pre-rotted in, 73, 74*f*, 75*f*
Agriculture
  ecological effect of antibiotics, 200
  pharmaceutical usage, 154
Alachlor
  conjugation to glutathione, 162*f*
  herbicides, 161
  LC-MS/MS ion spectrum of GSH conjugation product, 163*f*
Alfalfa, antibiotic exposure, 198
American Chemical Society (ACS), veterinary pharmaceutical symposium, 1
Aminoglycosides
  antimicrobial agent class, 106*t*
  fatty acid biosynthesis, 173*f*
  phytotoxicity, 177–178
  risk assessment for aquatic plants, 181, 182*f*
  translation and transcription processes in plastid as targets for, 175–176
Aminopenicillins, antimicrobial agent class, 106*t*
Androgens, possible exposure, 12–13
Animal agriculture, chlorate excretion, 35
Animal manure, reproductive hormones, 12–13
Animal pharmaceuticals
  residues in environment, 1–2, 3–4
  See also Veterinary medicines (VMs); Veterinary pharmaceuticals
Antibiotic-resistant bacteria, manure, 4
Antibiotics
  annual U.S. production, 171
  ecological effects in agriculture, 200
  environmental concern, 4
  environmental presence, 170–171
  glutathione *S*-transferase-mediated, detoxification in plants, 157–160
  resistance, 109–110
  resistance mechanism, 111–112
  tetracycline, 46
  tylosin, 94
  veterinary pharmaceuticals, 1, 93–94
  See also Aquatic plants; Fluoroquinolone antibiotics;

Sulfamethazine (SMZ); Tetracycline antibiotics
Antimicrobial drugs
aquaculture, 105
environmental fate of erythromycin A, 114–116
fate in aquaculture, 111, 112–114
food animal production, 105
therapy for pathogen infection, 116
See also Aquaculture industry
Aquaculture industry
antibiotic resistance, 109–110
antibiotics in European countries, 106–107
antimicrobial agents, 106*t*
antimicrobial drugs, 105
bacterial pathogens in aquatic environments, 107, 109
Center for Veterinary Medicine (CVM), 105
environmental fate of erythromycin A, 114–116
fate of antimicrobials, 112–114
lateral gene transfer mechanism, 111–112
mechanism of antibiotic resistance, 111–112
pathogenic bacteria transmitted by aquaculture, 108*t*
regulatory safety, 107
risk analysis of antimicrobial use, 112–113
risk ranking, 113*f*, 114
unapproved antimicrobial drugs, 107
United States Food and Drug Administration (US FDA), 105
Aquatic ecosystems
bioavailability of sulfamethazine, 125, 128–129
concentrated animal feeding operations (CAFOs), 209–210
concentrations of estrogens in water sources, 138*t*
estrogenic compound occurrence, 139–144
livestock estrogens occurrence, 138–139
proposed model for evaluating ecological hazard, 218*f*
veterinary medicines (VMs), 206
See also Ecological risk assessment (ERA); Livestock; Veterinary medicines (VMs)
Aquatic microcosms
pharmaceutical mixtures, 194
treatments, 197–198
Aquatic organisms, toxicity of antibiotics and pharmaceuticals, 196
Aquatic plants
aminoglycosides, 177–178
antibiotic classes targeting pathways, 173*f*
cephalexin, 180
chlorplastic fatty acid biosynthetic pathway and triclosan, 179–180
fatty acid biosynthesis, 172, 173*f*
folate biosynthetic pathway, 172, 173*f*
hazard quotients, 181, 182*f*
β-lactam antibiotics, 180
lincosamides, 177
macrolides, 177
metabolic pathways within mitochondria and plastid of plant cell, 173*f*
mitochondrial folate biosynthesis and sulfonamides, 178–179
monensin, 180
penicillin, 180
plastid replication and DNA gyrases as targets for fluoroquinolones, 172, 174–175

pleuromutilins, 177–178
receptors and pathways, 171–172
risk assessment, 181, 182*f*
sensitivity of blue-green algae, green algal and higher, 180–181
sulfonamides, 178–179
tetracyclines, 176–177
translation and transcription processes in plastid, 175–176
triclosan, 179–180
Aquatic worm, bioaccumulation of sulfamethazine, 128, 129*t*
Atorvastatin, EC50 values in *Glomus intraradices*, 198, 199*f*
Atrazine
conjugation to glutathione, 162*f*
herbicides, 161
time course for conjugate formation, 164*f*
Australian Commonwealth Scientific and Industrial Research Organisation (CSIRO), risk analysis, 113

B

Bacteria
chlorate degradation, 39–40
chlorate degradation in fecal, from cattle, 38, 39*f*
chlorate mechanism of lethal effect on *N*-respiring, 29, 30*f*
erythromycin A for treatment, 114
fluoroquinolone susceptibility to methanogenic, 82
Bacterial pathogens, aquatic environments, 107, 108*t*, 109
Basidiomycetous fungi, cultures on wheat straw, 70
Bioavailability
definition, 123
enrofloxacin degradation, 70–71
sulfamethazine, 128–129
veterinary pharmaceuticals, 4–5
Biodegradation, tylosin, 98
Biodegradation assay, fluoroquinolones, 88
Biosciences Laboratory, chlorate metabolism and residues, 32
Biotic process, tetracycline antibiotics, 47
Biotransformation kinetics, fluoroquinolones, 82, 83*f*
Birnessite
manganese oxide mineral, 47
reactivity toward oxytetracycline (OT), 58, 61
transformation of OT and chlortetracycline (CT), 58
See also Manganese oxide (MnO_2)
Blue-green algae
phytotoxic sensitivity, 180–181
risk assessment for antibiotics, 181, 182*f*
sensitivity to triclosan, 179
See also Aquatic plants
Broilers
chlorate elimination, 33*t*
chlorate salts against Gram-negative pathogens, 31*t*
concentrations of natural and synthetic estrogens from, 138*t*
natural and synthetic estrogens in streams from, 147*t*
See also Poultry

C

Campylobacter spp.
postharvest contamination, 108*t*
quinolones against, 106*t*
Carbamazepine
EC50 values in *Glomus intraradices*, 198, 199*f*
frequency and concentration in surface waters, 192, 193*f*
Carrot, antibiotic exposure, 198
Cation exchange capacity, tylosin leachability and, 100–101

Cattle
 chlorate degradation by fecal bacteria from, 38, 39*f*
 chlorate elimination, 33*t*
 chlorate salt activity against Gram-negative pathogens in, 31*t*
 concentrations of natural and synthetic estrogens from, 138*t*
 hormones in, 136–137
 land-use upstream and sampling sites, 140, 141*t*
 natural and synthetic estrogens in streams from, 147*t*
 pharmaceuticals and hormones in production, 3
 tylosin, 94
 See also Dairy cows
Cattle dung, biotransformation of [4-^{14}C]enrofloxacin, 84, 85*f*
Cattle dung pats
 biotransformation of fluoroquinolones, 76*t*
 $^{14}CO_2$ formation in fluoroquinolones, 75–77
Cattle waste pile
 biotransformation of fluoroquinolones, 76*t*
 $^{14}CO_2$ liberation from fluoroquinolones, 74, 75*f*
Center for Veterinary Medicine (CVM), antibiotics for aquaculture, 105
Cephalexin, phytotoxicity, 180
Chlamydia spp.
 macrolides against *C. pneumoniae*, 106*t*
 tetracyclines against, 106*t*
Chloramphenicol, fatty acid biosynthesis, 173*f*
Chlorate salts
 absorption and excretion of parent, 34
 activity against Gram-negative pathogens, 31*t*
 aerobic reduction, 39
 chemical composition variation of radiochlorine in soil columns, 37*f*
 chloride as sole metabolite, 32
 concentrations excreted in urine of treated animals, 37–38
 degradation by bacteria, 39–40
 degradation by *Salmonella enterica* and *Escherichia coli* O157:H7, 38, 39*f*
 elimination from cattle, swine, broilers and rats, 33*t*
 environmental fate, 34–40
 Gram-negative pathogens, 29–30
 mean transport through soil columns, 36*f*
 mechanism for lethal effect in nitrate-respiring bacteria, 29, 30*f*
 metabolism and disposition in food animals, 32, 34
 pre-harvest food safety tool, 29–30
 reducing Gram-negative pathogens, 30
 reduction to chloride, 34
 solubility and leaching, 35–37
 use as herbicide, 35
 waste waters containing, 35
Chloride
 chlorate metabolite, 32
 chlorate reduction to, 34
Chlorine dioxide, industrial use, 35
Chloroacetanilide
 herbicides, 161
 hypothesized MS/MS fragmentation, 163*f*
Chlorotriazine, herbicides, 161
Chlorplastic fatty acid biosynthesis, target for triclosan, 179–180
Chlortetracycline (CT)
 conjugation to glutathione, 162*f*
 degradation in pond water, 53–54
 EC50 values in *Glomus intraradices*, 199*f*
 environment, 60–61

glutathione *S*-transferase (GST) activity for maize roots after exposure, 157, 158*f*
mass spectrometry, 52, 60*f*
occurrence, 46
phytotoxicity, 176–177
preparation, 48
residues and detection limit, 192
structure, 49*f*
structures of transformation products and impurities, 56*t*, 57*t*
transformation in pond water and MnO_2-amended pond water, 52–53
transformation products, 54–55
ultraviolet absorbance data acquisition, 51–52
ultraviolet absorbing chromophores, 49*f*
See also Tetracycline antibiotics
Ciprofloxacin
adsorption by ion exchange, 208
$^{14}CO_2$ release from, in fresh cattle dung, 79*f*
$^{14}CO_2$ formation from, added to aging cattle dung, 77, 78*f*
$^{14}CO_2$ liberation from, in cattle waste pile, 74, 75*f*
phytotoxicity, 174, 175
structure, 69*f*
See also Fluoroquinolone antibiotics
Clay
erythromycin A, 115*f*
tylosin leachability and, 100–101
Clostridium botulinum, postharvest contamination, 108*t*
$^{14}CO_2$ formation
assessing biotransformation, 69–80
See also Fluoroquinolone antibiotics
Compost
biotransformation of fluoroquinolones, 76*t*
$^{14}CO_2$ release from fluoroquinolones in, 78–79
fate of parent fluoroquinolones in plant-derived, 81–82, 82, 84*f*
Concentrated animal feeding operations (CAFOs)
aquatic ecosystems, 209–210
hormone release into environment, 12–13, 137
non-point source nutrient pollution, 209
Confined animal feeding operations (CAFO), environmental concerns, 1, 154
Coumarins, fatty acid biosynthesis, 172, 173*f*
Cows. *See* Cattle; Dairy cows
Culture vessel, recording $^{14}CO_2$ formation, 71–72
Cyanobacteria, fluoroquinolones, 174

D

Dairy cows
concentrations of natural and synthetic estrogens from, 138*t*
natural and synthetic estrogens in streams from, 147*t*
See also Cattle
Danofloxacin
biotransformation kinetics, 82
structure, 69*f*
Degradation
bioavailability and, of enrofloxacin, 70–71
enrofloxacin, 86, 87*f*
erythromycin A, 115–116
pathways for tylosin, 102, 103*f*
Diatoms, sensitivity to triclosan, 179
Difloxacin, structure, 69*f*
DNA mutation, antibiotic resistance, 111
Doxycycline

EC50 values in *Glomus intraradices*, 198, 199*f*
frequency and concentration in surface waters, 192, 193*f*
phytotoxicity, 176–177

E

Ecological risk assessment (ERA)
considering veterinary medicine mechanism, 210–214
ecotoxicogenomics, 214–215
future perspectives, 217–219
higher tier approaches, 217–219
integrating target related endpoints, 217–219
probabilistic, of veterinary medicines (VMs), 215–217
proposed model for, of VM in aquatic ecosystems, 218*f*
proposed model for non-traditional response, 211*f*
quantitative structure-activity relationship (QSAR) predictions, 211–212
See also Microbiological risk assessment; Risk assessment
Ecology
antibiotics in agriculture, 200
fluoroquinolones, 72, 89
See also Ecological risk assessment (ERA)
Ecosystems. *See* Aquatic ecosystems
Ecotoxicology, veterinary pharmaceuticals, 5
Edwardsiella tarda, indigenous bacteria, 108*t*, 109
Endocrine disruption, natural and synthetic hormones, 136
Endocrine disruptor compounds (EDC), Federal Water Pollution Act, 12
Enrofloxacin (ENR)
bioavailability, 70–71
biodegradability, 70
biotransformation in cattle dung, 84, 85*f*
biotransformation kinetics in agricultural soils, 82, 83*f*
$^{14}CO_2$ formation from, added to aging cattle dung, 77*f*, 78*f*
$^{14}CO_2$ formation from [piperazine-2,3-^{14}C]ENR in soils, 72*f*
$^{14}CO_2$ liberation from, in cattle waste pile, 74, 75*f*
$^{14}CO_2$ release from core and amine labels of, in compost, 79*f*
$^{14}CO_2$ release from [4-^{14}C]ENR bound to agricultural soil, 71*f*
$^{14}CO_2$ release from, in fresh cattle dung, 79*f*
degradation scheme, 86, 87*f*
metabolites, 86, 87*f*
minimum inhibitory concentration (MIC), 86, 88*t*
radioactivity, 84, 85*f*
residual antibacterial activity of, and metabolites, 86, 88*t*
structure, 69*f*
See also Fluoroquinolone antibiotics
Enterobacteriaceae
lateral gene transfer, 112
postharvest contamination, 108*t*
quinolones against, 106*t*
sulfonamides against, 106*t*
Enterococcus spp., penicillins against, 106*t*
Environment
animal pharmaceutical residues, 1–2
antimicrobials and veterinary medicines in, 170–171
biological activity of pharmaceuticals and hormones, 4
concept for testing pharmaceutical mixtures, 194, 195*f*
estrogens in Two Rivers (Missouri), 143–144

hormone sources, 12–13
hormone transport and fate in, 137–138
occurrence of estrogenic compounds, 139–144
organic chemicals in Grand River Watershed (Missouri), 140–141, 142*f*
pharmaceuticals in, 155–156
regulatory framework for veterinary medicines, 206–207
reproductive hormones in, 12
sulfamethazine fate, 123–124, 126–127
tetracycline antibiotics, 60–61
tylosin, 94
tylosin degradation pathways, 102, 103*f*

Environmental estrogens
human exposure, 144–145, 147*t*
levels in human serum or plasma, 145, 146*t*

Environmental quality, North Bosque River, 209

Erysipelothrix rhusiopathiae, contact infection, 108*t*, 109

Erythromycin
environmental fate of, A, 114–116
frequency and concentration in surface waters, 192, 193*f*
glutathione *S*-transferase (GST) activity for maize roots after exposure, 158, 159*f*
resistance in chloroplasts, 176

Escherichia coli O157:H7
chlorate degradation, 38, 39*f*
prevalence in ground beef, 28, 29*f*
rate of infections, 28

17β-Estradiol
depth profile of sorbed, into soil, 18*f*
fractionation of radio labeled, 21*t*
incubation experiments, 19–20
mineralization, 19–20
redistribution in soil-water systems, 17–18
soil sorption, 14, 15, 17*t*
stream near manure application, 13
structure of radio-labeled, 15*f*
vitellogenin production, 12
See also Hormones

Estrogenic compounds
concentrations of medicinal estrogens in plasma or serum, 146*t*
detection and concentration in filtered stream water samples, 141*t*, 144*t*
human exposure to environmental, dietary and medicinal, 144–145
land-use upstream of sampling sites, 140, 141*t*
natural and synthetic, in streams from livestock, 147*t*
occurrence in Grand River Watershed, Missouri, 140–141, 142*f*
occurrence in Two Rivers, Missouri, 142*f*, 143–144
phytoestrogens and environmental, in human serum or plasma, 146*t*

Estrogens
cattle and swine farming operations, 140–141
human exposure, 145, 147*t*
livestock, 136
natural and synthetic, 200
possible exposure, 12–13
transport and fate in environment, 137

17-α-Ethinylestradiol, EC50 values in *Glomus intraradices*, 198, 199*f*

European Union (EU), pharmaceutical use, 192

Excretion
chlorate salts, 34
erythromycin A, 115*f*

Experimental chlorate products (ECP), drinking water or feed additives, 30

F

Fate
- antimicrobials in aquaculture, 112–114
- assessing, of parent fluoroquinolone, 80–84
- environmental, of erythromycin A, 114–116
- environmental, of sulfamethazine, 123–124, 126–127
- hormone, modeling, 20–22
- hormones in soil-water systems, 13–16, 17–19
- veterinary pharmaceuticals, 4
- *See also* Hormones

Fatty acid biosynthesis pathway
- plastid process, 172, 173*f*
- target for triclosan, 179–180

Fecal bacteria from cattle, chlorate degradation, 38, 39*f*

Feces, chlorate degradation in, 38, 39*f*

Federal Water Pollution Act, endocrine disruptor compound (EDC) screening, 12

Field application, manure, 3

Fish
- hormones, 1
- natural and synthetic estrogens in streams from, 147*t*

Florfenicol, approved antibiotic, 105

Fluoroquinolone antibiotics
- agricultural soils, 81–82
- antibacterial drugs, 68–69
- assessing biotransformation by $^{14}CO_2$ formation, 69–80
- assessing biotransformation by fate of parent drug, 80–84
- bioavailability, 70–71
- biotransformation in agricultural matrices, 76*t*
- biotransformation kinetics, 82, 83*f*
- cattle dung, 84, 85*f*
- cattle dung pats, 75–77
- cattle waste pile, 74, 75*f*
- chemical structures, 69*f*
- $^{14}CO_2$ formation from amine label positions in enro-, prado-, and ciprofloxacin in aging cattle dung, 78*f*
- $^{14}CO_2$ formation from core label positions in enro- and pradofloxacin in aging cattle dung, 77*f*
- $^{14}CO_2$ formation from [piperazine-2,3-^{14}C]ENR in soils, 72*f*
- $^{14}CO_2$ release from [4-^{14}C]enrofloxacin, 71*f*
- compost, 78–79
- cyanobacteria, 174
- degradation in organic matrices, 80
- ecology, 72, 89
- fatty acid biosynthesis, 172, 173*f*
- identified metabolites, 86, 87*f*
- metabolic scheme for enrofloxacin in *Gloeophyllum striatum*, 86, 87*f*
- phytotoxicity, 175
- plant-derived compost, 82, 84*f*
- plastid replication and DNA gyrases in plants, 172, 174–175
- pure cultures of basidiomycetous fungi on wheat straw, 70
- test system II for recording $^{14}CO_2$ formation, 71–72
- transformation kinetics, 88–89
- wheat straw pre-rotted in agricultural soil, 73, 74*f*, 75*f*
- *See also* Antibiotics; Aquatic plants

Folate biosynthetic pathway

mitochondria, 172, 173*f*
target for sulfonamides, 178–179
Food animals
chlorate metabolism and disposition in, 32, 34
human pathogens in, 28, 29*f*
See also Chlorate salts
Food safety
chlorate as pre-harvest safety tool, 29–30
pathogen contamination, 28

G

Gloeophyllum striatum
enrofloxacin biotransformation, 70, 71*f*
straw culture, 70
Glomus intraradices, EC50 values for pharmaceuticals and antibiotics, 198, 199*f*
Glutathione (GSH)
conjugation to alachlor, 162*f*
conjugation to herbicides, 161
enzymes glutathione *S*-transferases (GST), 160–161
Glutathione *S*-transferase (GST)
antibiotic detoxification in plants, 157–160
detoxification system in plants, 160–161, 164
GST activity of protein from maize after chlortetracycline exposure, 158*f*
GST activity of protein from maize after tylosin, erythromycin, and sulfadimethoxine exposure, 159*f*
Gram-negative pathogens
activity of chlorate salts against, 31*t*
chlorate salts, 29–30
Gram-positive bacteria, tylosin against, 94
Grand River Watershed
occurrence of organic chemicals, 140–141
study site, 142*f*
Green algae
phytotoxic sensitivity, 180–181
risk assessment for antibiotics, 181, 182*f*
sensitivity to triclosan, 179
See also Aquatic plants
Ground beef
Escherichia coli O157:H7 prevalence, 28, 29*f*
pathogen contamination, 28
See also Cattle

H

Hamburger
pathogen contamination, 28
See also Cattle; Ground beef
Hazard quotients, aquatic plants and antibiotics, 181, 182*f*
Hens. *See* Poultry
Herbicides
chlorate, 35
glutathione conjugation to, 161
Hogs
tetracycline antibiotics in, lagoons, 47
watershed and corporate farms, 140
See also Swine
Hormones
biological activity, 4
endocrine disruption, 136
environmental fate and transport in soil-water systems, 13–16
environmental sources, 12–13
17β-estradiol redistribution, 17–18
17β-estradiol sorption, 15, 17*t*
fate and transport in soil-water systems, 17–19
fate and transport process modeling, 20–22
fish and wildlife effects, 2

fractionation of radio-labeled 17β-estradiol and testosterone, 21*t*
future studies, 22–23
incubation experiments, 19–20
livestock, 136–137, 146–147
livestock production, 3
migration into ground water vs. drainage, 23
reproductive, in environment, 12
soil characteristic for sorption, miscible-displacement and incubation experiments, 16*t*
soil sorption of, 14, 16*t*
sorbed concentrations of 17β-estradiol to soil fractions, 15, 17*t*
structure of radio labeled 17β-estradiol and testosterone, 15*f*
synthetic, 1
testosterone redistribution, 18–19
testosterone soil sorption, 16, 17*t*
transport and fate in environment, 137–138
See also Synthetic hormones
Human pathogens, presence in food animals, 28, 29*f*
Humans
concentrations of medicinal estrogens in plasma or serum, 145, 146*t*
endocrine modulators, 136
erythromycin A, 114
exposure levels of estrogens, 147*t*
exposure levels of estrogens by source, 145, 147*t*
exposure to environmental, dietary and medicinal estrogens, 144–145
hormones from livestock, 139
phytoestrogens and environmental estrogens in human serum or plasma, 145, 146*t*

I

Incubation experiments, 17β-estradiol and testosterone, 19–20
Infectious bacterial diseases, erythromycin A, 114
International Cooperation on Harmonization of Technical Requirements for Registration of Veterinary Medicinal Products (VICH), 206–207
Iron oxides, tetracycline photodegradation, 47

K

Kinetics
biotransformation of fluoroquinolones, 82, 83*f*
$^{14}CO_2$ release from fluoroquinolones added to compost, 78–79
$^{14}CO_2$ release from fluoroquinolones in agricultural soil, 73, 74*f*, 75*f*
$^{14}CO_2$ release from fluoroquinolones in cattle dung pats, 76, 77*f*, 78*f*
$^{14}CO_2$ release from fluoroquinolones in cattle waste pile, 74, 75*f*
$^{14}CO_2$ release from fluoroquinolones in fresh cattle dung, 76, 79*f*
fluoroquinolones, 88–89
Klebsiella pneumoniae, phenicol resistance, 110

L

β-Lactam antibiotics
phytotoxicity, 180
risk assessment for aquatic plants, 181, 182*f*
Lateral gene transfer, mechanism, 111–112
Leachability, tylosin in soil columns, 100–101

Leaching, chlorate salts, 35–37
Leptospira interrogans, contact infection, 108*t*
Lettuce, antibiotic exposure, 198
Levofloxacin, EC50 values in *Glomus intraradices*, 198, 199*f*
Lincomysin, frequency and concentration in surface waters, 192, 193*f*
Lincosamides
fatty acid biosynthesis, 173*f*
pharmacological mode of action, 177
translation and transcription processes in plastid as targets for, 175–176
Listeria spp., penicillins against, 106*t*
Listeria monocytogenes, indigenous bacteria, 108*t*
Live animals, chlorate salts and Gram-negative pathogens, 30
Livestock
concentrations of natural and synthetic estrogens from, 138*t*
erythromycin A, 114
hormones in, 136–137, 146–147
human exposure levels of estrogens, 147*t*
natural and synthetic estrogens in streams from, 147*t*
occurrence and effects of, estrogens in environment, 138–139
waste for field application, 3
Lumbriculus variegatus, bioaccumulation of sulfamethazine, 125, 128, 129*t*

M

Macrolides
antimicrobial agent class, 106*t*
fatty acid biosynthesis, 173*f*
pharmacological mode of action, 177
risk assessment for aquatic plants, 181, 182*f*
translation and transcription processes in plastid as targets for, 175–176
tylosin, 94
Manganese oxide (MnO_2)
quenching of MnO_2-mediated reactions, 52–53
tetracycline transformation, 47–48
tetracycline transformation in presence of, 58–60
tetracycline transformation rates, 57–58
See also Tetracycline antibiotics
Manure
antibiotic-resistant bacteria, 4
field application, 3
reproductive hormones, 12–13
See also Hormones
Marbofloxacin, structure, 69*f*
Medicinal estrogens
concentrations in human serum or plasma, 145, 146*t*
human exposure, 144–145, 147*t*
Metabolism, chlorate, 32, 34
Metabolites, enrofloxacin, 86, 87*f*
Methanogenic bacteria, fluoroquinolone susceptibility, 82
Metolachlor, herbicides, 161
Microbiological risk assessment
antimicrobial agents in aquaculture, 111
See also Ecological risk assessment (ERA); Risk assessment
Mineralization
erythromycin A, 115–116
17β-estradiol and testosterone, 19–20
sulfamethazine, 124
Minimum inhibitory concentration, enrofloxacin, 86, 88*t*
Missouri
estrogens in Two Rivers, 142*f*, 143–144

organic chemicals in Grand River Watershed, 140–141, 142*f*

Mitochondria
  folate biosynthesis in plants as target for sulfonamides, 178–179
  folate biosynthetic pathway, 172, 173*f*

Monensin
  EC50 values in *Glomus intraradices*, 198, 199*f*
  frequency and concentration in surface waters, 192, 193*f*
  phytotoxicity, 180

Moxifloxacin (MFX)
  $^{14}CO_2$ liberation from, in cattle waste pile, 74, 75*f*
  $^{14}CO_2$ release from, in fresh cattle dung, 79*f*
  structure, 69*f*
  See also Fluoroquinolone antibiotics

Mycobacterium spp., macrolides against, 106*t*

Mycobacterium marinum, contact infection, 108*t*, 109

N

Nitrofurans, antimicrobial agent class, 106*t*

Non-target organisms
  target-specific responses in, 212–213
  veterinary pharmaceuticals and, 4

North America, pharmaceutical use, 191–192

O

Orbifloxacin, structure, 69*f*

Organic chemicals
  detection and concentration in filtered stream water samples, 140, 141*t*, 144*t*
  Grand River Watershed, Missouri, 140–141
  total ion chromatogram from GC/MS, 140, 142*f*

Organic matter, tylosin leachability and, 100–101

Oxytetracycline (OT)
  approved antibiotic, 105
  degradation in pond water, 53–54
  EC50 values in *Glomus intraradices*, 198, 199*f*
  environment, 60–61
  mass spectrometry, 52, 60*f*
  occurrence, 46
  phytotoxicity, 176–177
  preparation, 48
  residues and detection limit, 192
  resistance in catfish and goldfish, 109–110
  structure, 49*f*
  structures of transformation products and impurities, 56*t*, 57*t*
  transformation in pond water and MnO_2-amended pond water, 52–53
  transformation products, 54–55
  ultraviolet absorbance data acquisition, 51–52
  ultraviolet absorbing chromophores, 49*f*
  See also Tetracycline antibiotics

P

Pasteurella piscicida, phenicol resistance, 110

Penicillins
  antimicrobial agent class, 106*t*
  phytotoxicity, 180

Pharmaceuticals
  concept for testing mixtures, 195*f*
  ecological effects for antibiotics in agriculture, 200
  livestock production, 3

natural and synthetic estrogens, 200
presence in environment, 170–171
problem formulation, 193–196
residues in environment, 1–2, 4
responses to, in environment, 196–199
soils, 195–196, 198–199
surface water systems, 193–194, 196–198
usage in agriculture, 154
See also Veterinary pharmaceuticals

Phenicols
antimicrobial agent class, 106*t*
resistance, 110

Photobacterium damsela, contact infection, 108*t*, 109

Photolysis, veterinary medicines, 208

Photo-reaction products, tylosin, 97, 103*f*

Phytoestrogens
human exposure, 144–145, 147*t*
levels in human serum or plasma, 145, 146*t*

Phytotoxicity
antibiotic, 156
quinolones, 174, 175
See also Aquatic plants; Plants

Pigs. *See* Swine

Plants
antibiotic uptake into, 156–157
glutathione *S*-transferase (GST)-mediated antibiotic detoxification, 157–160
GST activity assay of protein from maize after chlortetracycline exposure, 158*f*
GST activity assay of protein from maize after tylosin, erythromycin, and sulfadimethoxine exposure, 159*f*
GSTs, 160–161, 164
LC-MS/MS ion spectrum of GSH conjugation product of alachlor, 163*f*
proposed MS/MS fragmentation of chloroacetanilide conjugated to GSH, 163*f*
schematic of conjugation of alachlor to glutathione by GST, 161, 162*f*
See also Aquatic plants

Plasmids, lateral gene transfer, 111–112

Plastid
fatty acid biosynthesis, 172, 173*f*
See also Aquatic plants

Plesiomonas shigelloides, indigenous bacteria, 108*t*

Pleuromutilins
phytotoxicity, 177–178
risk assessment for aquatic plants, 181, 182*f*
translation and transcription processes in plastid as targets for, 175–176

Pond water
composition, 51*t*
concentration of estrogens from livestock, 138*t*
degradation of tetracycline antibiotics, 53–54
dissipation profile of tylosin, 97, 98*t*
environmental fate of sulfamethazine, 123, 124, 126–127
MnO_2 influence on tetracycline transformation rates, 57–58
tetracycline antibiotics, 47
tetracycline antibiotic transformation products, 54–55, 56*t*, 57*t*
See also Tetracycline antibiotics

Poultry
chlorate elimination, 33*t*
chlorate salt activity against Gram-negative pathogens in, 31*t*

concentrations of natural and synthetic estrogens from, 138*t*
erythromycin A, 114
hormones in, 136–137
pharmaceuticals and hormones, 3
Salmonella, 28
tylosin, 94
Pradofloxacin (PRA)
$^{14}CO_2$ formation from, added to aging cattle dung, 77*f*, 78*f*
$^{14}CO_2$ liberation from, in cattle waste pile, 74, 75*f*
$^{14}CO_2$ release from, in fresh cattle dung, 79*f*
$^{14}CO_2$ release from core and amine labels of, in compost, 79*f*
development, 68
structure, 69*f*
See also Fluoroquinolone antibiotics
Propachlor
herbicides, 161
time course for conjugate formation, 164*f*
Pseudomonas spp., lateral gene transfer, 111–112

Q

Quality, environmental, of North Bosque River, 209
Quinolones
antimicrobial agent class, 106*t*
phytotoxicity, 174, 175
See also Fluoroquinolone antibiotics

R

Radioactivity, enrofloxacin (ENR), 84, 85*f*
Rats, chlorate elimination, 32, 33*t*
Regulatory framework, veterinary medicines in environment, 206–207
Reproductive hormones, environment, 12
Residual antibacterial activity, enrofloxacin and metabolites, 86, 88*t*
Resistance
antibiotic, 109–110
erythromycin, in chloroplasts, 176
mechanisms, 111–112
Rickettsia spp., tetracyclines against, 106*t*
Rifampicin, fatty acid biosynthesis, 173*f*
Risk assessment
antibiotics and aquatic plants, 181, 182*f*
antimicrobials in aquaculture, 111, 112–114
See also Ecological risk assessment (ERA); Microbiological risk assessment
Risk factors, *Salmonella* and *Escherichia coli* strains, 28
Risk ranking, management tool, 113*f*, 114
River water, concentration of estrogens from livestock, 138*t*
Row crops, land-use upstream and sampling sites, 140, 141*t*
Run-off, concentration of estrogens from livestock, 138*t*

S

Salmonella spp.
chlorate degradation, 38, 39*f*
food-animal production, 28
phenicol resistance, 110
postharvest contamination, 108*t*
quinolones against, 106*t*
rate of infections, 28
Sarafloxacin
addition to soils, 69
biotransformation kinetics, 82
structure, 69*f*

See also Fluoroquinolone antibiotics
Sediment
environmental fate of sulfamethazine, 124, 126–127
monitoring veterinary pharmaceuticals, 3–4
sulfamethazine bioavailability, 125, 128–129
Sedimentation, erythromycin A, 115–116
Self-mobilization, lateral gene transfer, 111
Sewage treatment plants (STPs)
human sources of pharmaceutical mixtures, 193–194
pharmaceuticals, 192
Sex steroid hormones, livestock, 136–137
Sheep, chlorate salt activity against Gram-negative pathogens in, 31*t*
Shigella spp., quinolones against, 106*t*
Soil. *See* Agricultural soils
Soil column
chemical composition variation of radiochlorine as chlorate, 36, 37*f*
chlorate solubility, 36
dissipation profile of tylosin, 97, 98*t*
mobility of tylosin, 99–101
tylosin study, 96
Soil dissipation
soil properties, 100*t*
sorption coefficient to tylosin, 100*t*
tylosin profile, 97–98
tylosin study, 95–96
Soil fate, sulfamethazine, 122
Soil mix, chlorate degradation in, 38, 39*f*
Soil sorption
17β-estradiol, 14, 15, 17*t*
hormones, 14
soil characteristics, 16*t*
testosterone, 14, 16, 17*t*
Soil-water systems
fate and transport of hormones, 13–16, 17–19
See also Hormones
Solubility, chlorate salts, 35–37
Springs, concentration of estrogens from livestock, 138*t*
Streams, concentration of estrogens from livestock, 138*t*
Streptococcus iniae, contact infection, 108*t*, 109
Streptococcus pneumoniae, macrolides against, 106*t*
Structural modifications, antibiotic resistance, 111
Sulfachlorpyridazine, frequency and concentration in surface waters, 192, 193*f*
Sulfadimethoxine
frequency and concentration in surface waters, 192, 193*f*
glutathione *S*-transferase (GST) activity for maize roots after exposure, 158, 159*f*
Sulfamerazine, approved antibiotic, 105
Sulfamethazine (SMZ)
bioaccumulation in aquatic worm, 129*t*
bioavailability, 123, 128–129
bioavailability study, 125
chemical structure, 122*f*
dissipation from pond water, 128*f*
environmental fate of, 126–127
environmental fate study, 123–124
frequency and concentration in surface waters, 192, 193 *f*
mass balance of SMZ-residues at day 7 and 63, 127*f*
materials and methods, 123–125
movement into sediment, 128*f*
soil fate studies, 122
sulfonamide antibiotic, 122
Sulfamethoxazole

EC50 values in *Glomus intraradices*, 198, 199*f*
frequency and concentration in surface waters, 192, 193*f*
Sulfathiozone, frequency and concentration in surface waters, 192, 193*f*
Sulfonamides
antimicrobial agent class, 106*t*
approved antibiotic, 105
folate biosynthetic pathway, 173*f*
mitochondrial folate biosynthesis in plants as target, 178–179
Surface waters
antibiotic and pharmaceutical toxicity to aquatic organisms, 196
frequency and concentrations of pharmaceuticals, 193*f*
microcosms treatments, 197–198
pharmaceutical problems, 193–194
pharmaceuticals, 192
tetracycline mixtures, 196–197
Swine
chlorate elimination, 33*t*
chlorate salt activity against Gram-negative pathogens in, 31*t*
hormone fate and transport in facility, 22
hormones in, 136–137
pharmaceuticals and hormones, 3
tylosin, 94
Synthetic hormones
endocrine disruption, 136
fish and wildlife effects, 2
natural and synthetic estrogens, 200
transport and fate in environment, 137–138
veterinary pharmaceuticals, 1
See also Hormones

T

Temperatures, chlorate as herbicide, 35
Testosterone
depth profile of sorbed, into soil, 18*f*
fractionation of radio labeled, 21*t*
incubation experiments, 19–20
mineralization, 19–20
redistribution in soil–water systems, 18–19
soil sorption, 14, 16, 17*t*
structure of radio labeled, 15*f*
See also Hormones
Tetracycline antibiotics
animal husbandry, 154
antimicrobial agent class, 106*t*
biotic and abiotic processes, 47
chlortetracycline (CT), 46
degradation in pond water, 53–54
EC50 values in *Glomus intraradices*, 198, 199*f*
environmental implications, 60–61
environmental occurrence, 155–156
fate, 46
fatty acid biosynthesis, 173*f*
formation of complexes with polyvalent cations, 46–47
inductively coupled plasma–optical emission spectroscopy (ICP–OES), 52
LC preparation, 50
manganese oxides, 47–48
mass fragmentation patterns, structure and fragmentation scheme for polyhydroxy product, 60*f*
mass spectrometry (MS), 52
materials and methods, 48–52
MnO_2 influence on transformation rates in pond water, 57–58
OT (oxytetracycline), 46

OT and CT transformation in pond water and MnO_2-amended pond water, 53*f*
OT and CT transformation products and impurities, 56*t*, 57*t*
phytotoxicity, 176–177
pond water composition, 51*t*
products of transformation in presence of MnO_2 in pond water, 58–60
quenching of MnO_2-mediated reactions, 52–53
reagents and preparation, 48
residues and detection limit, 192
risk assessment for aquatic plants, 181, 182*f*
sample processing and analysis scheme, 50*f*
structural moieties of OT and CT for ribosome binding and biological activity, 49*f*
surface water systems, 196–197
transformation in pond water, 48–49
transformation investigations, 47
transformation products in pond water, 54–55, 56*t*, 57*t*
translation and transcription processes in plastid as targets for, 175–176
ultraviolet absorbance data acquisition, 51–52
UV absorbing chromophores of tetracycline ring, 49*f*
See also Antibiotics
Thiolactomycin, fatty acid biosynthesis, 173*f*
Thiopeptides, fatty acid biosynthesis, 173*f*
Titanium oxides, tetracycline photodegradation, 47
Transcription, process in plastid as target for antibiotics, 175–176
Transformation
manganese oxide and tetracyclines, 47–48
pathways for tylosin, 102, 103*f*
See also Tetracycline antibiotics
Translation, process in plastid as target for antibiotics, 175–176
Transport
antibiotic resistance by, outside cell, 111
chlorate salts, 35, 36*f*
hormone, modeling, 20–22
hormones in soil-water systems, 13–16, 17–19
hypotheses of hormone, 23
Triclosan
chlorplastic fatty acid biosynthesis as target for, 179–180
EC50 values in *Glomus intraradices*, 198, 199*f*
fatty acid biosynthesis, 173*f*
risk assessment for aquatic plants, 181, 182*f*
Trimethoprim
folate biosynthetic pathway, 173*f*
frequency and concentration in surface waters, 192, 193*f*
Turkeys, chlorate salt activity against Gram-negative pathogens in, 31*t*
Two Rivers, Missouri
estrogens in monitoring study, 143–144
study site, 142*f*
Tylosin
analytical methods, 96–97
aquatic plants, 177
chemical structure, 94*f*
dissipation in surface water, 99*f*
dissipation profile in water and soil, 97–98
distribution of recovered, in packed soil column, 100*f*
EC50 values in *Glomus intraradices*, 198, 199*f*
glutathione *S*-transferase (GST) activity for maize roots after exposure, 158, 159*f*
leachability vs. cation exchange capacity, 101*f*

leachability vs. clay, 101*f*
leachability vs. organic matter, 101*f*
macrolide antibiotic, 94
materials and methods, 95–97
mobility in soil columns, 99–101
proposed degradation pathways, 102, 103*f*
residues and detection limit, 192
soil column study, 96
soil dissipation study, 95–96
soil properties and sorption coefficient to, 100*t*
transformation potential in environment, 102, 103*f*
water dissipation study, 95

U

United States, pharmaceutical use, 191–192
United States Food and Drug Administration (US FDA), drugs in aquaculture industry, 105
Urine
chlorate degradation in, 38, 39*f*
chlorate excretion, 37–38
US Department of Agriculture (USDA)
Biosciences Laboratory in Fargo, ND, 32
pre-harvest safety tools, 29

V

Veterinary medicines (VMs)
chemical toxicity distributions (CTDs), 216
ecotoxicogenomics in ecological risk assessment of, 214–215
evaluating ecological hazard in aquatic ecosystems, 217–218
integration of target related endpoints, 217–219
intra-species endpoint sensitivity distributions (IESDs), 216–217
mechanism and ecological risk assessment, 210–214
probabilistic ecological risk assessment, 215–217
quantitative structure-activity relationship (QSARs), 211–212
regulatory framework, 206–207
site-specific exposure, 208–210
species sensitivity distributions (SSDs), 215–217
target organism data and environmental decisions, 213*f*
target-specific responses in non-target organisms, 212–213
tiered approaches to ecological risk assessment, 218–219
See also Ecological risk assessment (ERA)
Veterinary pharmaceuticals
American Chemical Society national meeting, 1
antibiotics, 1, 93–94
bioavailability, 4–5
ecotoxicological studies, 5
fate, 4
livestock production, 3, 154
potential direct and indirect effects on non-target organisms, 4
presence in environment, 170–171
residues in environment, 3–4
sediment monitoring, 3–4
synthetic hormones, 1
Vibrio cholerae, indigenous bacteria, 108*t*
Vibrionaceae, lateral gene transfer, 112
Vibrio parahaemolyticus, indigenous bacteria, 108*t*
Vibrio vulnificus
contact infection, 108*t*, 109
indigenous bacteria, 108*t*

VICH (International Cooperation on Harmonization of Technical Requirements for Registration of Veterinary Medicinal Products), 206–207
Vitellogenin production, 17β-estradiol, 12

W

Water dissipation
- tylosin in surface water, 98, 99*f*
- tylosin profile, 97–98
- tylosin study, 95

Watershed
- estrogens in Two Rivers, 142*f*, 143–144
- organic chemicals in Grand River Watershed, 140–141, 142*f*

Wheat straw
- biotransformation of fluoroquinolones, 76*t*
- $^{14}CO_2$ formation from [4-^{14}C]enrofloxacin in pre-rotted, 73, 74*f*
- $^{14}CO_2$ formation from [piperazine-2,3-^{14}C]enrofloxacin in pre-rotted, 73, 75*f*
- cultures of basidiomycetous fungi on, 70
- pre-rotted in agricultural soil, 73

Wildlife
- endocrine modulators, 136
- hormones, 1
- hormones from livestock, 139

Z

Zoonotic pathogens, environmental habitats, 110